Chevrolet Sprint & Geo Metro Automotive Repair Manual

**by Larry Warren
and John H Haynes**

Member of the Guild of Motoring Writers

Models covered:
Chevrolet Sprint - 1985 through 1988
Geo Metro - 1989 through 2001

(24075-1T4)

Haynes Group Limited
Haynes North America, Inc.
www.haynes.com

Disclaimer

There are risks associated with automotive repairs. The ability to make repairs depends on the individual's skill, experience and proper tools. Individuals should act with due care and acknowledge and assume the risk of performing automotive repairs.

The purpose of this manual is to provide comprehensive, useful and accessible automotive repair information, to help you get the best value from your vehicle. However, this manual is not a substitute for a professional certified technician or mechanic.

This repair manual is produced by a third party and is not associated with an individual vehicle manufacturer. If there is any doubt or discrepancy between this manual and the owner's manual or the factory service manual, please refer to the factory service manual or seek assistance from a professional certified technician or mechanic.

Even though we have prepared this manual with extreme care and every attempt is made to ensure that the information in this manual is correct, neither the publisher nor the author can accept responsibility for loss, damage or injury caused by any errors in, or omissions from, the information given.

Acknowledgements

The publisher wishes to thank Tomco Industries, 1435 Woodson Road, St. Louis Missouri, 63132 for their gracious help. Wiring diagrams provided exclusively for Haynes North America by Valley Forge Technical Information Services. Technical consultants who contributed to this project include Mike Stubblefield, Ken Freund and Robert Maddox.

© **Haynes North America, Inc.** **1991, 1995, 1998, 2002**
With permission from Haynes Group Limited

A book in the Haynes Automotive Repair Manual Series

All rights reserved. No part of this book may be reproduced or transmitted in any form or by any means, electronic or mechanical, including photocopying, recording or by any information storage or retrieval system, without permission in writing from the copyright holder.

ISBN-10: 1-56392-453-6
ISBN-13: 978-1-56392-453-8

Library of Congress Control Number 2002100095

While every attempt is made to ensure that the information in this manual is correct, no liability can be accepted by the authors or publishers for loss, damage or injury caused by any errors in, or omissions from, the information given.

02-272

Contents

Haynes photographer, mechanic and author with Chevrolet Sprint

1990 Geo Metro

About this manual

Its purpose

The purpose of this manual is to help you get the best value from your vehicle. It can do so in several ways. It can help you decide what work must be done, even if you choose to have it done by a dealer service department or a repair shop; it provides information and procedures for routine maintenance and servicing; and it offers diagnostic and repair procedures to follow when trouble occurs.

We hope you use the manual to tackle the work yourself. For many simpler jobs, doing it yourself may be quicker than arranging an appointment to get the vehicle into a shop and making the trips to leave it and pick it up. More importantly, a lot of money can be saved by avoiding the expense the shop must pass on to you to cover its labor and overhead costs. An added benefit is the sense of satisfaction and accomplishment that you feel after doing the job yourself.

Using the manual

The manual is divided into Chapters. Each Chapter is divided into numbered Sections, which are headed in bold type between horizontal lines. Each Section consists of consecutively numbered paragraphs.

At the beginning of each numbered Section you will be referred to any illustrations which apply to the procedures in that Section. The reference numbers used in illustration captions pinpoint the pertinent Section and the Step within that Section. That is, illustration 3.2 means the illustration refers to Section 3 and Step (or paragraph) 2 within that Section.

Procedures, once described in the text, are not normally repeated. When it's necessary to refer to another Chapter, the reference will be given as Chapter and Section number. Cross references given without use of the word "Chapter" apply to Sections and/or paragraphs in the same Chapter. For example, "see Section 8" means in the same Chapter.

References to the left or right side of the vehicle assume you are sitting in the driver's seat, facing forward.

Even though we have prepared this manual with extreme care, neither the publisher nor the author can accept responsibility for any errors in, or omissions from, the information given.

NOTE

A **Note** provides information necessary to properly complete a procedure or information which will make the procedure easier to understand.

CAUTION

A **Caution** provides a special procedure or special steps which must be taken while completing the procedure where the Caution is found. Not heeding a Caution can result in damage to the assembly being worked on.

WARNING

A **Warning** provides a special procedure or special steps which must be taken while completing the procedure where the Warning is found. Not heeding a Warning can result in personal injury.

Introduction to the Chevrolet Sprint/Geo Metro

These models are available in two and four-door hatchback and four-door sedan body styles.

The transversely-mounted inline three and four-cylinder engines used in these models are equipped with either a carburetor or fuel injection.

The engine drives the front wheels through either a five-speed manual or three-speed automatic transaxle via independent driveaxles.

Independent suspension, featuring coil spring strut units is used at the front wheels. At the rear, a beam axle with leaf or coils springs and shock absorbers is used on Sprint models. Geo models have independent suspension with strut/coil spring units. The rack and pinion steering unit is mounted behind the engine.

The brakes are disc at the front and drums at the rear, with power assist standard.

Vehicle identification numbers

Modifications are a continuing and unpublicized process in vehicle manufacturing. Since spare parts manuals and lists are compiled on a numerical basis, the individual vehicle numbers are essential to correctly identify the component required.

Vehicle Identification Number (VIN)

This very important identification number is stamped on a plate attached to the left side of the dashboard on the driver's side of the vehicle, visible through the windshield (see illustration). The VIN also appears on the Vehicle Certificate of Title and Registration. It contains information such as where

and when the vehicle was manufactured, the model year and the body style.

Engine identification number

The engine ID number is stamped into the front side of the block at the rear (transaxle) end (see illustration).

Transaxle identification number

On manual transaxles, the ID number is stamped into the case near the transaxle-to-engine mounting surface and the clutch release lever. Automatic transaxle ID numbers are on the top of the case near the dipstick.

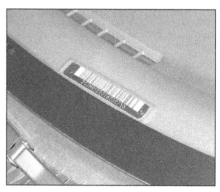

The Vehicle Identification Number (VIN) is stamped into a metal plate fastened to the dashboard on the driver's side - it is visible through the windshield

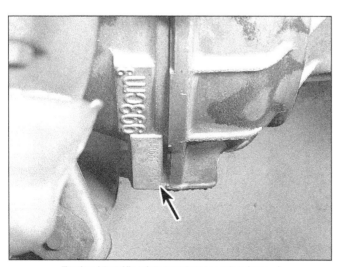

Engine identification number location (arrow)

Transaxle identification number location (arrow)

Buying parts

Replacement parts are available from many sources, which generally fall into one of two categories - authorized dealer parts departments and independent retail auto parts stores. Our advice concerning these parts is as follows:

Retail auto parts stores: Good auto parts stores will stock frequently needed components which wear out relatively fast, such as clutch components, exhaust systems, brake parts, tune-up parts, etc. These stores often supply new or reconditioned parts on an exchange basis, which can save a considerable amount of money. Discount auto parts stores are often very good places to buy materials and parts needed for general vehicle maintenance such as oil, grease, filters, spark plugs, belts, touch-up paint, bulbs, etc. They also usually sell tools and general accessories, have convenient hours, charge lower prices and can often be found not far from home.

Authorized dealer parts department: This is the best source for parts which are unique to the vehicle and not generally available elsewhere (such as major engine parts, transmission parts, trim pieces, etc.).

Warranty information: If the vehicle is still covered under warranty, be sure that any replacement parts purchased - regardless of the source - do not invalidate the warranty!

To be sure of obtaining the correct parts, have engine and chassis numbers available and, if possible, take the old parts along for positive identification.

Maintenance techniques, tools and working facilities

Maintenance techniques

There are a number of techniques involved in maintenance and repair that will be referred to throughout this manual. Application of these techniques will enable the home mechanic to be more efficient, better organized and capable of performing the various tasks properly, which will ensure that the repair job is thorough and complete.

Fasteners

Fasteners are nuts, bolts, studs and screws used to hold two or more parts together. There are a few things to keep in mind when working with fasteners. Almost all of them use a locking device of some type, either a lockwasher, locknut, locking tab or thread adhesive. All threaded fasteners should be clean and straight, with undamaged threads and undamaged corners on the hex head where the wrench fits. Develop the habit of replacing all damaged nuts and bolts with new ones. Special locknuts with nylon or fiber inserts can only be used once. If they are removed, they lose their locking ability and must be replaced with new ones.

Rusted nuts and bolts should be treated with a penetrating fluid to ease removal and prevent breakage. Some mechanics use turpentine in a spout-type oil can, which works quite well. After applying the rust penetrant, let it work for a few minutes before trying to loosen the nut or bolt. Badly rusted fasteners may have to be chiseled or sawed off or removed with a special nut breaker, available at tool stores.

If a bolt or stud breaks off in an assembly, it can be drilled and removed with a special tool commonly available for this purpose. Most automotive machine shops can perform this task, as well as other repair procedures, such as the repair of threaded holes that have been stripped out.

Flat washers and lockwashers, when removed from an assembly, should always be replaced exactly as removed. Replace any damaged washers with new ones. Never use a lockwasher on any soft metal surface (such as aluminum), thin sheet metal or plastic.

Fastener sizes

For a number of reasons, automobile manufacturers are making wider and wider use of metric fasteners. Therefore, it is important to be able to tell the difference between standard (sometimes called U.S. or SAE) and metric hardware, since they cannot be interchanged.

All bolts, whether standard or metric, are sized according to diameter, thread pitch and length. For example, a standard 1/2 - 13 x 1 bolt is 1/2 inch in diameter, has 13 threads per inch and is 1 inch long. An M12 - 1.75 x 25 metric bolt is 12 mm in diameter, has a thread pitch of 1.75 mm (the distance between threads) and is 25 mm long. The two bolts are nearly identical, and easily confused, but they are not interchangeable.

In addition to the differences in diameter, thread pitch and length, metric and standard bolts can also be distinguished by examining the bolt heads. To begin with, the distance across the flats on a standard bolt head is measured in inches, while the same dimension on a metric bolt is sized in millimeters (the same is true for nuts). As a result, a standard wrench should not be used on a metric bolt and a metric wrench should not be used on a standard bolt. Also, most stan-dard bolts have slashes radiating out from the center of the head to denote the grade or strength of the bolt, which is an indication of the amount of torque that can be applied to it. The greater the number of slashes, the greater the strength of the bolt. Grades 0 through 5 are commonly used on automobiles. Metric bolts have a property class (grade) number, rather than a slash, molded into their heads to indicate bolt strength. In this case, the higher the number, the stronger the bolt. Property class numbers 8.8, 9.8 and 10.9 are commonly used on automobiles.

Strength markings can also be used to distinguish standard hex nuts from metric hex nuts. Many standard nuts have dots stamped into one side, while metric nuts are marked with a number. The greater the number of dots, or the higher the number, the greater the strength of the nut.

Metric studs are also marked on their ends according to property class (grade). Larger studs are numbered (the same as metric bolts), while smaller studs carry a geometric code to denote grade.

It should be noted that many fasteners, especially Grades 0 through 2, have no distinguishing marks on them. When such is the case, the only way to determine whether it is standard or metric is to measure the thread pitch or compare it to a known fastener of the same size.

Standard fasteners are often referred to as SAE, as opposed to metric. However, it should be noted that SAE technically refers to a non-metric fine thread fastener only. Coarse thread non-metric fasteners are referred to as USS sizes.

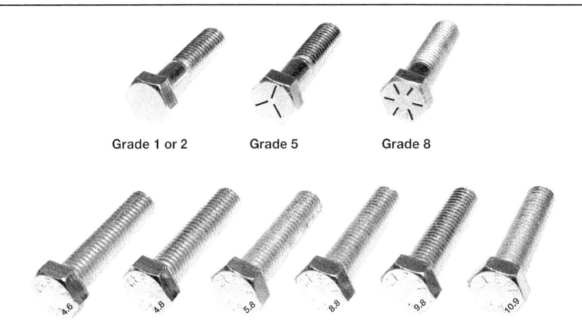

Grade 1 or 2   Grade 5   Grade 8

Bolt strength marking (standard/SAE/USS; bottom - metric)

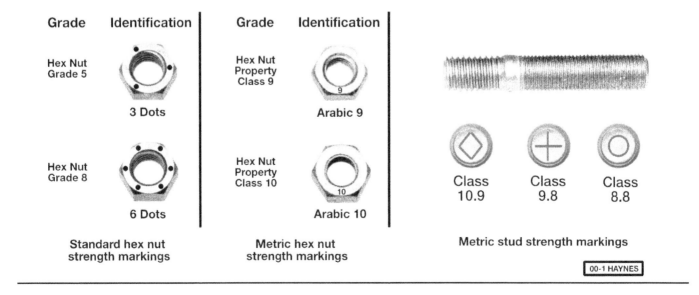

Grade	Identification
Hex Nut Grade 5	3 Dots
Hex Nut Grade 8	6 Dots

Standard hex nut strength markings

Grade	Identification
Hex Nut Property Class 9	Arabic 9
Hex Nut Property Class 10	Arabic 10

Metric hex nut strength markings

Class 10.9   Class 9.8   Class 8.8

Metric stud strength markings

Since fasteners of the same size (both standard and metric) may have different strength ratings, be sure to reinstall any bolts, studs or nuts removed from your vehicle in their original locations. Also, when replacing a fastener with a new one, make sure that the new one has a strength rating equal to or greater than the original.

Tightening sequences and procedures

Most threaded fasteners should be tightened to a specific torque value (torque is the twisting force applied to a threaded component such as a nut or bolt). Overtightening the fastener can weaken it and cause it to break, while undertightening can cause it to eventually come loose. Bolts, screws and studs, depending on the material they are made of and their thread diameters, have specific torque values, many of which are noted in the Specifications at the beginning of each Chapter. Be sure to follow the torque recommendations closely. For fasteners not assigned a specific torque, a general torque value chart is presented here as a guide. These torque values are for dry (unlubricated) fasteners threaded into steel or cast iron (not aluminum). As was previously mentioned, the size and grade of a fastener determine the amount of torque that can safely be applied to it. The figures listed here are approximate for Grade 2 and Grade 3 fasteners. Higher grades can tolerate higher torque values.

Fasteners laid out in a pattern, such as cylinder head bolts, oil pan bolts, differential cover bolts, etc., must be loosened or tightened in sequence to avoid warping the component. This sequence will normally be shown in the appropriate Chapter. If a specific pattern is not given, the following procedures can be used to prevent warping.

	Ft-lbs	Nm
Metric thread sizes		
M-6	6 to 9	9 to 12
M-8	14 to 21	19 to 28
M-10	28 to 40	38 to 54
M-12	50 to 71	68 to 96
M-14	80 to 140	109 to 154
Pipe thread sizes		
1/8	5 to 8	7 to 10
1/4	12 to 18	17 to 24
3/8	22 to 33	30 to 44
1/2	25 to 35	34 to 47
U.S. thread sizes		
1/4 - 20	6 to 9	9 to 12
5/16 - 18	12 to 18	17 to 24
5/16 - 24	14 to 20	19 to 27
3/8 - 16	22 to 32	30 to 43
3/8 - 24	27 to 38	37 to 51
7/16 - 14	40 to 55	55 to 74
7/16 - 20	40 to 60	55 to 81
1/2 - 13	55 to 80	75 to 108

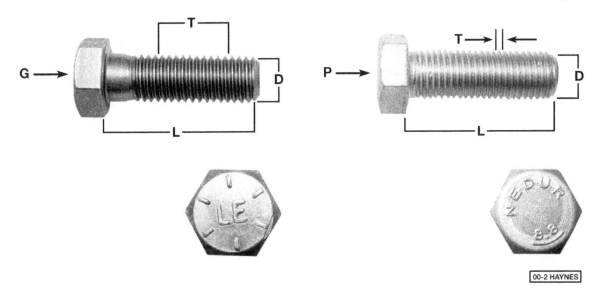

00-2 HAYNES

Standard (SAE and USS) bolt dimensions/grade marks

G   Grade marks (bolt strength)
L   Length (in inches)
T   Thread pitch (number of threads per inch)
D   Nominal diameter (in inches)

Metric bolt dimensions/grade marks

P   Property class (bolt strength)
L   Length (in millimeters)
T   Thread pitch (distance between threads in millimeters)
D   Diameter

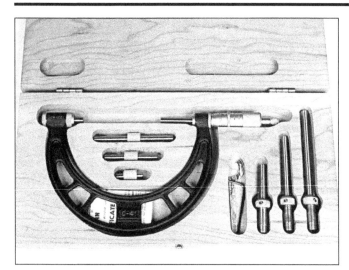

Micrometer set

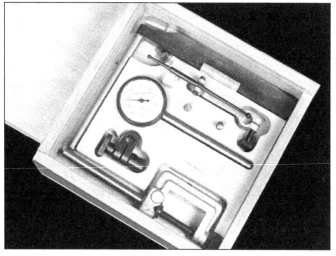

Dial indicator set

Initially, the bolts or nuts should be assembled finger-tight only. Next, they should be tightened one full turn each, in a criss-cross or diagonal pattern. After each one has been tightened one full turn, return to the first one and tighten them all one-half turn, following the same pattern. Finally, tighten each of them one-quarter turn at a time until each fastener has been tightened to the proper torque. To loosen and remove the fasteners, the procedure would be reversed.

Component disassembly

Component disassembly should be done with care and purpose to help ensure that the parts go back together properly. Always keep track of the sequence in which parts are removed. Make note of special characteristics or marks on parts that can be installed more than one way, such as a grooved thrust washer on a shaft. It is a good idea to lay the disassembled parts out on a clean surface in the order that they were removed. It may also be helpful to make sketches or take instant photos of components before removal.

When removing fasteners from a component, keep track of their locations. Sometimes threading a bolt back in a part, or putting the washers and nut back on a stud, can prevent mix-ups later. If nuts and bolts cannot be returned to their original locations, they should be kept in a compartmented box or a series of small boxes. A cupcake or muffin tin is ideal for this purpose, since each cavity can hold the bolts and nuts from a particular area (i.e. oil pan bolts, valve cover bolts, engine mount bolts, etc.). A pan of this type is especially helpful when working on assemblies with very small parts, such as the carburetor, alternator, valve train or interior dash and trim pieces. The cavities can be marked with paint or tape to identify the contents.

Whenever wiring looms, harnesses or connectors are separated, it is a good idea to identify the two halves with numbered pieces of masking tape so they can be easily reconnected.

Gasket sealing surfaces

Throughout any vehicle, gaskets are used to seal the mating surfaces between two parts and keep lubricants, fluids, vacuum or pressure contained in an assembly.

Many times these gaskets are coated with a liquid or paste-type gasket sealing compound before assembly. Age, heat and pressure can sometimes cause the two parts to stick together so tightly that they are very difficult to separate. Often, the assembly can be loosened by striking it with a soft-face hammer near the mating surfaces. A regular hammer can be used if a block of wood is placed between the hammer and the part. Do not hammer on cast parts or parts that could be easily damaged. With any particularly stubborn part, always recheck to make sure that every fastener has been removed.

Avoid using a screwdriver or bar to pry apart an assembly, as they can easily mar the gasket sealing surfaces of the parts, which must remain smooth. If prying is absolutely necessary, use an old broom handle, but keep in mind that extra clean up will be necessary if the wood splinters.

After the parts are separated, the old gasket must be carefully scraped off and the gasket surfaces cleaned. Stubborn gasket material can be soaked with rust penetrant or treated with a special chemical to soften it so it can be easily scraped off. A scraper can be fashioned from a piece of copper tubing by flattening and sharpening one end. Copper is recommended because it is usually softer than the surfaces to be scraped, which reduces the chance of gouging the part. Some gaskets can be removed with a wire brush, but regardless of the method used, the mating surfaces must be left clean and smooth. If for some reason the gasket surface is gouged, then a gasket sealer thick enough to fill scratches will have to be used during reassembly of the components. For most applications, a non-drying (or semi-drying) gasket sealer should be used.

Hose removal tips

Warning: *If the vehicle is equipped with air conditioning, do not disconnect any of the A/C hoses without first having the system depressurized by a dealer service department or a service station.*

Hose removal precautions closely parallel gasket removal precautions. Avoid scratching or gouging the surface that the hose mates against or the connection may leak. This is especially true for radiator hoses. Because of various chemical reactions, the rubber in hoses can bond itself to the metal spigot that the hose fits over. To remove a hose, first loosen the hose clamps that secure it to the spigot. Then, with slip-joint pliers, grab the hose at the clamp and rotate it around the spigot. Work it back and forth until it is completely free, then pull it off. Silicone or other lubricants will ease removal if they can be applied between the hose and the outside of the spigot. Apply the same lubricant to the inside of the hose and the outside of the spigot to simplify installation.

As a last resort (and if the hose is to be replaced with a new one anyway), the rubber can be slit with a knife and the hose peeled from the spigot. If this must be done, be careful that the metal connection is not damaged.

If a hose clamp is broken or damaged, do not reuse it. Wire-type clamps usually weaken with age, so it is a good idea to replace them with screw-type clamps whenever a hose is removed.

Tools

A selection of good tools is a basic requirement for anyone who plans to maintain and repair his or her own vehicle. For the owner who has few tools, the initial investment might seem high, but when compared to the spiraling costs of professional auto maintenance and repair, it is a wise one.

To help the owner decide which tools are needed to perform the tasks detailed in this manual, the following tool lists are offered: *Maintenance and minor repair,*

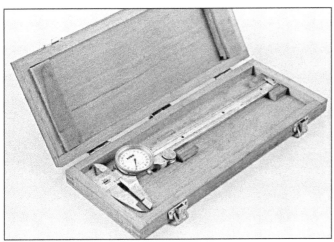

Dial caliper

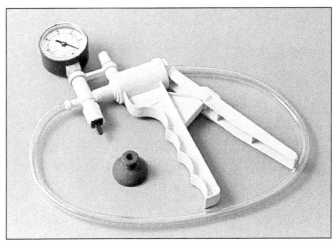

Hand-operated vacuum pump

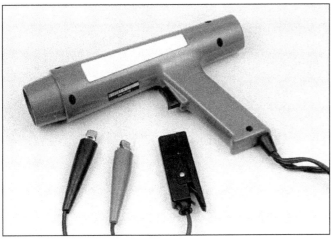

Timing light

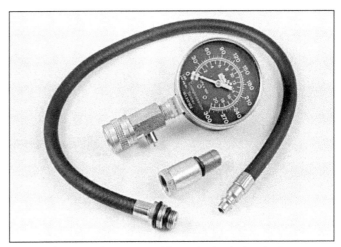

Compression gauge with spark plug hole adapter

Damper/steering wheel puller

General purpose puller

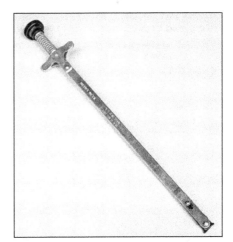

Hydraulic lifter removal tool

Repair/overhaul and *Special*.

The newcomer to practical mechanics should start off with the *maintenance and minor repair* tool kit, which is adequate for the simpler jobs performed on a vehicle. Then, as confidence and experience grow, the owner can tackle more difficult tasks, buying additional tools as they are needed.

Eventually the basic kit will be expanded into the *repair and overhaul* tool set. Over a period of time, the experienced do-it-yourselfer will assemble a tool set complete enough for most repair and overhaul procedures and will add tools from the special category when it is felt that the expense is justified by the frequency of use.

Maintenance and minor repair tool kit

The tools in this list should be considered the minimum required for performance of routine maintenance, servicing and minor repair work. We recommend the purchase of combination wrenches (box-end and open-

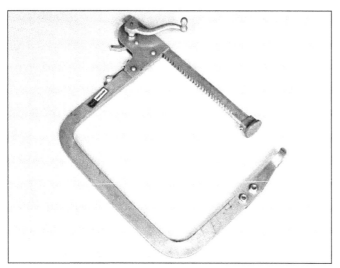

Valve spring compressor

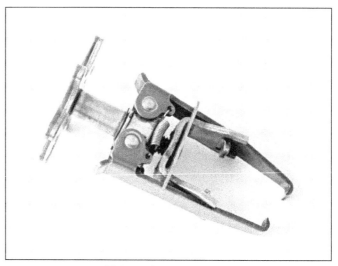

Valve spring compressor

Ridge reamer

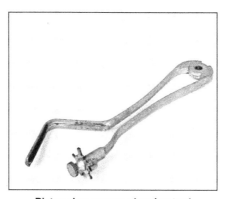

Piston ring groove cleaning tool

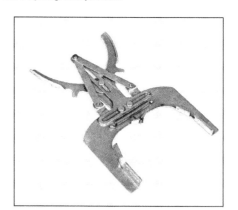

Ring removal/installation tool

end combined in one wrench). While more expensive than open end wrenches, they offer the advantages of both types of wrench.

> Combination wrench set (1/4-inch to
>   1 inch or 6 mm to 19 mm)
> Adjustable wrench, 8 inch
> Spark plug wrench with rubber insert
> Spark plug gap adjusting tool
> Feeler gauge set
> Brake bleeder wrench
> Standard screwdriver (5/16-inch x
>   6 inch)
> Phillips screwdriver (No. 2 x 6 inch)
> Combination pliers - 6 inch
> Hacksaw and assortment of blades
> Tire pressure gauge
> Grease gun
> Oil can
> Fine emery cloth
> Wire brush
> Battery post and cable cleaning tool
> Oil filter wrench
> Funnel (medium size)
> Safety goggles
> Jackstands (2)
> Drain pan

Note: *If basic tune-ups are going to be part of routine maintenance, it will be necessary to purchase a good quality stroboscopic timing*

light and combination tachometer/dwell meter. Although they are included in the list of special tools, it is mentioned here because they are absolutely necessary for tuning most vehicles properly.

Repair and overhaul tool set

These tools are essential for anyone who plans to perform major repairs and are in addition to those in the maintenance and minor repair tool kit. Included is a comprehensive set of sockets which, though expensive, are invaluable because of their versatility, especially when various extensions and drives are available. We recommend the 1/2-inch drive over the 3/8-inch drive. Although the larger drive is bulky and more expensive, it has the capacity of accepting a very wide range of large sockets. Ideally, however, the mechanic should have a 3/8-inch drive set and a 1/2-inch drive set.

> Socket set(s)
> Reversible ratchet
> Extension - 10 inch
> Universal joint
> Torque wrench (same size drive as
>   sockets)
> Ball peen hammer - 8 ounce
> Soft-face hammer (plastic/rubber)

Ring compressor

> Standard screwdriver (1/4-inch x 6 inch)
> Standard screwdriver (stubby -
>   5/16-inch)
> Phillips screwdriver (No. 3 x 8 inch)
> Phillips screwdriver (stubby - No. 2)
> Pliers - vise grip
> Pliers - lineman's
> Pliers - needle nose
> Pliers - snap-ring (internal and external)
> Cold chisel - 1/2-inch

Cylinder hone

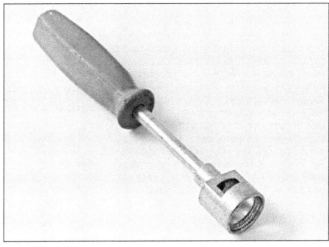

Brake hold-down spring tool

Scribe
Scraper (made from flattened copper tubing)
Centerpunch
Pin punches (1/16, 1/8, 3/16-inch)
Steel rule/straightedge - 12 inch
Allen wrench set (1/8 to 3/8-inch or 4 mm to 10 mm)
A selection of files

Torque angle gauge

Wire brush (large)
Jackstands (second set)
Jack (scissor or hydraulic type)

Note: *Another tool which is often useful is an electric drill with a chuck capacity of 3/8-inch and a set of good quality drill bits.*

Special tools

The tools in this list include those which are not used regularly, are expensive to buy, or which need to be used in accordance with their manufacturer's instructions. Unless these tools will be used frequently, it is not very economical to purchase many of them. A consideration would be to split the cost and use between yourself and a friend or friends. In addition, most of these tools can be obtained from a tool rental shop on a temporary basis.

This list primarily contains only those tools and instruments widely available to the public, and not those special tools produced by the vehicle manufacturer for distribution to dealer service departments. Occasionally, references to the manufacturer's special tools are included in the text of this manual. Generally, an alternative method of doing the job without the special tool is offered. However,

sometimes there is no alternative to their use. Where this is the case, and the tool cannot be purchased or borrowed, the work should be turned over to the dealer service department or an automotive repair shop.

Valve spring compressor
Piston ring groove cleaning tool
Piston ring compressor
Piston ring installation tool
Cylinder compression gauge
Cylinder ridge reamer
Cylinder surfacing hone
Cylinder bore gauge
Micrometers and/or dial calipers
Hydraulic lifter removal tool
Balljoint separator
Universal-type puller
Impact screwdriver
Dial indicator set
Stroboscopic timing light (inductive pick-up)
Hand operated vacuum/pressure pump
Tachometer/dwell meter
Universal electrical multimeter
Cable hoist
Brake spring removal and installation tools
Floor jack

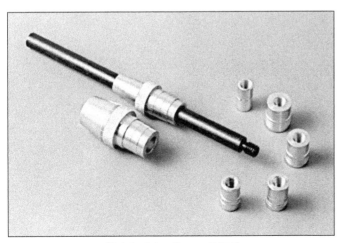

Clutch plate alignment tool

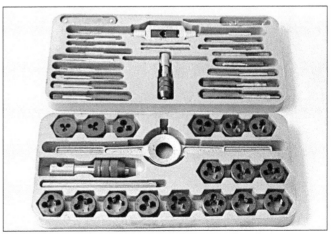

Tap and die set

Buying tools

For the do-it-yourselfer who is just starting to get involved in vehicle maintenance and repair, there are a number of options available when purchasing tools. If maintenance and minor repair is the extent of the work to be done, the purchase of individual tools is satisfactory. If, on the other hand, extensive work is planned, it would be a good idea to purchase a modest tool set from one of the large retail chain stores. A set can usually be bought at a substantial savings over the individual tool prices, and they often come with a tool box. As additional tools are needed, add-on sets, individual tools and a larger tool box can be purchased to expand the tool selection. Building a tool set gradually allows the cost of the tools to be spread over a longer period of time and gives the mechanic the freedom to choose only those tools that will actually be used.

Tool stores will often be the only source of some of the special tools that are needed, but regardless of where tools are bought, try to avoid cheap ones, especially when buying screwdrivers and sockets, because they won't last very long. The expense involved in replacing cheap tools will eventually be greater than the initial cost of quality tools.

Care and maintenance of tools

Good tools are expensive, so it makes sense to treat them with respect. Keep them clean and in usable condition and store them properly when not in use. Always wipe off any dirt, grease or metal chips before putting them away. Never leave tools lying around in the work area. Upon completion of a job, always check closely under the hood for tools that may have been left there so they won't get lost during a test drive.

Some tools, such as screwdrivers, pliers, wrenches and sockets, can be hung on a panel mounted on the garage or workshop wall, while others should be kept in a tool box or tray. Measuring instruments, gauges, meters, etc. must be carefully stored where they cannot be damaged by weather or impact from other tools.

When tools are used with care and stored properly, they will last a very long time. Even with the best of care, though, tools will wear out if used frequently. When a tool is damaged or worn out, replace it. Subsequent jobs will be safer and more enjoyable if you do.

How to repair damaged threads

Sometimes, the internal threads of a nut or bolt hole can become stripped, usually from overtightening. Stripping threads is an all-too-common occurrence, especially when working with aluminum parts, because aluminum is so soft that it easily strips out.

Usually, external or internal threads are only partially stripped. After they've been cleaned up with a tap or die, they'll still work. Sometimes, however, threads are badly damaged. When this happens, you've got three choices:

1) *Drill and tap the hole to the next suitable oversize and install a larger diameter bolt, screw or stud.*

2) *Drill and tap the hole to accept a threaded plug, then drill and tap the plug to the original screw size. You can also buy a plug already threaded to the original size. Then you simply drill a hole to the specified size, then run the threaded plug into the hole with a bolt and jam nut. Once the plug is fully seated, remove the jam nut and bolt.*

3) *The third method uses a patented thread repair kit like Heli-Coil or Slimsert. These easy-to-use kits are designed to repair damaged threads in straight-through holes and blind holes. Both are available as kits which can handle a variety of sizes and thread patterns. Drill the hole, then tap it with the special included tap. Install the Heli-Coil and the hole is back to its original diameter and thread pitch.*

Regardless of which method you use, be sure to proceed calmly and carefully. A little impatience or carelessness during one of these relatively simple procedures can ruin your whole day's work and cost you a bundle if you wreck an expensive part.

Working facilities

Not to be overlooked when discussing tools is the workshop. If anything more than routine maintenance is to be carried out, some sort of suitable work area is essential.

It is understood, and appreciated, that many home mechanics do not have a good workshop or garage available, and end up removing an engine or doing major repairs outside. It is recommended, however, that the overhaul or repair be completed under the cover of a roof.

A clean, flat workbench or table of comfortable working height is an absolute necessity. The workbench should be equipped with a vise that has a jaw opening of at least four inches.

As mentioned previously, some clean, dry storage space is also required for tools, as well as the lubricants, fluids, cleaning solvents, etc. which soon become necessary.

Sometimes waste oil and fluids, drained from the engine or cooling system during normal maintenance or repairs, present a disposal problem. To avoid pouring them on the ground or into a sewage system, pour the used fluids into large containers, seal them with caps and take them to an authorized disposal site or recycling center. Plastic jugs, such as old antifreeze containers, are ideal for this purpose.

Always keep a supply of old newspapers and clean rags available. Old towels are excellent for mopping up spills. Many mechanics use rolls of paper towels for most work because they are readily available and disposable. To help keep the area under the vehicle clean, a large cardboard box can be cut open and flattened to protect the garage or shop floor.

Whenever working over a painted surface, such as when leaning over a fender to service something under the hood, always cover it with an old blanket or bedspread to protect the finish. Vinyl covered pads, made especially for this purpose, are available at auto parts stores.

Booster battery (jump) starting

Observe these precautions when using a booster battery to start a vehicle:

a) *Before connecting the booster battery, make sure the ignition switch is in the Off position.*
b) *Turn off the lights, heater and other electrical loads.*
c) *Your eyes should be shielded. Safety goggles are a good idea.*
d) *Make sure the booster battery is the same voltage as the dead one in the vehicle.*
e) *The two vehicles MUST NOT TOUCH each other!*
f) *Make sure the transaxle is in Neutral (manual) or Park (automatic).*
g) *If the booster battery is not a maintenance-free type, remove the vent caps and lay a cloth over the vent holes.*

Connect the red jumper cable to the positive (+) terminals of each battery **(see illustration)**.

Connect one end of the black jumper cable to the negative (-) terminal of the booster battery. The other end of this cable should be connected to a good ground on the vehicle to be started, such as a bolt or bracket on the body.

Start the engine using the booster battery, then, with the engine running at idle speed, disconnect the jumper cables in the reverse order of connection.

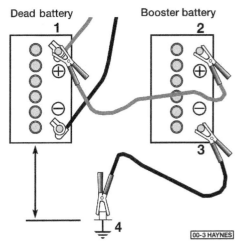

Make the booster battery cable connections in the numerical order shown (note that the negative cable of the booster battery is NOT attached to the negative terminal of the dead battery)

Jacking and towing

Jacking

The jack supplied with the vehicle should only be used for raising the vehicle when changing a tire or placing jackstands under the frame.

The vehicle should be on level ground with the wheels blocked and the transaxle in Park (automatic) or Reverse (manual). Remove the wheel covers (if equipped). **Warning:** *Wheel covers may have sharp edges - be very careful not to cut yourself.* If the wheel is being replaced, loosen the wheel nuts one-half turn and leave them in place until the wheel is raised off the ground.

Position the jack directly under the vehicle at the indicated jacking point. There is a front and rear jacking point **(see illustration)**. **Warning:** *Never work under the vehicle or start the engine while this jack is being used as the only means of support.* Loosen the lug nuts. Operate the jack with a slow, smooth motion until the wheel is raised off the ground. Remove the lug nuts, pull the tire off an replace it with the spare.

Install the wheel and lug nuts, tightening the nuts as securely as possible. Lower the vehicle, remove the jack and tighten the nuts (if loosened or removed) in a criss-cross sequence to the torque listed in the Chapter 1 Specifications Section. **Caution:** *The compact spare included with these vehicles is intended for temporary use only. Have the tire repaired and reinstall it on the vehicle the earliest opportunity.*

Towing

Do not tow the vehicle with all four wheels on the ground - transaxle damage may occur if you do. Use a towing dolly to keep the front wheels off the road. Make sure the parking brake is released and the ignition switch is in the ACC position.

Equipment specifically designed for towing should be used and should be attached to the main structural members of the vehicle, not the bumper or brackets.

Safety is a major consideration when towing and all applicable state and local laws must be obeyed. A safety chain system must be used for all towing.

The jack fits over the rocker panel flange (there are two jacking points on each side of the vehicle, indicated by notches in the rocker panel flange)

Automotive chemicals and lubricants

A number of automotive chemicals and lubricants are available for use during vehicle maintenance and repair. They include a wide variety of products ranging from cleaning solvents and degreasers to lubricants and protective sprays for rubber, plastic and vinyl.

Cleaners

Carburetor cleaner and choke cleaner is a strong solvent for gum, varnish and carbon. Most carburetor cleaners leave a dry-type lubricant film which will not harden or gum up. Because of this film it is not recommended for use on electrical components.

Brake system cleaner is used to remove grease and brake fluid from the brake system, where clean surfaces are absolutely necessary. It leaves no residue and often eliminates brake squeal caused by contaminants.

Electrical cleaner removes oxidation, corrosion and carbon deposits from electrical contacts, restoring full current flow. It can also be used to clean spark plugs, carburetor jets, voltage regulators and other parts where an oil-free surface is desired.

Demoisturants remove water and moisture from electrical components such as alternators, voltage regulators, electrical connectors and fuse blocks. They are non-conductive, non-corrosive and non-flammable.

Degreasers are heavy-duty solvents used to remove grease from the outside of the engine and from chassis components. They can be sprayed or brushed on and, depending on the type, are rinsed off either with water or solvent.

Lubricants

Motor oil is the lubricant formulated for use in engines. It normally contains a wide variety of additives to prevent corrosion and reduce foaming and wear. Motor oil comes in various weights (viscosity ratings) from 0 to 50. The recommended weight of the oil depends on the season, temperature and the demands on the engine. Light oil is used in cold climates and under light load conditions. Heavy oil is used in hot climates and where high loads are encountered. Multi-viscosity oils are designed to have characteristics of both light and heavy oils and are available in a number of weights from 5W-20 to 20W-50.

Gear oil is designed to be used in differentials, manual transmissions and other areas where high-temperature lubrication is required.

Chassis and wheel bearing grease is a heavy grease used where increased loads and friction are encountered, such as for wheel bearings, balljoints, tie-rod ends and universal joints.

High-temperature wheel bearing grease is designed to withstand the extreme temperatures encountered by wheel bearings in disc brake equipped vehicles. It usually contains molybdenum disulfide (moly), which is a dry-type lubricant.

White grease is a heavy grease for metal-to-metal applications where water is a problem. White grease stays soft under both low and high temperatures (usually from -100 to +190-degrees F), and will not wash off or dilute in the presence of water.

Assembly lube is a special extreme pressure lubricant, usually containing moly, used to lubricate high-load parts (such as main and rod bearings and cam lobes) for initial start-up of a new engine. The assembly lube lubricates the parts without being squeezed out or washed away until the engine oiling system begins to function.

Silicone lubricants are used to protect rubber, plastic, vinyl and nylon parts.

Graphite lubricants are used where oils cannot be used due to contamination problems, such as in locks. The dry graphite will lubricate metal parts while remaining uncontaminated by dirt, water, oil or acids. It is electrically conductive and will not foul electrical contacts in locks such as the ignition switch.

Moly penetrants loosen and lubricate frozen, rusted and corroded fasteners and prevent future rusting or freezing.

Heat-sink grease is a special electrically non-conductive grease that is used for mounting electronic ignition modules where it is essential that heat is transferred away from the module.

Sealants

RTV sealant is one of the most widely used gasket compounds. Made from silicone, RTV is air curing, it seals, bonds, waterproofs, fills surface irregularities, remains flexible, doesn't shrink, is relatively easy to remove, and is used as a supplementary sealer with almost all low and medium temperature gaskets.

Anaerobic sealant is much like RTV in that it can be used either to seal gaskets or to form gaskets by itself. It remains flexible, is solvent resistant and fills surface imperfections. The difference between an anaerobic sealant and an RTV-type sealant is in the curing. RTV cures when exposed to air, while an anaerobic sealant cures only in the absence of air. This means that an anaerobic sealant cures only after the assembly of parts, sealing them together.

Thread and pipe sealant is used for sealing hydraulic and pneumatic fittings and vacuum lines. It is usually made from a Teflon compound, and comes in a spray, a paint-on liquid and as a wrap-around tape.

Chemicals

Anti-seize compound prevents seizing, galling, cold welding, rust and corrosion in fasteners. High-temperature anti-seize, usually made with copper and graphite lubricants, is used for exhaust system and exhaust manifold bolts.

Anaerobic locking compounds are used to keep fasteners from vibrating or working loose and cure only after installation, in the absence of air. Medium strength locking compound is used for small nuts, bolts and screws that may be removed later. High-strength locking compound is for large nuts, bolts and studs which aren't removed on a regular basis.

Oil additives range from viscosity index improvers to chemical treatments that claim to reduce internal engine friction. It should be noted that most oil manufacturers caution against using additives with their oils.

Gas additives perform several functions, depending on their chemical makeup. They usually contain solvents that help dissolve gum and varnish that build up on carburetor, fuel injection and intake parts. They also serve to break down carbon deposits that form on the inside surfaces of the combustion chambers. Some additives contain upper cylinder lubricants for valves and piston rings, and others contain chemicals to remove condensation from the gas tank.

Miscellaneous

Brake fluid is specially formulated hydraulic fluid that can withstand the heat and pressure encountered in brake systems. Care must be taken so this fluid does not come in contact with painted surfaces or plastics. An opened container should always be resealed to prevent contamination by water or dirt.

Weatherstrip adhesive is used to bond weatherstripping around doors, windows and trunk lids. It is sometimes used to attach trim pieces.

Undercoating is a petroleum-based, tar-like substance that is designed to protect metal surfaces on the underside of the vehicle from corrosion. It also acts as a sound-deadening agent by insulating the bottom of the vehicle.

Waxes and polishes are used to help protect painted and plated surfaces from the weather. Different types of paint may require the use of different types of wax and polish. Some polishes utilize a chemical or abrasive cleaner to help remove the top layer of oxidized (dull) paint on older vehicles. In recent years many non-wax polishes that contain a wide variety of chemicals such as polymers and silicones have been introduced. These non-wax polishes are usually easier to apply and last longer than conventional waxes and polishes.

Conversion factors

Length (distance)
Inches (in)	X	25.4	= Millimeters (mm)	X	0.0394 = Inches (in)
Feet (ft)	X	0.305	= Meters (m)	X	3.281 = Feet (ft)
Miles	X	1.609	= Kilometers (km)	X	0.621 = Miles

Volume (capacity)
Cubic inches (cu in; in³)	X	16.387	= Cubic centimeters (cc; cm³)	X	0.061 = Cubic inches (cu in; in³)
Imperial pints (Imp pt)	X	0.568	= Liters (l)	X	1.76 = Imperial pints (Imp pt)
Imperial quarts (Imp qt)	X	1.137	= Liters (l)	X	0.88 = Imperial quarts (Imp qt)
Imperial quarts (Imp qt)	X	1.201	= US quarts (US qt)	X	0.833 = Imperial quarts (Imp qt)
US quarts (US qt)	X	0.946	= Liters (l)	X	1.057 = US quarts (US qt)
Imperial gallons (Imp gal)	X	4.546	= Liters (l)	X	0.22 = Imperial gallons (Imp gal)
Imperial gallons (Imp gal)	X	1.201	= US gallons (US gal)	X	0.833 = Imperial gallons (Imp gal)
US gallons (US gal)	X	3.785	= Liters (l)	X	0.264 = US gallons (US gal)

Mass (weight)
Ounces (oz)	X	28.35	= Grams (g)	X	0.035 = Ounces (oz)
Pounds (lb)	X	0.454	= Kilograms (kg)	X	2.205 = Pounds (lb)

Force
Ounces-force (ozf; oz)	X	0.278	= Newtons (N)	X	3.6 = Ounces-force (ozf; oz)
Pounds-force (lbf; lb)	X	4.448	= Newtons (N)	X	0.225 = Pounds-force (lbf; lb)
Newtons (N)	X	0.1	= Kilograms-force (kgf; kg)	X	9.81 = Newtons (N)

Pressure
Pounds-force per square inch (psi; lbf/in²; lb/in²)	X	0.070	= Kilograms-force per square centimeter (kgf/cm²; kg/cm²)	X	14.223 = Pounds-force per square inch (psi; lbf/in²; lb/in²)
Pounds-force per square inch (psi; lbf/in²; lb/in²)	X	0.068	= Atmospheres (atm)	X	14.696 = Pounds-force per square inch (psi; lbf/in²; lb/in²)
Pounds-force per square inch (psi; lbf/in²; lb/in²)	X	0.069	= Bars	X	14.5 = Pounds-force per square inch (psi; lbf/in²; lb/in²)
Pounds-force per square inch (psi; lbf/in²; lb/in²)	X	6.895	= Kilopascals (kPa)	X	0.145 = Pounds-force per square inch (psi; lbf/in²; lb/in²)
Kilopascals (kPa)	X	0.01	= Kilograms-force per square centimeter (kgf/cm²; kg/cm²)	X	98.1 = Kilopascals (kPa)

Torque (moment of force)
Pounds-force inches (lbf in; lb in)	X	1.152	= Kilograms-force centimeter (kgf cm; kg cm)	X	0.868 = Pounds-force inches (lbf in; lb in)
Pounds-force inches (lbf in; lb in)	X	0.113	= Newton meters (Nm)	X	8.85 = Pounds-force inches (lbf in; lb in)
Pounds-force inches (lbf in; lb in)	X	0.083	= Pounds-force feet (lbf ft; lb ft)	X	12 = Pounds-force inches (lbf in; lb in)
Pounds-force feet (lbf ft; lb ft)	X	0.138	= Kilograms-force meters (kgf m; kg m)	X	7.233 = Pounds-force feet (lbf ft; lb ft)
Pounds-force feet (lbf ft; lb ft)	X	1.356	= Newton meters (Nm)	X	0.738 = Pounds-force feet (lbf ft; lb ft)
Newton meters (Nm)	X	0.102	= Kilograms-force meters (kgf m; kg m)	X	9.804 = Newton meters (Nm)

Vacuum
Inches mercury (in. Hg)	X	3.377	= Kilopascals (kPa)	X	0.2961 = Inches mercury
Inches mercury (in. Hg)	X	25.4	= Millimeters mercury (mm Hg)	X	0.0394 = Inches mercury

Power
Horsepower (hp)	X	745.7	= Watts (W)	X	0.0013 = Horsepower (hp)

Velocity (speed)
Miles per hour (miles/hr; mph)	X	1.609	= Kilometers per hour (km/hr; kph)	X	0.621 = Miles per hour (miles/hr; mph)

Fuel consumption*
Miles per gallon, Imperial (mpg)	X	0.354	= Kilometers per liter (km/l)	X	2.825 = Miles per gallon, Imperial (mpg)
Miles per gallon, US (mpg)	X	0.425	= Kilometers per liter (km/l)	X	2.352 = Miles per gallon, US (mpg)

Temperature
Degrees Fahrenheit = (°C x 1.8) + 32

Degrees Celsius (Degrees Centigrade; °C) = (°F - 32) x 0.56

*It is common practice to convert from miles per gallon (mpg) to liters/100 kilometers (l/100km), where mpg (Imperial) x l/100 km = 282 and mpg (US) x l/100 km = 235

DECIMALS to MILLIMETERS

Decimal	mm	Decimal	mm
0.001	0.0254	0.500	12.7000
0.002	0.0508	0.510	12.9540
0.003	0.0762	0.520	13.2080
0.004	0.1016	0.530	13.4620
0.005	0.1270	0.540	13.7160
0.006	0.1524	0.550	13.9700
0.007	0.1778	0.560	14.2240
0.008	0.2032	0.570	14.4780
0.009	0.2286	0.580	14.7320
		0.590	14.9860
0.010	0.2540		
0.020	0.5080		
0.030	0.7620		
0.040	1.0160	0.600	15.2400
0.050	1.2700	0.610	15.4940
0.060	1.5240	0.620	15.7480
0.070	1.7780	0.630	16.0020
0.080	2.0320	0.640	16.2560
0.090	2.2860	0.650	16.5100
		0.660	16.7640
0.100	2.5400	0.670	17.0180
0.110	2.7940	0.680	17.2720
0.120	3.0480	0.690	17.5260
0.130	3.3020		
0.140	3.5560		
0.150	3.8100		
0.160	4.0640	0.700	17.7800
0.170	4.3180	0.710	18.0340
0.180	4.5720	0.720	18.2880
0.190	4.8260	0.730	18.5420
		0.740	18.7960
0.200	5.0800	0.750	19.0500
0.210	5.3340	0.760	19.3040
0.220	5.5880	0.770	19.5580
0.230	5.8420	0.780	19.8120
0.240	6.0960	0.790	20.0660
0.250	6.3500		
0.260	6.6040		
0.270	6.8580	0.800	20.3200
0.280	7.1120	0.810	20.5740
0.290	7.3660	0.820	21.8280
		0.830	21.0820
0.300	7.6200	0.840	21.3360
0.310	7.8740	0.850	21.5900
0.320	8.1280	0.860	21.8440
0.330	8.3820	0.870	22.0980
0.340	8.6360	0.880	22.3520
0.350	8.8900	0.890	22.6060
0.360	9.1440		
0.370	9.3980		
0.380	9.6520		
0.390	9.9060		
		0.900	22.8600
0.400	10.1600	0.910	23.1140
0.410	10.4140	0.920	23.3680
0.420	10.6680	0.930	23.6220
0.430	10.9220	0.940	23.8760
0.440	11.1760	0.950	24.1300
0.450	11.4300	0.960	24.3840
0.460	11.6840	0.970	24.6380
0.470	11.9380	0.980	24.8920
0.480	12.1920	0.990	25.1460
0.490	12.4460	1.000	25.4000

FRACTIONS to DECIMALS to MILLIMETERS

Fraction	Decimal	mm	Fraction	Decimal	mm
1/64	0.0156	0.3969	33/64	0.5156	13.0969
1/32	0.0312	0.7938	17/32	0.5312	13.4938
3/64	0.0469	1.1906	35/64	0.5469	13.8906
1/16	0.0625	1.5875	9/16	0.5625	14.2875
5/64	0.0781	1.9844	37/64	0.5781	14.6844
3/32	0.0938	2.3812	19/32	0.5938	15.0812
7/64	0.1094	2.7781	39/64	0.6094	15.4781
1/8	0.1250	3.1750	5/8	0.6250	15.8750
9/64	0.1406	3.5719	41/64	0.6406	16.2719
5/32	0.1562	3.9688	21/32	0.6562	16.6688
11/64	0.1719	4.3656	43/64	0.6719	17.0656
3/16	0.1875	4.7625	11/16	0.6875	17.4625
13/64	0.2031	5.1594	45/64	0.7031	17.8594
7/32	0.2188	5.5562	23/32	0.7188	18.2562
15/64	0.2344	5.9531	47/64	0.7344	18.6531
1/4	0.2500	6.3500	3/4	0.7500	19.0500
17/64	0.2656	6.7469	49/64	0.7656	19.4469
9/32	0.2812	7.1438	25/32	0.7812	19.8438
19/64	0.2969	7.5406	51/64	0.7969	20.2406
5/16	0.3125	7.9375	13/16	0.8125	20.6375
21/64	0.3281	8.3344	53/64	0.8281	21.0344
11/32	0.3438	8.7312	27/32	0.8438	21.4312
23/64	0.3594	9.1281	55/64	0.8594	21.8281
3/8	0.3750	9.5250	7/8	0.8750	22.2250
25/64	0.3906	9.9219	57/64	0.8906	22.6219
13/32	0.4062	10.3188	29/32	0.9062	23.0188
27/64	0.4219	10.7156	59/64	0.9219	23.4156
7/16	0.4375	11.1125	15/16	0.9375	23.8125
29/64	0.4531	11.5094	61/64	0.9531	24.2094
15/32	0.4688	11.9062	31/32	0.9688	24.6062
31/64	0.4844	12.3031	63/64	0.9844	25.0031
1/2	0.5000	12.7000	1	1.0000	25.4000

Safety first!

Regardless of how enthusiastic you may be about getting on with the job at hand, take the time to ensure that your safety is not jeopardized. A moment's lack of attention can result in an accident, as can failure to observe certain simple safety precautions. The possibility of an accident will always exist, and the following points should not be considered a comprehensive list of all dangers. Rather, they are intended to make you aware of the risks and to encourage a safety conscious approach to all work you carry out on your vehicle.

Essential DOs and DON'Ts

DON'T rely on a jack when working under the vehicle. Always use approved jackstands to support the weight of the vehicle and place them under the recommended lift or support points.

DON'T attempt to loosen extremely tight fasteners (i.e. wheel lug nuts) while the vehicle is on a jack - it may fall.

DON'T start the engine without first making sure that the transmission is in Neutral (or Park where applicable) and the parking brake is set.

DON'T remove the radiator cap from a hot cooling system - let it cool or cover it with a cloth and release the pressure gradually.

DON'T attempt to drain the engine oil until you are sure it has cooled to the point that it will not burn you.

DON'T touch any part of the engine or exhaust system until it has cooled sufficiently to avoid burns.

DON'T siphon toxic liquids such as gasoline, antifreeze and brake fluid by mouth, or allow them to remain on your skin.

DON'T inhale brake lining dust - it is potentially hazardous (see *Asbestos* below).

DON'T allow spilled oil or grease to remain on the floor - wipe it up before someone slips on it.

DON'T use loose fitting wrenches or other tools which may slip and cause injury.

DON'T push on wrenches when loosening or tightening nuts or bolts. Always try to pull the wrench toward you. If the situation calls for pushing the wrench away, push with an open hand to avoid scraped knuckles if the wrench should slip.

DON'T attempt to lift a heavy component alone - get someone to help you.

DON'T rush or take unsafe shortcuts to finish a job.

DON'T allow children or animals in or around the vehicle while you are working on it.

DO wear eye protection when using power tools such as a drill, sander, bench grinder, etc. and when working under a vehicle.

DO keep loose clothing and long hair well out of the way of moving parts.

DO make sure that any hoist used has a safe working load rating adequate for the job.

DO get someone to check on you periodically when working alone on a vehicle.

DO carry out work in a logical sequence and make sure that everything is correctly assembled and tightened.

DO keep chemicals and fluids tightly capped and out of the reach of children and pets.

DO remember that your vehicle's safety affects that of yourself and others. If in doubt on any point, get professional advice.

Asbestos

Certain friction, insulating, sealing, and other products - such as brake linings, brake bands, clutch linings, torque converters, gaskets, etc. - may contain asbestos. Extreme care must be taken to avoid inhalation of dust from such products, since it is hazardous to health. If in doubt, assume that they do contain asbestos.

Fire

Remember at all times that gasoline is highly flammable. Never smoke or have any kind of open flame around when working on a vehicle. But the risk does not end there. A spark caused by an electrical short circuit, by two metal surfaces contacting each other, or even by static electricity built up in your body under certain conditions, can ignite gasoline vapors, which in a confined space are highly explosive. Do not, under any circumstances, use gasoline for cleaning parts. Use an approved safety solvent.

Always disconnect the battery ground (-) cable at the battery before working on any part of the fuel system or electrical system. Never risk spilling fuel on a hot engine or exhaust component. It is strongly recommended that a fire extinguisher suitable for use on fuel and electrical fires be kept handy in the garage or workshop at all times. Never try to extinguish a fuel or electrical fire with water.

Fumes

Certain fumes are highly toxic and can quickly cause unconsciousness and even death if inhaled to any extent. Gasoline vapor falls into this category, as do the vapors from some cleaning solvents. Any draining or pouring of such volatile fluids should be done in a well ventilated area.

When using cleaning fluids and solvents, read the instructions on the container carefully. Never use materials from unmarked containers.

Never run the engine in an enclosed space, such as a garage. Exhaust fumes contain carbon monoxide, which is extremely poisonous. If you need to run the engine, always do so in the open air, or at least have the rear of the vehicle outside the work area.

If you are fortunate enough to have the use of an inspection pit, never drain or pour gasoline and never run the engine while the vehicle is over the pit. The fumes, being heavier than air, will concentrate in the pit with possibly lethal results.

The battery

Never create a spark or allow a bare light bulb near a battery. They normally give off a certain amount of hydrogen gas, which is highly explosive.

Always disconnect the battery ground (-) cable at the battery before working on the fuel or electrical systems.

If possible, loosen the filler caps or cover when charging the battery from an external source (this does not apply to sealed or maintenance-free batteries). Do not charge at an excessive rate or the battery may burst.

Take care when adding water to a non maintenance-free battery and when carrying a battery. The electrolyte, even when diluted, is very corrosive and should not be allowed to contact clothing or skin.

Always wear eye protection when cleaning the battery to prevent the caustic deposits from entering your eyes.

Household current

When using an electric power tool, inspection light, etc., which operates on household current, always make sure that the tool is correctly connected to its plug and that, where necessary, it is properly grounded. Do not use such items in damp conditions and, again, do not create a spark or apply excessive heat in the vicinity of fuel or fuel vapor.

Secondary ignition system voltage

A severe electric shock can result from touching certain parts of the ignition system (such as the spark plug wires) when the engine is running or being cranked, particularly if components are damp or the insulation is defective. In the case of an electronic ignition system, the secondary system voltage is much higher and could prove fatal.

Troubleshooting

Contents

This section provides an easy reference guide to the more common problems that may occur during the operation of your vehicle. These problems and their possible causes are grouped under headings denoting various components or systems, such as Engine, Cooling system, etc. They also refer you to the chapter and/or section that deals with the problem.

Remember that successful troubleshooting is not a mysterious black art practiced only by professional mechanics. It is simply the result of the right knowledge combined with an intelligent, systematic approach to the problem. Always work by a process of elimination, starting with the simplest solution and working through to the most complex - and never overlook the obvious. Anyone can run the gas tank dry or leave the lights on overnight, so don't assume that you are exempt from such oversights.

Finally, always establish a clear idea of why a problem has occurred and take steps to ensure that it doesn't happen again. If the electrical system fails because of a poor connection, check the other connections in the system to make sure that they don't fail as well. If a particular fuse continues to blow, find out why - don't just replace one fuse after another. Remember, failure of a small component can often be indicative of potential failure or incorrect functioning of a more important component or system.

Engine

1 Engine will not rotate when attempting to start

1 Battery terminal connections loose or corroded (Chapter 1).
2 Battery discharged or faulty (Chapter 1).
3 Automatic transmission not completely engaged in Park (Chapter 7B) or clutch not completely depressed (Chapter 8).
4 Broken, loose or disconnected wiring in the starting circuit (Chapters 5 and 12).
5 Starter motor pinion jammed in flywheel ring gear (Chapter 5).
6 Starter solenoid faulty (Chapter 5).
7 Starter motor faulty (Chapter 5).
8 Ignition switch faulty (Chapter 12).
9 Starter pinion or flywheel teeth worn or broken (Chapter 5).

2 Engine rotates but will not start

1 Fuel tank empty.
2 Battery discharged (engine rotates slowly) (Chapter 5).
3 Battery terminal connections loose or corroded (Chapter 1).
4 Leaking fuel injector(s), faulty fuel pump, pressure regulator, etc. (Chapter 4).
5 Fuel not reaching carburetor, fuel injec-

tion throttle body or fuel rail (Chapter 4).
6 Ignition components damp or damaged (Chapter 5).
7 Worn, faulty or incorrectly gapped spark plugs (Chapter 1).
8 Broken, loose or disconnected wiring in the starting circuit (Chapter 5).
9 Loose distributor is changing ignition timing (Chapter 5).
10 Broken, loose or disconnected wires at the ignition coil or faulty coil (Chapter 5).
11 Broken or stripped timing belt (Chapter 2).
12 Defective fuel pump relay and/or harness at relay (Chapter 4)

3 Engine hard to start when cold

1 Battery discharged or low (Chapter 1).
2 Malfunctioning fuel system (Chapter 4).
3 Carburetor or fuel injector(s) leaking (Chapter 4).
4 Distributor rotor carbon tracked (Chapter 5).

4 Engine hard to start when hot

1 Air filter clogged (Chapter 1).
2 Fuel not reaching the fuel injection system (Chapter 4).
3 Corroded battery connections, especially ground (Chapter 1).
4 Malfunctioning EVAP system (Chapter 6)

5 Starter motor noisy or excessively rough in engagement

1 Pinion or flywheel gear teeth worn or broken (Chapter 5).
2 Starter motor mounting bolts loose or missing (Chapter 5).

6 Engine starts but stops immediately

1 Loose or faulty electrical connections at distributor, coil or alternator (Chapter 5).
2 Insufficient fuel reaching the carburetor or fuel injector(s) (Chapters 1 and 4).
3 Vacuum leak at the gasket between the intake manifold and carburetor or fuel injection throttle body (Chapters 1 and 4).

7 Oil puddle under engine

1 Oil pan gasket and/or oil pan drain bolt washer leaking (Chapter 2).
2 Oil pressure sending unit leaking (Chapter 2).
3 Valve covers leaking (Chapter 2).
4 Engine oil seals leaking (Chapter 2).

8 Engine lopes while idling or idles erratically

1 Vacuum leakage (Chapters 2 and 4).
2 Defective EGR valve (Chapter 6).
3 Air filter clogged (Chapter 1).
4 Fuel pump not delivering sufficient fuel to the carburetor or fuel injection system (Chapter 4).
5 Leaking head gasket (Chapter 2).
6 Timing belt and/or pulleys worn (Chapter 2).
7 Camshaft lobes worn (Chapter 2).

9 Engine misses at idle speed

1 Spark plugs worn or not gapped properly (Chapter 1).
2 Faulty spark plug wires (Chapter 1).
3 Vacuum leaks (Chapter 1).
4 Incorrect ignition timing (Chapter 1).
5 Uneven or low compression (Chapter 2).

10 Engine misses throughout the driving speed range

1 Fuel filter clogged and/or impurities in the fuel system (Chapter 1).
2 Low fuel pressure (Chapter 4).
3 Faulty or incorrectly gapped spark plugs (Chapter 1).
4 Incorrect ignition timing (Chapter 5).
5 Cracked distributor cap, disconnected distributor wires or damaged distributor components (Chapters 1 and 5).
6 Spark plug wires damaged and shorting out (Chapters 1 or 5).
7 Faulty emission system components (Chapter 6).
8 Low or uneven cylinder compression pressures (Chapter 2).
9 Weak or faulty ignition system (Chapter 5).
10 Vacuum leak in carburetor or fuel injection system (Chapter 4), intake manifold (Chapters 2A/2B), fuel injection air control valve (Chapter 6) or vacuum hoses.

11 Engine stumbles on acceleration

1 Spark plugs fouled (Chapter 1).
2 Carburetor or fuel injection system faulty (Chapter 4).
3 Fuel filter clogged (Chapters 1 and 4).
4 Incorrect ignition timing (Chapter 5).
5 Intake air leak (Chapters 2 and 4).

12 Engine surges while holding accelerator steady

1 Intake air leak (Chapter 4).
2 Fuel pump faulty (Chapter 4).

3 Loose fuel injector wire harness connectors (Chapter 4).
4 Defective ECU or information sensor (Chapter 6).

13 Engine stalls

1 Idle speed incorrect (Chapter 1).
2 Fuel filter clogged and/or water and impurities in the fuel system (Chapters 1 and 4).
3 Distributor components damp or damaged (Chapter 5).
4 Faulty emissions system components (Chapter 6).
5 Faulty or incorrectly gapped spark plugs (Chapter 1).
6 Faulty spark plug wires (Chapter 1).
7 Vacuum leak in the intake manifold or vacuum hoses (Chapters 2 and 4).
8 Valve clearances incorrectly set (Chapter 1).

14 Engine lacks power

1 Incorrect ignition timing (Chapter 5).
2 Excessive play in distributor shaft (Chapter 5).
3 Worn rotor, distributor cap or wires (Chapters 1 and 5).
4 Faulty or incorrectly gapped spark plugs (Chapter 1).
5 Carburetor or fuel injection system malfunction (Chapter 4).
6 Faulty coil (Chapter 5).
7 Brakes binding (Chapter 9).
8 Automatic transaxle fluid level incorrect (Chapter 1).
9 Clutch slipping (Chapter 8).
10 Fuel filter clogged and/or impurities in the fuel system (Chapters 1 and 4).
11 Emission control system not functioning properly (Chapter 6).
12 Low or uneven cylinder compression pressures (Chapter 2).
13 Obstructed exhaust system (Chapter 4).

15 Engine backfires

1 Emission control system not functioning properly (Chapter 6).
2 Ignition timing incorrect (Chapter 5).
3 Faulty secondary ignition system (cracked spark plug insulator, faulty plug wires, distributor cap and/or rotor) (Chapters 1 and 5).
4 Fuel injection system malfunctioning (Chapter 4).
5 Vacuum leak at carburetor or fuel injector(s), intake manifold, air control valve or vacuum hoses (Chapters 2 and 4).
6 Valve clearances incorrectly set and/or valves sticking (Chapter 1).

16 Pinging or knocking engine sounds during acceleration or uphill

1 Incorrect grade of fuel.
2 Ignition timing incorrect (Chapter 5).
3 Carburetor or fuel injection system faulty (Chapter 4).
4 Improper or damaged spark plugs or wires (Chapter 1).
5 Worn or damaged distributor components (Chapter 5).
6 EGR valve not functioning (Chapter 6).
7 Vacuum leak (Chapters 2 and 4).
8 Knock sensor malfunctioning (Chapter 6).

17 Engine runs with oil pressure light on

1 Low oil level (Chapter 1).
2 Short in wiring circuit (Chapter 12).
3 Faulty oil pressure sender (Chapter 2).
4 Worn engine bearings and/or oil pump (Chapter 2).

18 Engine diesels (continues to run) after switching off

1 Leaking fuel injector(s).
2 Excessive engine operating temperature (Section 24).
3 Excessive carbon build-up on piston crowns (Chapter 2).

Engine electrical system

19 Battery will not hold a charge

1 Alternator drivebelt defective or not adjusted properly (Chapter 1).
2 Battery electrolyte level low (Chapter 1).
3 Battery terminals loose or corroded (Chapter 1).
4 Alternator not charging properly (Chapter 5).
5 Loose, broken or faulty wiring in the charging circuit (Chapter 5).
6 Short in vehicle wiring (Chapter 12).
7 Internally defective battery (Chapters 1 and 5).

20 Alternator light fails to go out

1 Faulty alternator or charging circuit (Chapter 5).
2 Alternator drivebelt defective or out of adjustment (Chapter 1).
3 Alternator voltage regulator inoperative (Chapter 5).

21 Alternator light fails to come on when key is turned on

1 Warning light bulb defective (Chapter 5).
2 Fault in the printed circuit, dash wiring or bulb holder (Chapter 12).

Fuel system

22 Excessive fuel consumption

1 Dirty or clogged air filter element (Chapter 1).
2 Incorrectly set ignition timing (Chapters 1 and 5).
3 Emissions system not functioning properly (Chapter 6).
4 Carburetor or fuel injection system malfunctioning (Chapter 4).
5 Low tire pressure or incorrect tire size (Chapter 1).

23 Fuel leakage and/or fuel odor

1 Leaking fuel feed or return line (Chapters 1 and 4).
2 Tank overfilled.
3 Evaporative canister filter clogged (Chapters 1 and 6).
4 Carburetor or fuel injectors faulty (Chapter 4).

Cooling system

24 Overheating

1 Insufficient coolant in system (Chapter 1).
2 Radiator core blocked or grille restricted (Chapter 3).
3 Thermostat faulty (Chapter 3).
4 Electric cooling fan circuit problem (Chapter 3).
5 Radiator cap not maintaining proper pressure (Chapter 3).
6 Ignition timing incorrect (Chapters 1 and 5).

25 Overcooling

1 Faulty thermostat (Chapter 3).
2 Inaccurate temperature gauge sending unit (Chapter 3).
3 Electric cooling fan circuit problem (Chapter 3).

26 External coolant leakage

1 Deteriorated/damaged hoses; loose clamps (Chapters 1 and 3).

2 Water pump defective (Chapter 3).
3 Leakage from radiator core or coolant reservoir bottle (Chapter 3).
4 Engine drain or water jacket core plugs leaking (Chapter 2).

27 Internal coolant leakage

1 Leaking cylinder head gasket (Chapter 2).
2 Cracked cylinder bore or cylinder head (Chapter 2).

28 Coolant loss

1 Too much coolant in system (Chapter 1).
2 Coolant boiling away because of overheating (Chapter 3).
3 Internal or external leakage (Chapter 3).
4 Faulty radiator cap (Chapter 3).

29 Poor coolant circulation

1 Inoperative water pump (Chapter 3).
2 Restriction in cooling system (Chapters 1 and 3).
3 Thermostat sticking (Chapter 3).

Clutch

30 Pedal travels to floor - no pressure or very little resistance

1 Damaged or misadjusted clutch cable.
2 Faulty clutch master cylinder, release cylinder or hydraulic line (Chapter 8).
3 Broken release bearing or fork (Chapter 8).

31 Unable to select gears

1 Faulty transaxle (Chapter 7).
2 Faulty clutch disc (Chapter 8).
3 Release lever and bearing not assembled properly (Chapter 8).
4 Faulty pressure plate (Chapter 8).
5 Pressure plate-to-flywheel bolts loose (Chapter 8).

32 Clutch slips (engine speed increases with no increase in vehicle speed)

1 Clutch plate worn (Chapter 8).
2 Clutch plate is oil soaked by leaking rear main seal (Chapter 8).
3 Clutch plate not seated. It may take 30 or 40 normal starts for a new one to seat.

4 Warped pressure plate or flywheel (Chapter 8).
5 Weak diaphragm spring (Chapter 8).
6 Clutch plate overheated. Allow to cool.

33 Grabbing (chattering) as clutch is engaged

1 Oil on clutch plate lining, burned or glazed facings (Chapter 8).
2 Worn or loose engine or transaxle mounts (Chapters 2 and 7).
3 Worn splines on clutch plate hub (Chapter 8).
4 Warped pressure plate or flywheel (Chapter 8).
5 Burned or smeared resin on flywheel or pressure plate (Chapter 8).

34 Transaxle rattling (clicking)

1 Release lever loose (Chapter 8).
2 Low engine idle speed (Chapters 1 and 5).

35 Noise in clutch area

1 Fork shaft improperly installed (Chapter 8).
2 Faulty bearing (Chapter 8).

36 Clutch pedal stays on floor

1 Faulty clutch master or release cylinder (Chapter 8).
2 Broken release bearing or fork (Chapter 8).

37 High pedal effort

1 Piston binding in bore of clutch master or release cylinder (Chapter 8).
2 Pressure plate faulty (Chapter 8).

Manual transaxle

38 Knocking noise at low speeds

1 Worn driveaxle constant velocity (CV) joints (Chapter 8).
2 Worn driveaxle bore in differential case (Chapter 7A).*

39 Noise most pronounced when turning

Differential gear noise (Chapter 7A).*

40 Clunk on acceleration or deceleration

1 Loose engine or transaxle mounts (Chapters 2 and 7A).
2 Worn differential pinion shaft in case.*
3 Worn driveaxle bore in differential case (Chapter 7A).*
4 Worn or damaged driveaxle inboard CV joints (Chapter 8).

41 Clicking noise in turns

Worn or damaged outboard CV joint (Chapter 8).

42 Vibration

1 Rough wheel bearing (Chapters 1 and 10).
2 Damaged driveaxle (Chapter 8).
3 Out-of-round tires (Chapter 1).
4 Tire out of balance (Chapters 1 and 10).
5 Worn CV joint (Chapter 8).

43 Noisy in neutral with engine running

1 Damaged input gear bearing (Chapter 7A).*
2 Damaged clutch release bearing (Chapter 8).

44 Noisy in one particular gear

1 Damaged or worn constant mesh gears (Chapter 7A).*
2 Damaged or worn synchronizers (Chapter 7A).*
3 Bent reverse fork (Chapter 7A).*
4 Damaged fourth speed gear or output gear (Chapter 7A).*
5 Worn or damaged reverse idler gear or idler bushing (Chapter 7A).*

45 Noisy in all gears

1 Insufficient lubricant (Chapter 7A).
2 Damaged or worn bearings (Chapter 7A).*
3 Worn or damaged input gear shaft and/or output gear shaft (Chapter 7A).*

46 Slips out of gear

1 Worn or improperly adjusted linkage (Chapter 7A).
2 Transaxle loose on engine (Chapter 7A).
3 Shift linkage does not work freely, binds (Chapter 7A).

4　Input gear bearing retainer broken or loose (Chapter 7A).*
5　Dirt between clutch cover and engine block (Chapter 7A).
6　Worn shift fork (Chapter 7A).*

47　Leaks lubricant

1　Driveaxle oil seals worn (Chapter 7).
2　Excessive amount of lubricant in transaxle (Chapters 1 and 7A).
3　Loose or broken input gear shaft bearing retainer (Chapter 7A).*
4　Input gear bearing retainer O-ring and/or lip seal damaged (Chapter 7A).*

48　Locked in gear

Lock pin or interlock pin missing (Chapter 7A).*

Although the corrective action necessary to remedy the symptoms described is beyond the scope of the home mechanic, the above information should be helpful in isolating the cause of the condition so that the owner can communicate clearly with a professional mechanic.

Automatic transaxle

Note: *Due to the complexity of the automatic transaxle, it is difficult for the home mechanic to properly diagnose and service this component. For problems other than the following, the vehicle should be taken to a dealer or transmission shop.*

49　Fluid leakage

1　Automatic transmission fluid is a deep red color. Fluid leaks should not be confused with engine oil, which can easily be blown onto the transaxle by air flow.
2　To pinpoint a leak, first remove all built-up dirt and grime from the transaxle housing with degreasing agents and/or steam cleaning. Then drive the vehicle at low speeds so air flow will not blow the leak far from its source. Raise the vehicle and determine where the leak is coming from. Common areas of leakage are:
a) *Pan (Chapters 1 and 7)*
b) *Dipstick tube (Chapters 1 and 7)*
c) *Transaxle oil lines (Chapter 7)*
d) *Speed sensor (Chapter 7)*

50　Transaxle fluid brown or has a burned smell

Transaxle fluid overheated (Chapter 1).

51　General shift mechanism problems

1　Chapter 7, Part B, deals with checking and adjusting the shift linkage on automatic transaxles. Common problems which may be attributed to poorly adjusted linkage are:
a) *Engine starting in gears other than Park or Neutral.*
b) *Indicator on shifter pointing to a gear other than the one actually being used.*
c) *Vehicle moves when in Park.*
2　Refer to Chapter 7B for the shift linkage adjustment procedure.

52　Transaxle will not downshift with accelerator pedal pressed to the floor

TV cable out of adjustment (Chapter 7B).

53　Engine will start in gears other than Park or Neutral

Neutral start switch malfunctioning (Chapter 7B).

54　Transaxle slips, shifts roughly, is noisy or has no drive in forward or reverse gears

There are many probable causes for the above problems, but the home mechanic should be concerned with only fluid level and fluid and filter condition. Before taking the vehicle to a repair shop, check the level and condition of the fluid as described in Chapter 1. Correct the fluid level as necessary or change the fluid and filter if needed. If the problem persists, have a professional diagnose the cause.

Driveaxles

55　Clicking noise in turns

Worn or damaged outboard CV joint (Chapter 8).

56　Shudder or vibration during acceleration

1　Excessive toe-in (Chapter 10).
2　Incorrect spring heights (Chapter 10).
3　Worn or damaged inboard or outboard CV joints (Chapter 8).
4　Sticking inboard CV joint assembly (Chapter 8).

57　Vibration at highway speeds

1　Out of balance front wheels and/or tires (Chapters 1 and 10).
2　Out of round front tires (Chapters 1 and 10).
3　Worn CV joint(s) (Chapter 8).

Brakes

Note: *Before assuming that a brake problem exists, make sure that:*
a) *The tires are in good condition and properly inflated (Chapter 1).*
b) *The front end alignment is correct (Chapter 10).*
c) *The vehicle is not loaded with weight in an unequal manner.*

58　Vehicle pulls to one side during braking

1　Incorrect tire pressures (Chapter 1).
2　Front end out of line (have the front end aligned).
3　Front, or rear, tires not matched to one another.
4　Restricted brake lines or hoses (Chapter 9).
5　Malfunctioning drum brake or caliper assembly (Chapter 9).
6　Loose suspension parts (Chapter 10).
7　Loose calipers (Chapter 9).
8　Excessive wear of brake shoe or pad material or disc/drum on one side.

59　Noise (high-pitched squeal when the brakes are applied)

Disc brake pads worn out (Chapter 9).

60　Brake roughness or chatter (pedal pulsates)

1　Excessive lateral runout (Chapter 9).
2　Uneven pad wear (Chapter 9).
3　Defective disc (Chapter 9).

61　Excessive brake pedal effort required to stop vehicle

1　Malfunctioning power brake booster (Chapter 9).
2　Partial system failure (Chapter 9).
3　Excessively worn pads or shoes (Chapter 9).
4　Piston in caliper or wheel cylinder stuck or sluggish (Chapter 9).
5　Brake pads or shoes contaminated with oil or grease (Chapter 9).
6　New pads or shoes installed and not yet

seated. It will take a while for the new material to seat against the disc or drum.

62 Excessive brake pedal travel

1 Partial brake system failure (Chapter 9).
2 Insufficient fluid in master cylinder (Chapters 1 and 9).
3 Air trapped in system (Chapters 1 and 9).

63 Dragging brakes

1 Incorrect adjustment of brake light switch (Chapter 9).
2 Master cylinder pistons not returning correctly (Chapter 9).
3 Restricted brakes lines or hoses (Chapters 1 and 9).
4 Incorrect parking brake adjustment (Chapter 9).

64 Grabbing or uneven braking action

1 Brake pads or shoes worn out (Chapter 9).
2 Malfunction of proportioning valve (Chapter 9).
3 Binding brake pedal mechanism (Chapter 9).

65 Brake pedal feels spongy when depressed

1 Air in hydraulic lines (Chapter 9).
2 Master cylinder mounting bolts loose (Chapter 9).
3 Master cylinder defective (Chapter 9).

66 Brake pedal travels to the floor with little resistance

1 Little or no fluid in the master cylinder reservoir caused by a leak in the system (Chapter 9).
2 Loose, damaged or disconnected brake lines (Chapter 9).
3 Defective master cylinder (Chapter 9).

67 Parking brake does not hold

Parking brake linkage improperly adjusted (Chapters 1 and 9).

Suspension and steering systems

Note: *Before attempting to diagnose the suspension and steering systems, perform the following preliminary checks:*

a) *Tires for wrong pressure and uneven wear.*
b) *Steering universal joints from the column to the steering gear for loose connectors or wear.*
c) *Front and rear suspension and the steering gear assembly for loose or damaged parts.*
d) *Out-of-round or out-of-balance tires, bent rims and loose and/or rough wheel bearings.*

68 Vehicle pulls to one side during braking

1 Mismatched or uneven tires (Chapter 10).
2 Broken or sagging springs (Chapter 10).
3 Wheel alignment (Chapter 10).
4 Front brake dragging (Chapter 9).

69 Abnormal or excessive tire wear

1 Wheel alignment (Chapter 10).
2 Sagging or broken springs (Chapter 10).
3 Tire out of balance (Chapter 10).
4 Worn strut damper (Chapter 10).
5 Overloaded vehicle.
6 Tires not rotated regularly.

70 Wheel makes a thumping noise

1 Blister or bump on tire (Chapter 10).
2 Worn strut damper (Chapter 10).

71 Shimmy, shake or vibration

1 Tire or wheel out-of-balance or out-of-round (Chapter 10).
2 Loose or worn front hub or wheel bearings (Chapters 1, 8 and 10).
3 Worn tie-rod ends (Chapter 10).
4 Worn lower balljoints (Chapters 1 and 10).
5 Excessive wheel runout (Chapter 10).
6 Blister or bump on tire (Chapter 10).

72 Hard steering

1 Lack of lubrication at balljoints and tie-rod ends (Chapters 1 and 10).
2 Front wheel alignment (Chapter 10).
3 Low tire pressure(s) (Chapters 1 and 10).

73 Poor returnability of steering to center

1 Lack of lubrication at balljoints and tie-rod ends (Chapters 1 and 10).

2 Binding in balljoints (Chapter 10).
3 Binding in steering column (Chapter 10).
4 Lack of lubricant in steering gear assembly (Chapter 10).
5 Front wheel alignment (Chapter 10).

74 Abnormal noise at the front end

1 Lack of lubrication at balljoints and tie-rod ends (Chapters 1 and 10).
2 Damaged shock absorber mount (Chapter 10).
3 Worn control arm bushings or tie-rod ends (Chapter 10).
4 Loose stabilizer bar (Chapter 10).
5 Loose wheel nuts (Chapters 1 and 10).
6 Loose suspension bolts (Chapter 10)

75 Wander or poor steering stability

1 Mismatched or uneven tires (Chapter 10).
2 Lack of lubrication at balljoints and tie-rod ends (Chapters 1 and 10).
3 Worn strut assemblies (Chapter 10).
4 Loose stabilizer bar (Chapter 10).
5 Broken or sagging springs (Chapter 10).
6 Wheels out of alignment (Chapter 10).

76 Erratic steering when braking

1 Front hub bearings worn (Chapter 10).
2 Broken or sagging springs (Chapter 10).
3 Leaking wheel cylinder or caliper (Chapter 10).
4 Warped discs or drums (Chapter 10).

77 Excessive pitching and/or rolling around corners or during braking

1 Loose stabilizer bar (Chapter 10).
2 Worn strut assemblies or mountings (Chapter 10).
3 Broken or sagging springs (Chapter 10).
4 Overloaded vehicle.

78 Suspension bottoms

1 Overloaded vehicle.
2 Worn strut assemblies (Chapter 10).
3 Incorrect, broken or sagging springs (Chapter 10).

79 Cupped tires

1 Front wheel or rear wheel alignment (Chapter 10).
2 Worn strut assemblies (Chapter 10).

3   Wheel bearings worn (Chapter 10).
4   Excessive tire or wheel runout (Chapter 10).
5   Worn balljoints (Chapter 10).

80   Excessive tire wear on outside edge

1   Inflation pressures incorrect (Chapter 1).
2   Excessive speed in turns.
3   Front end alignment incorrect (excessive toe-in). Have professionally aligned.
4   Suspension arm bent or twisted (Chapter 10).

81   Excessive tire wear on inside edge

1   Inflation pressures incorrect (Chapter 1).
2   Front end alignment incorrect (toe-out). Have professionally aligned.
3   Loose or damaged steering or suspension components (Chapter 10).

82   Tire tread worn in one place

1   Tires out of balance.
2   Damaged or buckled wheel. Inspect and replace if necessary.
3   Defective tire (Chapter 1).

83   Excessive play or looseness in steering system

1   Front hub bearing(s) worn (Chapter 10).
2   Tie-rod end loose (Chapter 10).
3   Steering gear loose or worn (Chapter 10).
4   Worn or loose steering intermediate shaft (Chapter 10).

84   Rattling or clicking noise in steering gear

1   Steering gear loose (Chapter 10).
2   Steering gear defective.

Chapter 1
Tune-up and routine maintenance

Contents

Specifications

Recommended lubricants and fluids

Note: *Manufacturers occasionally upgrade their fluid and lubricant specifications. Check with your local auto parts store for the most current recommendations.*

Engine oil	
Type	API grade SG multigrade fuel efficient oil
Viscosity	5W-30
Automatic transmission fluid	Dexron III automatic transmission fluid
Manual transaxle lubricant	SAE 75W/85 GL-5 gear oil
Brake fluid type	DOT 3 brake fluid
Power steering system fluid	Dexron III automatic transmission fluid
Fuel type	Unleaded gasoline, 87 octane or higher
Coolant	50/50 mixture of ethylene glycol and water

Engine oil viscosity chart - for best fuel economy and cold starting, select the lowest SAE viscosity grade for the expected temperature range

Capacities*

Engine oil, with filter change	3.5 qts
Automatic transaxle	
With fluid and filter change	1.6 qts
Overhaul	5.2 qts
Manual transaxle	2.5 qts
Coolant	6.4 qts

All capacities approximate. Fill below total capacity, then add as necessary to bring to appropriate level.

Ignition system

Spark plug type and gap	
1991 and earlier	
Type	AC R43CSLX or NGK BPR6ES-11
Gap	0.039 to 0.043-inch (1.0 to 1.1 mm)
1992 through 1998	
Type	AC R42XLS or NGK BPR6ES-11
Gap	0.039 to 0.045-inch (1.0 to 1.1 mm)
1999 and later	
1.0L engine	
Type	AC R42XLS
Gap	0.039 to 0.045-inch (1.0 to 1.1 mm)
1.3L engine	
Type	NGK BKR6E-11 or Denso K20PR-U11
Gap	0.039 to 0.045-inch (1.0 to 1.1 mm)
Spark plug wire resistance	10 K-ohms per 12 inches of length
Engine firing order	
1.0L engine	1-3-2
1.3L engine	1-3-4-2
Valve clearances	
Intake valve	
Engine hot	0.010-inch (0.27 mm
Engine cold	0.006-inch (0.17 mm
Exhaust valve	
Engine hot	0.012-inch (0.32 mm
Engine cold	0.008-inch (0.22 mm
Radiator pressure cap rating	13 psi
Clutch pedal freeplay (minimum)	9/16 to 13/16-inch (15 to 20 mm)
Clutch release arm freeplay	3/16-inch 10 19/32- inch (2 to 4 mm)
Parking brake adjustment	3 to 6 clicks

Firing order 1-3-2

2A-00 Specs HAYNES

Cylinder location and distributor rotation (three-cylinder models)
The blackened terminal shown on the distributor cap indicates the Number One spark plug wire position

Torque specifications

	Ft-lbs (unless otherwise indicated)	Nm
Oil drain plug	33 to 35	44 to 47
Automatic transaxle		
Drain plug	17 to 23	23 to 31
Fluid pan bolts	53 in-lb	6
Filter	53 in-lb	6
Carburetor/TBI nut/bolts	14 to 18	18 to 24
Manual transaxle drain and fill plugs	29	39
Spark plugs	21	28
Wheel lug nuts	45	60

Cylinder location and distributor rotation (1997 and earlier four-cylinder models)
The blackened terminal shown on the distributor cap indicates the Number One spark plug wire position

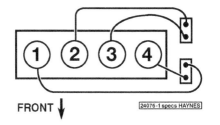

FRONT

24075-1 specs HAYNES

Cylinder location and spark-plug wire routing (1998 four-cylinder models)

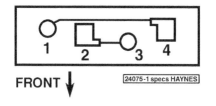

FRONT

24075-1 specs HAYNES

Cylinder location and spark-plug wire routing (1999 and later four-cylinder models)

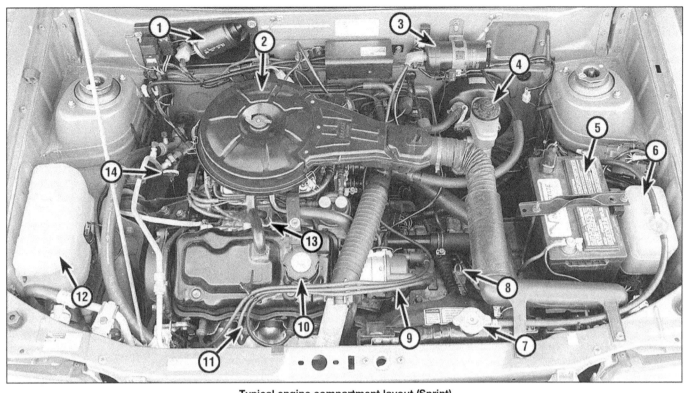

Typical engine compartment layout (Sprint)

1	Wiper motor	6	Engine coolant reservoir	11	Spark plugs
2	Air filter housing	7	Radiator cap	12	Windshield washer fluid reservoir
3	Ignition coil	8	Automatic transaxle fluid dipstick	13	PCV valve
4	Brake master cylinder	9	Spark plug wires	14	Engine oil dipstick
5	Battery	10	Engine oil filler cap		

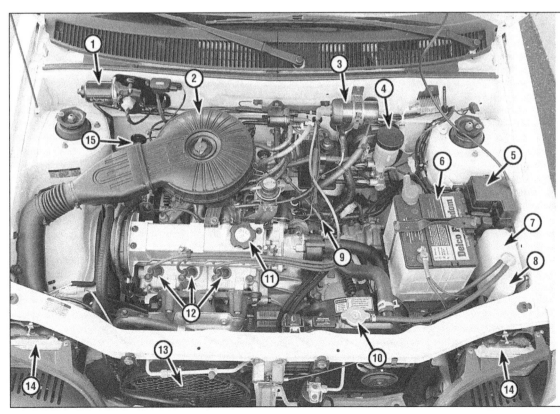

Typical engine compartment layout (1994 and earlier Metro)

1 Wiper motor
2 Air filter housing
3 Ignition coil
4 Brake master cylinder
5 Fuse/relay panel
6 Battery
7 Engine coolant reservoir
8 Windshield washer fluid reservoir
9 Automatic transaxle dipstick
10 Radiator cap
11 Engine oil filler cap
12 Spark plug and wire boot cover
13 Air conditioning cooling fan (not all models)
14 Headlights
15 Engine oil dipstick

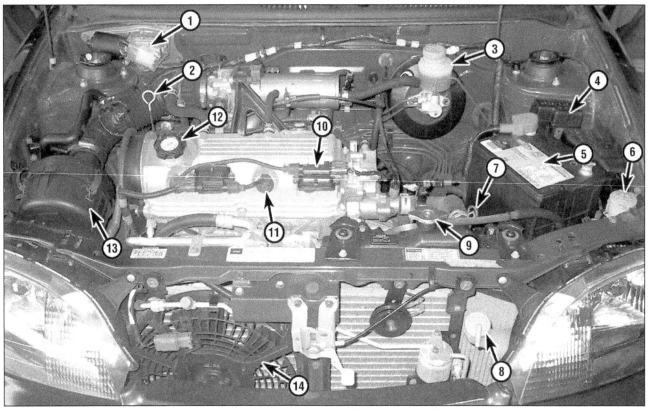

Typical engine compartment layout (1995 and later Metro)

1 Wiper motor	6 Windshield washer fluid reservoir	11 Spark plug
2 Engine oil dipstick	7 Automatic transaxle fluid dipstick	12 Engine oil filler cap
3 Brake master cylinder	8 Engine coolant reservoir	13 Air filter housing
4 Fuse/relay panel	9 Radiator cap	14 Air conditioning cooling fan (not all
5 Battery	10 Ignition coil pack/spark plug	models)

Typical engine compartment underside components (Sprint shown)

1   Engine oil filter
2   Radiator drain plug
3   Automatic transaxle drain plug
4   Front disc brake
5   Driveaxle CV joint boot
6   Steering gear boot
7   Exhaust pipe
8   Engine oil drain plug
9   Engine drivebelt

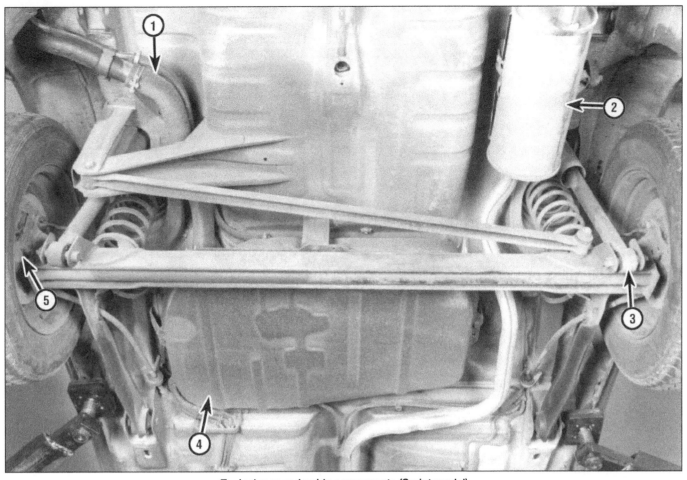

Typical rear underside components (Sprint model)

| 1 | *Fuel tank filler hose* | 2 | *Muffler* | 3 | *Shock absorber* | 4 | *Fuel tank* | 5 | *Drum brake* |

1 Chevrolet Sprint and Geo Metro Maintenance schedule

The maintenance intervals in this manual are provided with the assumption that you, not the dealer, will be doing the work. These are the minimum maintenance intervals recommended by the factory for vehicles that are driven daily. If you wish to keep your vehicle in peak condition at all times, you may wish to perform some of these procedures even more often. Because frequent maintenance enhances the efficiency, performance and resale value of your car, we encourage you to do so. If you drive in dusty areas, tow a trailer, idle or drive at low speeds for extended periods or drive for short distances (less than four miles) in below freezing temperatures, shorter intervals are also recommended.

When your vehicle is new, follow the maintenance schedule to the letter, record the maintenance performed in your owners manual and keep all receipts to protect the new vehicle warranty. In many cases, the initial maintenance check is done at no cost to the owner.

Every 250 miles or weekly, whichever comes first

Check the engine oil level (Section 4)
Check the engine coolant level (Section 4)
Check the windshield washer fluid level (Section 4)
Check the brake fluid level (Section 4)
Check the tires and tire pressures (Section 5)

Every 3000 miles or 3 months, whichever comes first

All items listed above plus:

Check the automatic transaxle fluid level (Section 6)
Check the power steering fluid level (Section 7)
Change the engine oil and oil filter (Section 8)

Every 7500 miles or 6 months, whichever comes first

All items listed above plus:

Check seat belt operation (Section 9)
Inspect and replace, if necessary, the windshield wiper blades (Section 10)
Check and adjust, if necessary, the clutch pedal freeplay (Section 11)
Check and service the battery (Section 12)
Check and adjust, if necessary, the engine drivebelts (Section 13)
Inspect and replace, if necessary, all underhood hoses (Section 14)
Check the cooling system (Section 15)
Rotate the tires (Section 16)

Every 15,000 miles or 12 months, whichever comes first

All items listed above plus:

Inspect the brake system (Section 17)*
Check the manual transaxle lubricant level (Section 18)
Check and adjust, if necessary, the valve clearances (Sprint and 1998 and later four-cylinder Metro models) (Section 19)
Check the carburetor/TBI bolt/nut torque (Section 20)*

Every 30,000 miles or 24 months, whichever comes first

All items listed above plus:

Replace the air filter (Section 21)*
Replace the spark plugs (Section 22)
Inspect and replace, if necessary, the spark plug wires (Section 23)
Inspect/replace the distributor cap and rotor (models so equipped) (Section 24)
Check the carburetor choke (Section 25)
Check the thermostatically controlled air cleaner (Section 26)
Check/adjust the engine idle speed (1994 and earlier models only) (Section 27)

Check and, if necessary, replace the PCV valve (Section 28)
Service the cooling system (drain, flush and refill) (Section 29)
Inspect the suspension, steering components and driveaxle boots (Section 30)*
Change the automatic transaxle fluid (Section 31)**
Change the manual transaxle lubricant (Section 32)
Check the exhaust system (Section 33)
Inspect the fuel system (Section 34)
Replace the fuel filter (Section 35)
Check the fuel cutoff system (carbureted models only) (Section 36)
Check the Pulse Air System operation (Section 37)
Check and reset the oxygen sensor light (1985 and 1986 models only) (Section 38)

Every 60,000 miles or 48 months, whichever comes first

Check/adjust the ignition timing (Chapter 5)
Replace the timing belt (Chapter 2A and 2B)

* This item is affected by "severe" operating conditions as described below. If your vehicle is operated under "severe" conditions, perform all maintenance indicated with a * at 7500 mile/6 month intervals. Severe conditions are indicated if you mainly operate your vehicle under one or more of the following conditions:

Operating in dusty areas
Towing a trailer
Idling for extended periods and/or low speed operation
Operating when outside temperatures remain below freezing and when most trips are less than five miles

** If operated under one or more of the following conditions, change the automatic transaxle fluid every 15,000 miles:

In heavy city traffic where the outside temperature regularly reaches 90-degrees F (32-degrees C) or higher
In hilly or mountainous terrain
Operating in dusty areas
Towing a trailer
Idling for extended periods and/or low speed operation
Operating when outside temperatures remain below freezing and when most trips are less than five miles
In heavy city traffic or where the outside temperature regularly reaches 90-degrees F (32-degrees C) or higher

2   Introduction

This Chapter is designed to help the home mechanic maintain his/her car for peak performance, economy, safety and long life. The following Sections deal specifically with each item on the maintenance schedule. Visual checks, adjustments, component replacement and other helpful items are included. Refer to the accompanying photos of the engine compartment and the underside of the vehicle for the location of various components.

Servicing your vehicle in accordance with the mileage/time maintenance schedule and the following Sections will provide it with a planned maintenance program that should result in a long and reliable service life. This is a comprehensive plan, so maintaining some items but not others at the specified service intervals will not produce the same results.

As you service your car, you will discover that many of the procedures can - and should - be grouped together because of the nature of the particular procedure you're performing or because of the close proximity of two otherwise unrelated components to one

another. For example, if the vehicle is raised for chassis lubrication, you should inspect the exhaust, suspension, steering and fuel systems while you're under the vehicle. When you're rotating the tires, it makes good sense to check the brakes and wheel bearings since the wheels are already removed.

Finally, let's suppose you have to borrow or rent a torque wrench. Even if you only need to tighten the spark plugs, you might as well check the torque of as many critical fasteners as time allows.

The first step of this maintenance program is to prepare you before the actual work

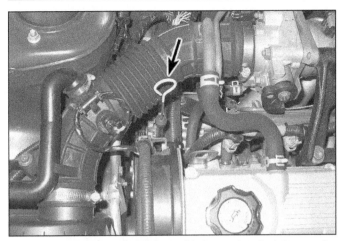

4.2 The engine oil dipstick (arrow) is located at the left rear of the engine compartment

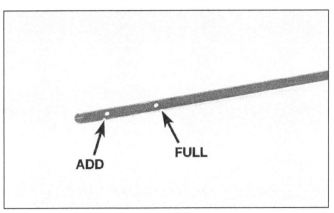

4.4 The oil level should be between the two marks on the dipstick - if it isn't, add enough oil to bring the level to or near the upper line (it takes one quart to raise the level from the lower line to the upper line)

begins. Read through all Sections pertinent to the procedures you're planning to do, then make a list of and gather together all the parts and tools you will need to do the job. If it looks as if you might run into problems during a particular segment of some procedure, seek advice from your local parts man or dealer service department.

Owner's Manual and VECI label information

Your vehicle owner's manual was written for your year and model and contains very specific information on component locations, specifications, fuse ratings, part numbers, etc. The Owner's Manual is an important resource for the do-it-yourselfer to have; if one was not supplied with your vehicle, it can generally be ordered from a dealer parts department.

Among other important information, the Vehicle Emissions Control Information (VECI) label contains specifications and procedures for tune-up adjustments (if applicable) and sparks plugs (see Chapter 6 for more information on the VECI label). The information on this label is the exact maintenance data recommended by the manufacturer. This data often varies by intended altitude, local emissions regulations, month of manufacture, etc.

This Chapter contains procedural details, safety information and more ambitious maintenance intervals than you might find in manufacturer's literature. However, you may find procedures or specifications in your Owner's Manual or VECI label can be considered correct, since it is specific to your particular vehicle.

3   Tune-up general information

The term tune-up is used in this manual to represent a combination of individual operations rather than one specific procedure.

If, from the time the vehicle is new, the

routine maintenance schedule is followed closely and frequent checks are made of fluid levels and high wear items, as suggested throughout this manual, the engine will be kept in relatively good running condition and the need for additional work will be minimized.

More likely than not, however, there will be times when the engine is running poorly due to lack of regular maintenance. This is even more likely if a used vehicle, which has not received regular and frequent maintenance checks, is purchased. In such cases, an engine tune-up will be needed outside of the regular routine maintenance intervals.

The first step in any tune-up or engine diagnosis to help correct a poor running engine would be a cylinder compression check. A check of the engine compression (see Chapter 2 Part B) will give valuable information regarding the overall performance of many internal components and should be used as a basis for tune-up and repair procedures. If, for instance, a compression check indicates serious internal engine wear, a conventional tune-up will not help the running condition of the engine and would be a waste of time and money. Because of its importance, someone should perform compression checking with the proper compression testing gauge and the knowledge to use it properly.

The following series of operations are those most often needed to bring a generally poor running engine back into a proper state of tune.

Minor tune-up

Check all engine related fluids (Section 4)
Clean, inspect and test the battery (Section 12)
Check and adjust the drivebelts (Section 13)
Check all underhood hoses (Section 14)
Check the cooling system (Section 15)
Check the air filter (Section 21)
Inspect the spark plug wires, distributor cap and rotor (Sections 23 and 24)

Major tune-up

All items listed under minor tune-up, plus . . .
Replace the air filter (Section 21)
Replace the spark plugs (Section 22)
Replace the spark plug wires (Section 23)
Check the fuel system (Section 34)

4   Fluid level checks (every 250 miles or weekly)

1   Fluids are an essential part of the lubrication, cooling, brake, clutch and other systems. Because these fluids gradually become depleted and/or contaminated during normal operation of the vehicle, they must be periodically replenished. See *Recommended lubricants, fluids and capacities* at the beginning of this Chapter before adding fluid to any of the following components. **Note:** *The vehicle must be on level ground before fluid levels can be checked.*

Engine oil

Refer to illustrations 4.2, 4.4 and 4.6
2   The engine oil level is checked with a dipstick located at the left rear of the engine **(see illustration)**. The dipstick extends through a metal tube from which it protrudes down into the engine oil pan.
3   The oil level should be checked before the vehicle has been driven, or about 15 minutes after the engine has been shut off. If the oil is checked immediately after driving the vehicle, some of the oil will remain in the upper engine components, producing an inaccurate reading on the dipstick.
4   Pull the dipstick from the tube and wipe all the oil from the end with a clean rag or paper towel. Insert the clean dipstick all the way back into its metal tube and pull it out again. Observe the oil at the end of the dipstick. At its highest point, the level should be between the upper and lower holes **(see illustration)**.
5   It takes one quart of oil to raise the level

4.6 The threaded oil filler cap is located on the valve cover - to
prevent dirt from contaminating the engine, always make
sure the area around this opening is clean before
unscrewing the cap counterclockwise

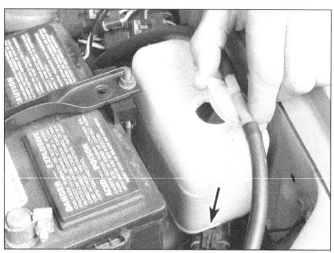

4.8a Make sure the coolant level in the reservoir doesn't drop
below the LOW mark (arrow) - if it's near the mark, add more
antifreeze/water mixture (earlier model)

from the lower line to the upper line on the dipstick. Do not allow the level to drop below the lower line or oil starvation may cause engine damage. Conversely, overfilling the engine (adding oil above the upper line) may cause oil-fouled spark plugs, oil leaks or oil seal failures.

6   Remove the threaded cap from the valve cover to add oil **(see illustration)**. Use an oil can spout or funnel to prevent spills. After adding the oil, install the filler cap hand tight. Start the engine and look carefully for any small leaks around the oil filter or drain plug. Stop the engine and check the oil level again after it has had sufficient time to drain from the upper block and cylinder head galleys.

7   Checking the oil level is an important preventive maintenance step. A continually dropping oil level indicates oil leakage through damaged seals, from loose connections, or past worn rings or valve guides. If the oil looks milky in color or has water droplets in it, a cylinder head gasket may be blown or the oil cooler could be leaking. The engine should be checked immediately. The condition of the oil should also be checked. Each time you check the oil level, slide your thumb and index finger up the dipstick before wiping off the oil. If you see small dirt or metal particles clinging to the dipstick, the oil should be changed (see Section 8).

Engine coolant

Refer to illustrations 4.8a and 4.8b
Warning: *Do not allow antifreeze to come in contact with your skin or painted surfaces of the vehicle. Rinse off spills immediately with plenty of water. Antifreeze is highly toxic if ingested. Never leave antifreeze lying around in an open container or in puddles on the floor; children and pets are attracted by its sweet smell and may drink it. Check with local authorities on disposing of used anti-freeze. Many communities have collection centers,*

which will see that antifreeze is disposed of safely.
Note: *Non-toxic antifreeze is now manufactured and available at local auto parts stores, but even this type should be disposed of properly.*

8   All vehicles covered by this manual are equipped with a pressurized coolant-recovery system. A coolant reservoir located in the front corner of the engine compartment or in front of the radiator and is connected by a hose to the base of the radiator filler neck **(see illustrations)**. If the coolant heats up during engine operation, coolant can escape through the pressurized filler cap, then through the connecting hose into the reservoir. As the engine cools, the coolant is automatically drawn back into the cooling system to maintain the correct level.

9   The coolant level in the reservoir should be checked regularly. The level must be between the MAX and MIN lines on the reservoir tank. The level will vary with the temperature of the engine. When the engine is cold, the coolant level should be at or slightly above the lower mark on the dipstick or tank. Once the engine has warmed up, the level should be at or near the upper mark. If it isn't, allow the fluid in the tank to cool, then remove the dipstick or cap from the reservoir and add coolant to bring the level up to the upper line. **Caution:** *Use only ethylene glycol type coolant and water in the mixture ratio recommended by your owner's manual. Do not use supplemental inhibitors or additives. If only a small amount of coolant is required to bring the system up to the proper level, water can be used. However, repeated additions of water will dilute the recommended antifreeze and water solution. In order to maintain the proper ratio of antifreeze and water, it is advisable to top up the coolant level with the correct mixture. Refer to your owner's manual for the recommended ratio.*

10   If the coolant level drops within a short

time after replenishment, there may be a leak in the system. Inspect the radiator, hoses, engine coolant filler cap, drain plugs and water pump. If no leak is evident, have the radiator cap pressure tested. **Warning:** *Never remove the radiator cap or the coolant recovery reservoir cap when the engine is running or has just been shut down, because the cooling system is hot. Escaping steam and scalding liquid could cause serious injury.*

11   If it is necessary to open the radiator cap, wait until the system has cooled completely, then wrap a thick cloth around the cap and turn it to the first stop. If any steam escapes, wait until the system has cooled further, then remove the cap.

12   When checking the coolant level, always note its condition. It should be relatively clear. If it is brown or rust colored, the system should be drained, flushed and refilled. Even if the coolant appears to be normal, the corrosion inhibitors wear out with use, so it must be replaced at the specified intervals.

13   Do not allow antifreeze to come in contact with your skin or painted surfaces of the vehicle. Flush contacted areas immediately with plenty of water.

4.8b On later models the coolant
reservoir is located in front of the radiator

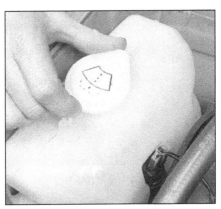

4.14a The windshield washer reservoir is located on the passenger side of the engine compartment on Sprint models - flip up the cap to add fluid to the washer reservoir

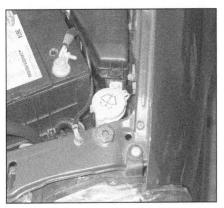

4.14b On later models, the windshield washer fluid reservoir is located in the front corner of the engine compartment

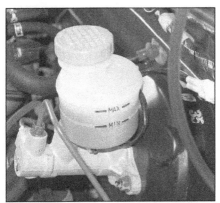

4.16 The brake fluid should be kept between the MIN and MAX marks on the reservoir - turn and lift up the cap to add fluid

Washer fluid

Refer to illustration 4.14a and 4.14b

14   Fluid for the windshield washer system is stored in a plastic reservoir, which is located in the engine compartment **(see illustrations)**. Check the windshield washer fluid reservoir filler neck to make sure it is near the top of the neck. The rear window

4.23 Check the water level in maintenance-type batteries

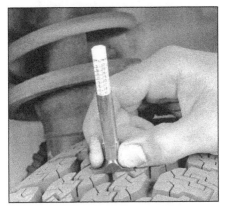

5.2 Use a tire tread depth gauge to monitor tire wear; they are available at auto parts stores and service stations and cost very little

fluid reservoir is located on the left side of the luggage compartment. Make sure the level is near the top of the filler neck. In milder climates, plain water can be used to top up the reservoir, but the reservoir should be kept no more than 2/3 full to allow for expansion should the water freeze. In colder climates, the use of a specially designed windshield washer fluid, available at your dealer and any auto parts store, will help lower the freezing point of the fluid. Mix the solution with water in accordance with the manufacturer's directions on the container. Do not use regular antifreeze. It will damage the vehicle's paint.

Brake fluid

Refer to illustration 4.16

15   The brake master cylinder is mounted on the front of the power booster unit in the engine compartment.
16   To check the fluid level of the brake master cylinder, simply look at the MAX and MIN marks on the reservoir **(see illustration)**. The level should be between the two marks.
17   If the level is low, wipe the top of the reservoir cover with a clean rag to prevent contamination of the brake system before lifting the cap.
18   Add only the specified brake fluid to the brake reservoir (refer to *Recommended lubricants and fluids* at the front of this Chapter or to your owner's manual). Mixing different types of brake fluid can damage the system. Fill the brake master cylinder reservoir to the MAX line. **Warning:** *Use caution when filling the reservoir - brake fluid can harm your eyes and damage painted surfaces. Do not use brake fluid that has been opened for more than one year or has been left open. Brake fluid absorbs moisture from the air. Excess moisture can cause a dangerous loss of braking.*
19   While the reservoir cap is removed, inspect the master cylinder reservoir for contamination. If deposits, dirt particles or water droplets are present, the system should be drained and refilled (see Chapters 8 or 9).
20   After filling the reservoir to the proper level, make sure the lid is properly seated to

prevent fluid leakage and/or system pressure loss.
21   The brake fluid in the master cylinder will drop slightly as the brake pads at each wheel wear down during normal operation. If the master cylinder requires repeated replenishing to keep it at the proper level, this is an indication of leakage in the brake system, which should be corrected immediately. Check all brake lines and connections, along with the wheel cylinders and booster (see Section 15 for more information). A drop in the clutch reservoir level indicates a leak in the clutch hydraulic system (see Chapter 8).
22   If, upon checking the brake master cylinder fluid level, you discover an empty or nearly empty reservoir, the brake system should be bled (see Chapter 9).

Battery electrolyte

Refer to illustration 4.23

23   All vehicles with which this manual is concerned are equipped with a battery that is permanently sealed (except for vent holes) and has no filler caps. Water doesn't have to be added to these batteries at any time. If an aftermarket maintenance-type battery is installed, the caps on top of the battery should be removed periodically to check for a low water level **(see illustration)**. This check is most critical during warm summer months.

---

5   Tire and tire pressure checks (every 250 miles or weekly)

Refer to illustrations 5.2, 5.3, 5.4a, 5.4b and 5.8

1   Periodic inspection of the tires may spare you from the inconvenience of being stranded with a flat tire. It can also provide you with vital information regarding possible problems in the steering and suspension systems before major damage occurs.
2   Normal tread wear can be monitored with a simple, inexpensive device known as a tread depth indicator **(see illustration)**. When the tread depth reaches the specified minimum, replace the tire(s).

UNDERINFLATION

CUPPING

Cupping may be caused by:
• Underinflation and/or mechanical irregularities such as out-of-balance condition of wheel and/or tire, and bent or damaged wheel.
• Loose or worn steering tie-rod or steering idler arm.
• Loose, damaged or worn front suspension parts.

OVERINFLATION

INCORRECT TOE-IN
OR EXTREME CAMBER

FEATHERING DUE
TO MISALIGNMENT

5.3 This chart will help you determine the condition of the tires, the probable cause(s) of abnormal wear and the corrective action necessary

3   Note any abnormal tread wear **(see illustration)**. Tread pattern irregularities such as cupping, flat spots and more wear on one side than the other are indications of front end alignment and/or balance problems. If any of these conditions are noted, take the vehicle to a tire shop or service station to correct the problem.

4   Look closely for cuts, punctures and embedded nails or tacks. Sometimes a tire will hold its air pressure for a short time or leak down very slowly even after a nail has embedded itself into the tread. If a slow leak persists, check the valve core to make sure it

is tight **(see illustration)**. Examine the tread for an object that may have embedded itself into the tire or for a "plug" that may have begun to leak (radial tire punctures are repaired with a plug that is installed in a puncture). If a puncture is suspected, it can be easily verified by spraying a solution of soapy water onto the puncture area **(see illustration)**. The soapy solution will bubble if there is a leak. Unless the puncture is inordinately large, a tire shop or gas station can usually repair the punctured tire.

5   Carefully inspect the inner side of each tire for evidence of brake fluid leakage. If you

see any, inspect the brakes immediately.

6   Correct tire air pressure adds miles to the lifespan of the tires, improves mileage and enhances overall ride quality. Tire pressure cannot be accurately estimated by looking at a tire, particularly if it is a radial. A tire pressure gauge is therefore essential. Keep an accurate gauge in the glovebox. The pressure gauges fitted to the nozzles of air hoses at gas stations are often inaccurate.

7   Always check tire pressure when the tires are cold. "Cold," in this case, means the vehicle has not been driven over a mile in the three hours preceding a tire pressure check.

5.4a If a tire loses air on a steady basis, check the valve core first to make sure it's snug (special inexpensive wrenches are commonly available at auto parts stores)

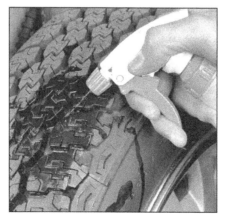

5.4b If the valve core is tight, raise the corner of the vehicle with the low tire and spray a soapy water solution onto the tread as the tire is turned slowly - leaks will cause small bubbles to appear

5.8 To extend the life of the tires, check the air pressure at least once a week with an accurate gauge (don't forget the spare)

6.3 The automatic transaxle dipstick (arrow) is located next to the battery

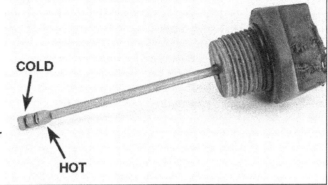

6.5 The automatic transaxle fluid level should be between the two upper marks on the dipstick when the fluid is hot (earlier model shown)

COLD

HOT

A pressure rise of four to eight pounds is not uncommon once the tires are warm.

8   Unscrew the valve cap protruding from the wheel or hubcap and push the gauge firmly onto the valve **(see illustration)**. Note the reading on the gauge and compare this figure to the recommended tire pressure shown on the tire placard on the left door jamb. Be sure to reinstall the valve cap to keep dirt and moisture out of the valve stem mechanism. Check all four tires and, if necessary, add enough air to bring them up to the recommended pressure levels.

9   Don't forget to keep the spare tire inflated to the specified pressure (consult your owner's manual). Note that the air pressure specified for the compact spare is significantly higher than the pressure of the regular tires.

6   Automatic transaxle fluid level check (every 3000 miles or 3 months)

Refer to illustrations 6.3 and 6.5

1   The level of the automatic transaxle fluid should be carefully maintained. Low fluid level can lead to slipping or loss of drive, while overfilling can cause foaming, loss of fluid and transaxle damage.

2   The fluid level should only be checked on level ground with the engine idling and the transaxle in Park.

3   Remove the dipstick - it's located next to the battery **(see illustration)**. Check the level of the fluid on the dipstick and note its condition.

4   Wipe the fluid from the dipstick with a clean rag and reinsert it.

5   Pull the dipstick out again and note the fluid level **(see illustration)**. The level should be between the upper and lower marks on the dipstick. If the level is low, add the specified automatic transmission fluid. Add the fluid through the dipstick opening with a funnel.

6   Add just enough of the specified fluid to fill the transaxle to the proper level. It takes

about one pint to raise the level from the lower mark to the upper mark, so add the fluid a little at a time and keep checking the level until it is correct.

7   The condition of the fluid should also be checked along with the level. If the fluid at the end of the dipstick is black or a dark reddish brown color, or if it emits a burned smell, the fluid should be changed (see Section 27). If you are in doubt about the condition of the fluid, purchase some new fluid and compare the two for color and smell.

7   Power steering fluid level check (every 3000 miles or 3 months)

1   The power steering system relies on fluid, which may, over a period of time, require replenishing.

2   The fluid reservoir for the power steering pump is located on the right inner fender panel near the front of the engine compartment.

3   For the check, the front wheels should be pointed straight ahead and the engine should be off. The fluid should be cold when checking the level.

4   On earlier models, remove the cap and make sure the High and Low marks on the dipstick. On later models, make sure the fluid level is between the MAX and MIN marks on the reservoir.

5   If additional fluid is required, pour the specified type directly into the reservoir, using a funnel to prevent spills.

6   If the reservoir requires frequent fluid additions, all power steering hoses, hose connections, the power steering pump and the steering gear should be carefully checked for leaks.

8   Engine oil and oil filter change (every 3000 miles or 3 months)

Refer to illustrations 8.2, 8.7, 8.12 and 8.14

1   Frequent oil changes are the best preventive maintenance the home mechanic can give the engine, because aging oil becomes diluted and contaminated, which leads to premature engine wear.

2   Make sure you have all the necessary

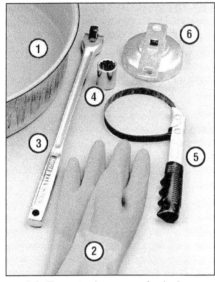

8.2 These tools are required when changing the engine oil and filter

1   *Drain pan - It should be fairly shallow in depth, but wide to prevent spills*

2   *Rubber gloves - When removing the drain plug and filter, you will get oil on your hands (the gloves will prevent burns)*

3   *Breaker bar - Sometimes the oil drain plug is tight, and a long breaker bar is needed to loosen it*

4   *Socket - To be used with the breaker bar or a ratchet (must be the correct size to fit the drain plug)*

5   *Filter wrench - This is a metal band-type wrench, which requires clearance around the filter to be effective*

6   *Filter wrench - This type fits on the bottom of the filter and can be turned with a ratchet or breaker bar (different-size wrenches are available for different types of filters)*

tools before you begin this procedure **(see illustration)**. You should also have plenty of rags or newspapers handy for mopping up any spills.

3   Access to the underside of the vehicle is greatly improved if the vehicle can be lifted on a hoist, driven onto ramps or supported by jackstands. **Warning:** *Do not work under a*

8.7 Use the proper size box-end wrench or socket to remove the oil drain plug without rounding off the corners

8.12 The oil filter is usually on very tight and will require a special wrench for removal - DO NOT use the wrench to tighten the new filter

8.14 Lubricate the oil filter gasket with clean engine oil before installing the filter on the engine

vehicle which is supported only by a bumper, hydraulic or scissors-type jack.

4   If this is your first oil change, familiarize yourself with the locations of the oil drain plug and the oil filter. The engine and exhaust components will be warm during the actual work, so try to anticipate any potential problems before the engine and accessories are hot.

5   Park the vehicle on a level spot. Start the engine and allow it to reach its normal operating temperature. Warm oil and sludge will flow out more easily. Turn off the engine when it's warmed up. Remove the filler cap from the valve cover.

6   Raise the vehicle and support it securely on jackstands. **Warning:** *Never get beneath the vehicle when it is supported only by a jack. The jack provided with your vehicle is designed solely for raising the vehicle to remove and replace the wheels. Always use jackstands to support the vehicle when it becomes necessary to place your body underneath the vehicle.*

7   Being careful not to touch the hot exhaust components, place the drain pan under the drain plug in the bottom of the pan and remove the plug **(see illustration)**. You may want to wear gloves while unscrewing the plug the final few turns if the engine is hot.

8   Allow the old oil to drain into the pan. It may be necessary to move the pan farther under the engine as the oil flow slows to a trickle. Inspect the old oil for the presence of metal shavings and chips.

9   After all the oil has drained, wipe off the drain plug with a clean rag. Even minute metal particles clinging to the plug would immediately contaminate the new oil.

10   Clean the area around the drain plug opening, reinstall the plug and tighten it securely, but do not strip the threads.

11   Move the drain pan into position under the oil filter.

12   Loosen the oil filter **(see illustration)** by turning it counterclockwise with an oil filter wrench. Once the filter is loose, use your

hands to unscrew it from the block. Just as the filter is detached from the block, immediately tilt the open end up to prevent the oil inside the filter from spilling out. **Warning:** *The exhaust system may still be hot, so be careful.*

13   With a clean rag, wipe off the mounting surface on the block. If a residue of old oil is allowed to remain, it will smoke when the block is heated up. Also make sure that none of the old gasket remains stuck to the mounting surface. It can be removed with a scraper if necessary.

14   Compare the old filter with the new one to make sure they are the same type. Smear some clean engine oil on the rubber gasket of the new filter and screw it into place **(see illustration)**. Because overtightening the filter will damage the gasket, do not use a filter wrench to tighten the filter. Tighten it by hand until the gasket contacts the seating surface. Then seat the filter by giving it an additional 3/4-turn.

15   Remove all tools, rags, etc. from under the vehicle, being careful not to spill the oil in the drain pan, then lower the vehicle.

16   Add new oil to the engine through the oil filler cap in the valve cover. Use a funnel, if necessary, to prevent oil from spilling onto the top of the engine. Pour the specified amount of fresh oil into the engine. Wait a few minutes to allow the oil to drain into the pan, then check the level on the oil dipstick (see Section 4 if necessary). If the oil level is at or near the upper hole on the dipstick, install the filler cap hand tight, start the engine and allow the new oil to circulate.

17   Allow the engine to run for about a minute. While the engine is running, look under the vehicle and check for leaks at the oil pan drain plug and around the oil filter. If either is leaking, stop the engine and tighten the plug or filter.

18   Wait a few minutes to allow the oil to trickle down into the pan, then recheck the level on the dipstick and, if necessary, add enough oil to bring the level to the upper hole.

19   During the first few trips after an oil change, make it a point to check frequently for leaks and proper oil level.

20   Used motor oil cannot be re-used in its present state and should be recycled. Oil reclamation centers, auto repair shops and gas stations will normally accept the oil, which can be refined and used again. After the oil has cooled, it can be drained into a suitable container (capped plastic jugs, topped bottles, milk cartons, etc.) for transport to one of these recycling sites. New or used oil should never be allowed to go into street drains or into the ground.

9   Seat belt check (every 7500 miles or 6 months)

1   Check seat belts, buckles, latch plates and guide loops for obvious damage and signs of wear.

2   See if the seat belt reminder light comes on when the key is turned to the Run or Start position. A chime should also sound.

3   Seat belts are designed to lock up during a sudden stop or impact, yet allow free movement during normal driving. Make sure the retractors return the belt against your chest while driving and rewind the belt fully when the buckle is unlatched.

4   If any of the above checks reveal problems with the seat belt system, replace parts as necessary.

10   Wiper blade inspection and replacement (every 7500 miles or 6 months)

Refer to illustrations 10.6, 10.8 and 10.9

1   The windshield and rear (if equipped) wiper and blade assemblies should be inspected periodically for damage, loose components and cracked or worn blade elements.

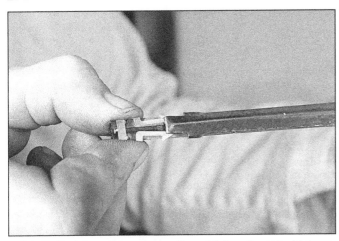

10.6 On Sprint models, squeeze the retaining clip and withdraw the blade element from the frame

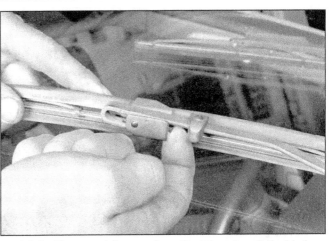

10.8 On Metro models press in on the tab and push the blade assembly out of the hook at the end to remove it

2   Road film can build up on the wiper blades and affect their efficiency, so they should be washed regularly with a mild detergent solution.
3   The action of the wiping mechanism can loosen bolts, nuts and fasteners, so they should be checked and tightened, as necessary, at the same time the wiper blades are checked.
4   If the wiper blade elements are cracked, worn or warped, or no longer clean adequately, they should be replaced with new ones.
5   Lift the arm assembly away from the glass for clearance.

Sprint

6   Squeeze the release clip and slide the element out of the blade frame (see illustration).
7   Installation is the reverse of removal.

Metro

8   Press in on the lock tab and push the blade assembly down the wiper arm, out of the arm hook. (see illustration).

9   Squeeze the blade element tabs tightly and pull the element out of the metal frame (see illustration).
10   Remove the metal retainers from the element and install them in the new element.
11   Insert the element into the frame and push it until the element tabs lock.
12   Place the metal arm assembly in the hook on the wiper arm and press it into place until the lock tab snaps into place.

11   Clutch pedal height and freeplay check and adjustment (every 7500 miles or 6 months)

Refer illustrations 11.1, 11.3 and 11.4
1   The clutch pedal must be at the same height as the brake pedal. If the pedal is more than 5/16-inch higher or lower than the brake pedal, loosen the locknut and turn the adjusting bolt until the height is as listed in specification section of this Chapter (see illustration).
2   Tighten the locknut and recheck the pedal height.

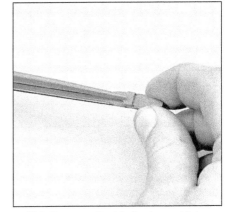

10.9 Squeeze the blade element tabs, then pull the element out of the metal frame and remove it

3   Push down on the clutch pedal and use a ruler to measure the distance it moves freely before resistance is felt (see illustration). The freeplay should be within the limits listed in this Chapter's specifications. If it isn't it must be adjusted.

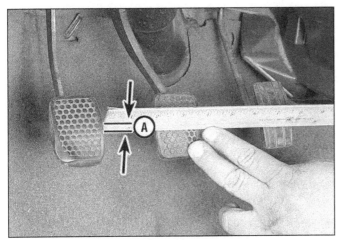

11.1 Use a ruler to determine if the clutch pedal is the same height (A - no more than 5/16-inch) as the brake pedal - adjust the bolt on the pedal bracket if necessary

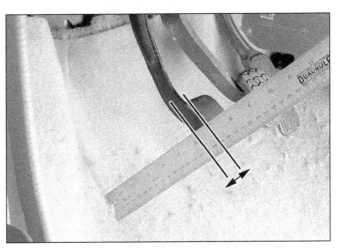

11.3 Freeplay is the distance the clutch pedal moves without resistance

11.4 After loosening the locknut, turn the adjusting nut (arrow) until the release lever is within limits

4   Turn the adjusting nut at the end of the clutch cable in the engine compartment until the release lever freeplay is within the limits listed in this Chapter's specifications. **(see illustration)**.

12 Battery check, maintenance and charging (every 7500 miles or 6 months)

Warning: *Certain precautions must be followed when checking and servicing the battery. Hydrogen gas, which is highly flammable, is always present in the battery cells, so keep lighted tobacco and all other open flames and sparks away from the battery. The electrolyte inside the battery is actually dilute sulfuric acid, which will cause injury if splashed on your skin or in your eyes. It will also ruin clothes and painted surfaces. When removing the battery cables, always detach the negative cable first and hook it up last!*

Check

Refer to illustration 12.1

1   A routine preventive maintenance program for the battery in your vehicle is the only way to ensure quick and reliable starts. But before performing any battery maintenance, make sure that you have the proper equipment necessary to work safely around the battery **(see illustration)**.

2   There are also several precautions that should be taken whenever battery maintenance is performed. Before servicing the battery, always turn the engine and all accessories off and disconnect the cable from the negative terminal of the battery.

3   The battery produces hydrogen gas, which is both flammable and explosive. Never create a spark, smoke or light a match around the battery. Always charge the battery in a ventilated area.

4   Electrolyte contains poisonous and corrosive sulfuric acid. Do not allow it to get in your eyes, on your skin on your clothes. Never ingest it. Wear protective safety glasses when working near the battery. Keep children away from the battery.

5   Note the external condition of the battery. If the positive terminal and cable clamp on your vehicle's battery is equipped with a rubber protector, make sure that it's not torn or damaged. It should completely cover the terminal. Look for any corroded or loose connections, cracks in the case or cover or loose hold-down clamps. Also check the entire length of each cable for cracks and frayed conductors.

6   Some models with sealed batteries have a battery condition indicator on top of the battery. Compare the color showing in the window to the condition color chart on the battery. You may catch a low-charge battery condition before it strands you on the road-

side. If the color indicates a low state of charge, charge the battery and examine the charging system (see Chapter 5 and this Section).

Maintenance

Refer to illustrations 12.7a, 12.7b, 12.8a and 12.8b

7   If corrosion, which looks like white, fluffy deposits **(see illustration)** is evident, particularly around the terminals, the battery should be removed for cleaning. Loosen the cable clamp bolts with a wrench, being careful to remove the ground cable first, and slide them off the terminals **(see illustration)**. Then disconnect the hold-down clamp bolt and nut, remove the clamp and lift the battery from the engine compartment.

8   Clean the cable clamps thoroughly with a battery brush or a terminal cleaner and a solution of warm water and baking soda **(see illustration)**. Wash the terminals and the top of the battery case with the same solution but make sure that the solution doesn't get into the battery. When cleaning the cables, terminals and battery top, wear safety goggles and rubber gloves to prevent any solution from coming in contact with your eyes or hands. Wear old clothes too - even diluted, sulfuric acid splashed onto clothes will burn holes in them. If the terminals have been extensively corroded, clean them up with a terminal cleaner **(see illustration)**. Thoroughly wash all cleaned areas with plain water.

9   Whenever the battery is removed for cleaning or charging, inspect the battery carrier before reinstalling the battery in the engine compartment. If the carrier is dirty or covered with corrosion, clean it in the same solution of warm water and baking soda. Inspect the metal brackets that support the carrier to make sure that they are not covered with corrosion. If they are, wash them off. If

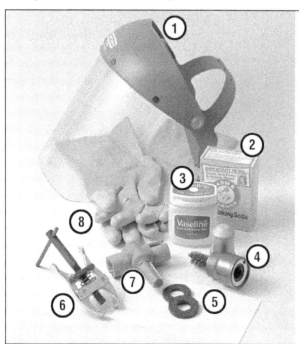

12.1 Tools and materials required for battery maintenance

1   *Face shield/safety goggles* - When removing corrosion with a brush, the acidic particles can easily fly up into your eyes

2   *Baking soda* - A solution of baking soda and water can be used to neutralize corrosion

3   *Petroleum jelly* - A layer of this on the battery posts will help prevent corrosion

4   *Battery post/cable cleaner* - This wire brush cleaning tool will remove all traces of corrosion from the battery posts and cable clamps

5   *Treated felt washers* - Placing one of these on each post, directly under the cable clamps, will help prevent corrosion

6   *Puller* - Sometimes the cable clamps are very difficult to pull off the posts, even after the nut/bolt has been completely loosened. This tool pulls the clamp straight up and off the post without damage

7   *Battery post/cable cleaner* - Here is another cleaning tool which is a slightly different version of Number 4 above, but it does the same thing

8   *Rubber gloves* - Another safety item to consider when servicing the battery; remember that's acid inside the battery!

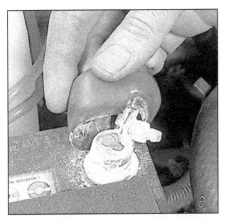

12.7a Battery terminal corrosion usually appears as light, fluffy powder

12.7b Removing the cable from a battery post with a wrench - sometimes special battery pliers are required for this procedure if corrosion has caused deterioration of the nut hex (always remove the ground cable first and hook it up last!)

corrosion is extensive, sand the brackets down to bare metal and spray them with a zinc-based primer (available in spray cans at auto paint and body supply stores).

10   Reinstall the battery back into the engine compartment. Make sure that no parts or wires are laying on the carrier during installation of the battery. Information on removing and installing the battery can be found in Chapter 5. Information on jump starting can be found at the front of this manual. For more detailed battery checking procedures, refer to the *Haynes Automotive Electrical Manual.*

11   Install a pair of specially-treated felt washers around the terminals (available at auto parts stores), then coat the terminals and the cable clamps with petroleum jelly or grease to prevent further corrosion. Install the cable clamps and tighten the nuts, being careful to install the negative cable last.

12   Install the hold-down clamp and nuts. Tighten the nuts only enough to hold the battery firmly in place. Overtightening these nuts can crack the battery case.

13   Make sure that the battery tray is in good condition and the hold-down clamp

bolts are tight. If the battery is removed from the tray, make sure no parts remain in the bottom of the tray when the battery is reinstalled. When reinstalling the hold-down clamp bolts, do not overtighten them.

Charging

Warning: *When batteries are being charged, hydrogen gas, which is very explosive and flammable, is produced. Do not smoke or allow open flames near a charging or a recently charged battery. Wear eye protection when near the battery during charging. Also, make sure the charger is unplugged before connecting or disconnecting the battery from the charger.*

14   Slow-rate charging is the best way to restore a battery that's discharged to the point where it will not start the engine. It's also a good way to maintain the battery charge in a vehicle that's only driven a few miles between starts. Maintaining the battery

charge is particularly important in the winter when the battery must work harder to start the engine and electrical accessories that drain the battery are in greater use.

15   It's best to use a one or two-amp battery charger (sometimes called a "trickle" charger). They are the safest and put the least strain on the battery. They are also the least expensive. For a faster charge, you can use a higher amperage charger, but don't use one rated more than 1/10th the amp/hour rating of the battery. Rapid boost charges that claim to restore the power of the battery in one to two hours are hardest on the battery and can damage batteries not in good condition. This type of charging should only be used in emergency situations.

16   The average time necessary to charge a battery should be listed in the instructions that come with the charger. As a general rule, a trickle charger will charge a battery in 12 to 16 hours.

13 Drivebelt check, adjustment and replacement (every 7500 miles or 6 months)

Check

Refer to illustrations 13.3a, 13.3b and 13.4

1   The drivebelts are located at the front of the engine and play an important role in the overall operation of the vehicle and its components. Due to their function and material make-up, the belts are prone to failure after a period of time and should be inspected and adjusted periodically to prevent major engine damage.

2   The number of belts used on a particular vehicle depends on the accessories installed. Drivebelts are used to turn the alternator, power steering pump, water pump and air conditioning compressor. Depending on the pulley arrangement, more than one of these components may be driven by a single belt.

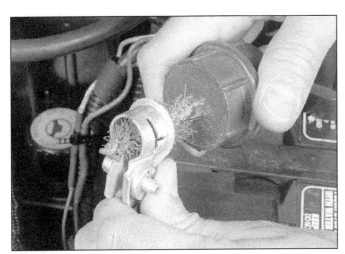

12.8a When cleaning the cable clamps, all corrosion must be removed

12.8b Regardless of the type of tool used on the battery posts, a clean, shiny surface should be the result (the inside of the clamp is tapered to match the taper on the post, so don't remove too much material)

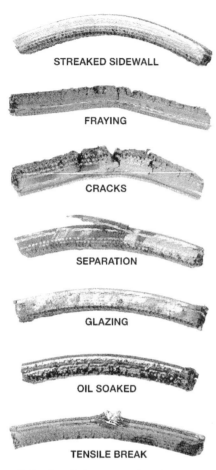

STREAKED SIDEWALL

FRAYING

CRACKS

SEPARATION

GLAZING

OIL SOAKED

TENSILE BREAK

13.3a Check the conventional belts for signs of wear like these - if the belt looks worn, replace it

3   With the engine off, open the hood and locate the various belts at the front of the engine. Using your fingers (and a flashlight, if necessary), move along the belts checking for cracks and separation of the belt plies. Also check for fraying and glazing, which gives the belt a shiny appearance **(see illustrations)**. Both sides of each belt should be inspected, which means you will have to twist the belt to check the underside.

4   The tension of each belt is checked by pushing on the belt at a distance halfway between the pulleys. Push firmly with your thumb and see how much the belt moves (deflects) **(see illustration)**. As rule of thumb, if the distance from pulley center-to-pulley center is between 7 and 11 inches, the belt should deflect 1/4-inch. If the belt travels between pulleys spaced 12 to 16 inches apart, the belt should deflect 1/2-inch.

Adjustment

Refer to illustration 13.5a and 13.5b

5   If the air conditioner compressor belt tension must be adjusted, locate the idler pulley bolt at the left front corner of the engine and turn the idler pulley adjuster bolt **(see illustrations)**. To adjust the power steering pump belt, loosen the adjuster lock bolt, then move the adjuster in-or-out to loosen or tighten the belt. The water pump and alternator belt tension is adjusted by loosening the lock bolt and turning the adjuster bolt.

6   After the belts have been adjusted, measure the belt tension in accordance with one of the above methods. Repeat the adjustment procedure until the drivebelt is tensioned properly.

Replacement

7   Follow the above adjustment procedures to loosen the belt, slip the belt off the pulleys and remove it. Since belts tend to wear out more or less at the same time, it's a good idea to replace all of them at the same time.

8   Take the old belts with you when purchasing new ones in order to make a direct comparison for length, width and design. Keep in mind that your old belt may have stretched, and the correct new belt may be slightly shorter. When installing a new ribbed belt, make sure it is centered on its drive pulley.

9   Install the belt by reversing the removal procedures. When installing a ribbed belt, make sure it is centered on the pulleys, it must not overlap either edge of the pulleys. Adjust the belt as described earlier in this Section.

14   Underhood hose check and replacement (every 7500 miles or 6 months)

Caution: *Replacement of air conditioning hoses must be left to a dealer service department or air conditioning shop that has the equipment to evacuate the system safely. Never remove air conditioning components or hoses until the system has been evacuated and the refrigerant recovered by an air conditioning shop.*

General

1   High temperatures in the engine compartment can cause the deterioration of the rubber and plastic hoses used for engine, accessory and emission systems operation. Periodic inspection should be made for cracks, loose clamps, material hardening and leaks.

2   Information specific to the cooling system hoses can be found in Section 15

3   Some, but not all, hoses are secured to the fittings with clamps. Where clamps are used, check to be sure they haven't lost their tension, allowing the hose to leak. If clamps aren't used, make sure the hose has not expanded and/or hardened where it slips over the fitting, allowing it to leak.

Vacuum hoses

4   Its quite common for vacuum hoses, especially those in the emissions system, to be color coded or identified by colored

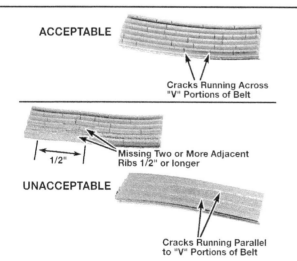

ACCEPTABLE

Cracks Running Across
"V" Portions of Belt

1/2"

Missing Two or More Adjacent
Ribs 1/2" or longer

UNACCEPTABLE

Cracks Running Parallel
to "V" Portions of Belt

13.3b On ribbed belts, checks for signs of wear like these - if the belt looks worn, replace it

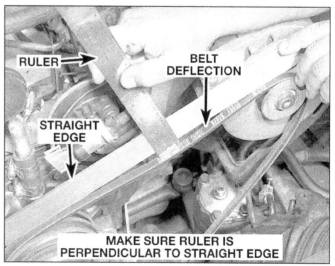

RULER

BELT DEFLECTION

STRAIGHT EDGE

MAKE SURE RULER IS PERPENDICULAR TO STRAIGHT EDGE

13.4 Measuring drivebelt deflection with a straightedge and ruler

13.5a On earlier model air conditioning equipped models, turn the idler pulley bolt (arrow) to adjust the drivebelt tension

13.5b On later model air conditioning equipped models, adjust the drivebelt tension by loosening the idler pulley nut (arrow) - move the idler pulley and tighten the nut

Check for a chafed area that could fail prematurely.

Check for a soft area indicating the hose has deteriorated inside.

Overtightening the clamp on a hardened hose will damage the hose and cause a leak.

Check each hose for swelling and oil-soaked ends. Cracks and breaks can be located by squeezing the hose.

15.4 Hoses, like drivebelts, have a habit of failing at the worst possible time - to prevent the inconvenience of a blown radiator or heater hose, inspect them carefully as shown here

stripes molded into them. Various systems require hoses with different wall thickness, collapse resistance and temperature resistance. When replacing hoses, be sure the new ones are made of the same material.

5   Often the only effective way to check a hose is to remove it completely from the vehicle. If more than one hose is removed, be sure to label the hoses and fittings to ensure correct installation.

6   When checking vacuum hoses, be sure to include any plastic T-fittings in the check. Inspect the fittings for cracks and the hose where it fits over the fitting for distortion, which could cause leakage.

7   A small piece of vacuum hose (1/4-inch inside diameter) can be used as a stethoscope to detect vacuum leaks. Hold one end of the hose to your ear and probe around vacuum hoses and fittings, listening for the "hissing" sound characteristic of a vacuum leak. **Warning:** *When probing with the vacuum hose stethoscope, be very careful not to come into contact with moving engine components such as the drivebelts, cooling fan, etc.*

Fuel hose

Warning: *Gasoline is extremely flammable, so take extra precautions when you work on any part of the fuel system. Don't smoke or allow open flames or bare light bulbs near the work area, and don't work in a garage where a natural gas-type appliance (such as a water heater or clothes dryer) with a pilot light is present. Since gasoline is carcinogenic, wear latex gloves when there's a possibility of being exposed to fuel, and, if you spill any fuel on your skin, rinse it off immediately with soap and water. Mop up any spills immediately and do not store fuel-soaked rags where they could ignite. The fuel system is under constant pressure, so, if any fuel lines are to be disconnected, the fuel pressure in the system must be relieved first (see Chapter 4 for more information). When you perform any kind of work on the fuel system, wear safety glasses and have a Class B type fire extinguisher on hand.*

8   Check all rubber fuel lines for deterioration and chafing. Check especially for cracks in areas where the hose bends and just before fittings, such as where a hose attaches to the fuel filter.

9   When replacing hose, use only hose that is specifically designed for your fuel injection system.

Metal lines

10   Sections of metal line are often used for fuel line between the fuel pump and fuel injection system. Check carefully to be sure the line has not been bent or crimped and that cracks have not started in the line.

11   If a section of metal fuel line must be replaced, only seamless steel tubing should be used, since copper and aluminum tubing don't have the strength necessary to withstand normal engine vibration.

12   Check the metal brake lines where they enter the master cylinder and brake proportioning unit (if used) for cracks in the lines or loose fittings. Any sign of brake fluid leakage calls for an immediate, thorough inspection of the brake system.

15   Cooling system check (every 7500 miles or 6 months)

Refer to illustration 15.4

1   Many major engine failures can be attributed to a faulty cooling system. If the vehicle is equipped with an automatic transaxle, the cooling system also cools the transmission fluid and thus plays an important role in prolonging transaxle life.

2   The cooling system should be checked with the engine cold. Do this before the vehicle is driven for the day or after the engine has been shut off for at least three hours.

3   Remove the radiator cap by turning it to the left until it reaches a stop. Refer to the underhood photographs at the front of this Chapter to locate the radiator cap. If you hear a hissing sound (indicating there is still pressure in the system), wait until it stops. Now press down on the cap with the palm of your hand and continue turning to the left until the cap can be removed. Thoroughly clean the cap, inside and out, with clean water. Also clean the filler neck on the radiator. All traces of corrosion should be removed. The coolant inside the radiator should be relatively transparent. If it's rust colored, the system should be drained and refilled (see Section 29). If the coolant level isn't up to the top, add additional antifreeze/coolant mixture (see Section 4).

4   Carefully check the large upper and lower radiator hoses along with the smaller diameter heater hoses which run from the engine to the firewall. Inspect each hose along its entire length, replacing any hose that is cracked, swollen or shows signs of deterioration. Cracks may become more apparent if the hose is squeezed **(see illustration)**.

5   Make sure that all hose connections are tight. A leak in the cooling system will usually show up as white or rust colored deposits on the areas adjoining the leak. If wire-type clamps are used at the ends of the hoses, it may be a good idea to replace them with more secure screw-type clamps.

6   Use compressed air or a soft brush to remove bugs, leaves, etc. from the front of the radiator or air conditioning condenser. Be careful not to damage the delicate cooling fins or cut yourself on them.

7   Every other inspection, or at the first indication of cooling system problems, have the cap and system pressure tested. If you don't have a pressure tester, most gas stations and repair shops will do this for a minimal charge.

16   Tire rotation (every 7500 miles or 6 months)

Refer to illustration 16.2

1   The tires should be rotated at the specified intervals and whenever uneven wear is noticed. Since the vehicle will be raised and the tires removed anyway, check the brakes (see Section 16) at this time.

2   Radial tires must be rotated in a specific pattern **(see illustration)**.

3   Refer to the information in *Jacking and towing* at the front of this manual for the proper procedures to follow when raising the vehicle and changing a tire. If the brakes are to be checked, do not apply the parking brake as stated. Make sure the tires are blocked to prevent the vehicle from rolling.

4   Preferably, the entire vehicle should be raised at the same time. This can be done on a hoist or by jacking up each corner and then lowering the vehicle onto jackstands placed under the frame rails. Always use four jackstands and make sure the vehicle is firmly supported.

5   After rotation, check and adjust the tire pressures as necessary and be sure to check the lug nut tightness. Ideally, lug nuts should

17.7 You will find an inspection window in each caliper - placing a ruler across the hole should enable you to determine the thickness of the remaining pad material for both the inner and outer pads

be torqued to Specifications with a torque wrench, and rechecked after 25 miles of driving.

6   For further information on the wheels and tires, refer to Chapter 10.

17   Brake system check (every 15,000 miles or 12 months)

Warning: *The dust created by the brake system may contain asbestos, which is harmful to your health. Never blow it out with compressed air and don't inhale any of it. An approved filtering mask should be worn when working on the brakes. Do not, under any circumstances, use petroleum-based solvents to clean brake parts. Use brake system cleaner only! Try to use non-asbestos replacement parts whenever possible.*
Note: *For detailed photographs of the brake system, refer to Chapter 9.*

1   In addition to the specified intervals, the brakes should be inspected every time the wheels are removed or whenever a defect is suspected.

2   Any of the following symptoms could indicate a potential brake system defect: The vehicle pulls to one side when the brake pedal is depressed; the brakes make squealing or dragging noises when applied; brake pedal travel is excessive; the pedal pulsates; or brake fluid leaks, usually onto the inside of the tire or wheel.

3   Loosen the wheel lug nuts.

4   Raise the vehicle and place it securely on jackstands.

5   Remove the wheels (see *Jacking and towing* at the front of this book, or your owner's manual, if necessary).

Front disc brakes

Refer to illustrations 17.7, 17.9 and 17.11

6   There are two pads (an outer and an inner) in each caliper. The pads are visible with the wheels removed.

7   Check the pad thickness by looking at each end of the caliper and through the inspection window in the caliper body **(see illustration)**. If the lining material is less than the thickness listed in this Chapter's Specifications, replace the pads. **Note:** *Keep in mind that the lining material is riveted or bonded to*

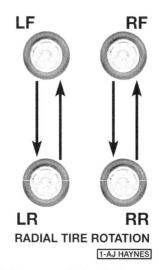

16.2 Recommended radial tire rotation pattern for these models

a metal backing plate and the metal portion is not included in this measurement.

8   If it is difficult to determine the exact thickness of the remaining pad material by the above method, or if you are at all concerned about the condition of the pads, remove the caliper(s), then remove the pads from the calipers for further inspection (refer to Chapter 9).

9   Once the pads are removed from the calipers, clean them with brake cleaner and re-measure them with a ruler or a vernier caliper **(see illustration)**.

10   Measure the disc thickness with a micrometer to make sure that it still has service life remaining. If any disc is thinner than the specified minimum thickness, replace it (refer to Chapter 9). Even if the disc has service life remaining, check its condition. Look for scoring, gouging and burned spots. If these conditions exist, remove the disc and have it resurfaced (see Chapter 9).

11   Before installing the wheels, check all brake lines and hoses for damage, wear, deformation, cracks, corrosion, leakage, bends and twists, particularly in the vicinity of the rubber hoses at the calipers **(see illustration)**. Check the clamps for tightness and the connections for leakage. Make sure that all hoses and lines are clear of sharp edges,

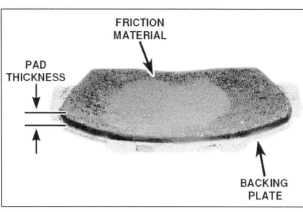

17.9 If a more precise measurement of pad thickness is necessary, remove the pads and measure the remaining friction material

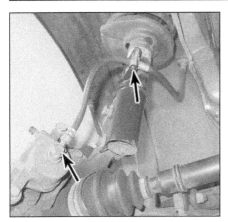

17.11 Check along the brake hoses and at each fitting (arrows) for deterioration and cracks

17.12a You can check the thickness of the remaining brake lining material by removing the rubber plug (arrow) in the backing plate with a small screwdriver

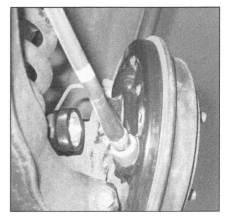

17.12b Use a small flashlight to inspect the lining

moving parts and the exhaust system. If any of the above conditions are noted, repair, reroute or replace the lines and/or fittings as necessary (see Chapter 9).

Rear drum brakes

Refer to illustrations 17.12a, 17.12b, 17.15 and 17.17

12   On Sprint models, the rear brake lining can be checked without removing the brake drum. Remove the rubber plugs in the backing plates with a small screwdriver and use a flashlight to check the lining thickness **(see illustrations)**.

13   For a more complete brake inspection, refer to Chapter 9 and remove the brake drums.

14   With the drums removed, carefully clean the brake assembly with brake system cleaner. **Warning:** *Don't blow the dust out with compressed air and don't inhale any of it (it may contain asbestos, which is harmful to your health).*

15   Note the thickness of the lining material on both front and rear brake shoes. If the material has worn away to within 1/16-inch of

the recessed rivets or 1/8-inch of the metal backing on bonded type shoes, the shoes should be replaced **(see illustration)**. The shoes should also be replaced if they're cracked, glazed (shiny areas), or covered with brake fluid.

16   Make sure all the brake assembly springs are connected and in good condition, referring to the photographs in Chapter 9, if necessary.

17   Check the brake components for signs of fluid leakage. With your finger or a small screwdriver, carefully pry back the rubber cups on the wheel cylinder located at the top of the brake shoes **(see illustration)**. Any leakage here is an indication that the wheel cylinders should be replaced immediately (see Chapter 9). Also, check all hoses and connections for signs of leakage.

18   Wipe the inside of the drum with a clean rag and denatured alcohol or brake cleaner. Again, be careful not to breathe the dangerous asbestos dust.

19   Check the inside of the drum for cracks, score marks, deep scratches and "hard

spots" which will appear as small discolored areas. If imperfections cannot be removed with fine emery cloth, the drum must be taken to an automotive machine shop for resurfacing.

20   Repeat the procedure for the remaining wheel. If the inspection reveals that all parts are in good condition, reinstall the brake drums, install the wheels and lower the vehicle to the ground.

Brake booster check

21   Sit in the driver's seat and perform the following sequence of tests.

22   With the brake fully depressed, start the engine - the pedal should move down a little when the engine starts.

23   With the engine running, depress the brake pedal several times - the travel distance should not change.

24   Depress the brake, stop the engine and hold the pedal in for about 30 seconds - the pedal should neither sink nor rise.

25   Restart the engine, run it for about a minute and turn it off. Then firmly depress the brake several times - the pedal travel should decrease with each application.

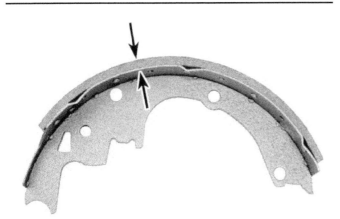

17.15 If the lining is bonded to the brake shoe, measure the lining thickness from the outer surface to the metal shoe, as shown here; if the lining is riveted to the shoe, measure from the lining outer surface to the rivet head

17.17 Check the wheel cylinder boots for leaking fluid indicating that the cylinder must be replaced or rebuilt

26 If your brakes do not operate as described, the brake booster has failed. Refer to Chapter 9 for the replacement procedure.

Parking brake

27 Slowly pull up on the parking brake and count the number of clicks you hear until the handle is up as far as it will go. The adjustment is correct if you hear the specified number of clicks (see this Chapter's Specifications). If you hear more or fewer clicks, it's time to adjust the parking brake (see Chapter 9).
28 An alternative method of checking the parking brake is to park the vehicle on a steep hill with the parking brake set and the transaxle in Neutral. If the parking brake cannot prevent the vehicle from rolling, it is in need of adjustment (see Chapter 9).

---

18 Manual transaxle lubricant level check (every 15,000 miles or 12 months)

Refer to illustration 18.1
1 On Sprint models, the lubricant level can be checked by unscrewing the dipstick **(see illustration)**. On later models the manual transaxle does not have a dipstick. To check the fluid level, raise the vehicle and support it securely on jackstands. The vehicle must be level. The check/fill plug is on the back side of the transaxle, and the drain plug is on the bottom edge of the housing. Remove the check/fill plug with a socket or box-end wrench. If the lubricant level is correct, it should be up to the lower edge of the hole. Often, lubricant will leak out when the plug is removed, indicating the level is correct. If lubricant does not leak out when the plug is removed, use your finger as a dipstick to check that it is up to the level of the hole.
2 If the transaxle needs more lubricant (if the level is not up to the hole), use a funnel to add more. Handy lubricant containers with built-in pumps are available from auto parts stores; these provide a still easier method for

a moderate extra expense. Stop filling the transaxle when the lubricant begins to run out the hole.
3 Install the plug and tighten it securely. Drive the vehicle a short distance, then check for leaks.

---

19 Valve clearance - check and adjustment (not all models) (every 15,000 miles or 12 months)

Refer to illustrations 19.4, 19.6, 19.7, 19.8 and 19.10
Warning: *The electric cooling fan can activate at any time. Disconnect the fan motor or the negative battery cable when working in the vicinity of the fan.*
Note: *This procedure applies only to Sprint and 1998 and later four-cylinder Metro models.*
1 The valve clearances are checked and adjusted with the engine at normal operating temperature.
2 Remove the air cleaner assembly
3 Remove the valve cover (see Chapter 2, Part A).

1.0L three-cylinder engine

4 Place the number one piston at Top Dead Center (TDC) on the compression stroke (see Chapter 2, Part A). The number one cylinder rocker arms (closest to the timing belt end of the engine) should be loose (able to move up-and-down slightly) and the notch in the crankshaft pulley should line up with the 0 on the timing tag **(see illustration)**.
5 Check/adjust only the valves indicated by arrows in illustration 19.4. The valve clearances can be found in the specifications section at the beginning of this Chapter.
6 The clearance is measured by inserting the specified size feeler gauge between the end of the valve stem and the adjusting screw **(see illustration)**. You should feel a slight amount of drag when the feeler gauge is moved back-and-forth **(see illustration)**.

18.1 On Sprint models, the manual transaxle dipstick is located adjacent to the distributor - the lubricant level should be kept between the Full and Add lines

7 If the gap is too large or too small, loosen the locknut and turn the adjusting screw to obtain the correct gap **(see illustration)**. Recheck the clearance to make sure it hasn't changed.
8 Rotate the crankshaft 240 -degrees until the piston closest to the transaxle is at TDC on the compression stroke. The number 3 cylinder rocker arms should be loose and the notch in the crankshaft pulley should be lined up with the timing cover bolt on the lower right timing cover bolt at the lower left of the engine **(see illustration)**.
9 Adjust the valves indicated by arrows in **illustration 19.8** as described in Steps 6 and 7.
10 Rotate the crankshaft 240 -degrees until the number 2 piston is at TDC on the compression stroke. The number 2 cylinder rocker arms should be loose and the notch in the crankshaft pulley should be lined up with the timing cover bolt on the lower right timing cover bolt at the lower left of the engine **(see illustration)**.
11 Adjust the valves indicated by arrows in **illustration 19.10**.

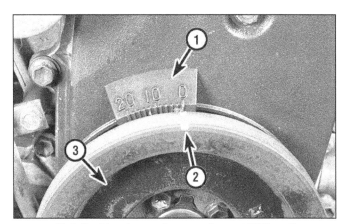 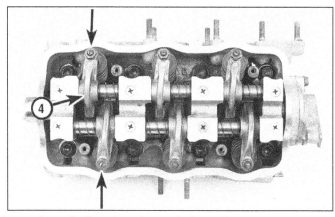

19.4 Position the number one piston at TDC on the compression stroke and adjust the valves as indicated by the arrows (three cylinder models)

| 1 | Timing tag | 2 | Pulley notch | 3 | Crankshaft pulley | 4 | Number one cylinder rocker arms (arrows) |

19.6 You should feel a slight amount of drag when the feeler gauge is moved back-and-forth

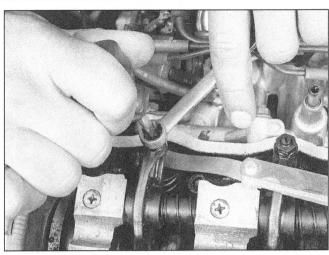

19.7 To adjust the valve clearance, use a box-end wrench to loosen the locknut slightly and a screwdriver to turn the adjusting screw (three cylinder models)

19.8 Number three cylinder valve adjustment details (three cylinder models)

1	Rotate the crankshaft 240-degrees	3	Bolt	4 Number three cylinder rocker arms (arrows)
2	Pulley notch			

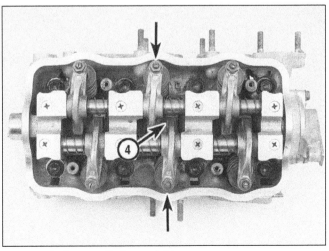

19.10 Number two cylinder valve adjustment details (three cylinder models)

1	Rotate the crankshaft 240-degrees	3	Bolt	4 Number two cylinder rocker arms (arrows)
2	Pulley notch			

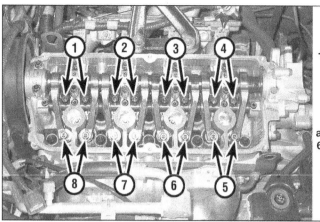

19.14 Valve numbering on 1.3L four-cylinder engines - with the number one piston at TDC on the compression stroke, adjust valves 1, 2, 8 and 6; rotate the crankshaft 360 degrees (one revolution) and adjust valves 3, 4, 7 and 5

1.3L four-cylinder engine

Refer to illustration 19.14

12   Position the number one piston at TDC on the compression stroke (see Chapter 2, Part A).

13   Make sure the rocker arms for the number one cylinder valves are loose and number four are tight. If they aren't, the number one piston is not at TDC on the compression stroke. If the rocker arms are not loose, rotate the crankshaft one complete revolution (360-degrees). The rocker arms should now be loose.

14   Check and adjust the valves numbered 1, 2, 8 and 6 in the accompanying illustration **(see illustration)**. The valve clearances can be found in the Specifications at the beginning of this Chapter.

15   The clearance is measured by inserting the specified size feeler gauge between the end of the valve stem and the adjusting screw **(see illustration 19.6)**. You should feel a slight amount of drag when the feeler gauge is moved back-and-forth.

16   If the gap is too large or too small, loosen the locknut and turn the adjusting screw to obtain the correct gap **(see illustration 19.7)**.

17   Once the gap has been set, hold the screw in position with a screwdriver and retighten the locknut. Recheck the valve

clearance - sometimes it will change slightly when the locknut is tightened. If so, readjust it until it's correct.

18   Repeat the procedure for the remaining valves **(see illustration 19.14)**, then turn the crankshaft one complete revolution (360-degrees) and realign the notch in the pulley with the zero on the engine.

19   Adjust valves 3, 4, 7 and 5 **(see illustration 19.14)**.

All models

20   Installation of the spark plugs, valve cover, spark plug wires and boots, accelerator cable bracket, etc. is the reverse of removal.

---

20   Carburetor/TBI mounting nut torque check (every 15,000 miles or 12 months)

---

1   The carburetor or Throttle Body Injection (TBI) body is attached to the top of the intake manifold by several bolts or nuts. These fasteners can sometimes work loose from vibration and temperature changes during normal engine operation and cause a vacuum leak.

2   If you suspect a vacuum leak exists at the bottom of the carburetor or throttle body,

obtain a length of hose. Start the engine and place one end of the hose next to your ear as you probe around the base with the other end. You'll hear a hissing sound if a leak exists (be careful of hot or moving engine components).

3   Remove the air cleaner assembly (see Chapter 4), tagging each hose to be disconnected with a piece of numbered tape to make reassembly easier.

4   Locate the mounting bolts at the base of the carburetor or top of the throttle body. Decide what special tools or adapters will be necessary, if any, to tighten the fasteners.

5   Tighten the bolts or nuts to the torque listed in this Chapter's specifications. Don't overtighten them as the threads could strip.

6   If, after the bolts are properly tightened, a vacuum leak still exists, the carburetor or throttle body must be removed and a new gasket installed. See Chapter 4 for more information.

7   After tightening the fasteners, reinstall the air cleaner and return all hoses to their original positions.

---

21   Air filter replacement (every 30,000 miles or 24 months)

---

1   At the specified intervals, the air filter should be replaced with a new one.

Early models

Refer to illustrations 21.2 and 21.3

2   Release the air cleaner cover clips and unscrew the wing nut **(see illustration)**.

3   Lift the air filter element out of the housing and wipe out the inside of the air cleaner housing with a clean rag **(see illustration)**.

4   While the air cleaner cover is off, be careful not to drop anything down into the air cleaner assembly.

5   Place the new filter in the air cleaner housing. Make sure it seats properly in the lower half of the housing.

6   Install the air cleaner cover and tighten the wing nut securely.

21.2 After removing the wing nut, release the clips around the edge and lift off the air cleaner cover

21.3 Hold the top out of the way and lift out the air filter element, then wipe out the inside of the housing with a clean rag

21.7 Detach the air cleaner housing clips

21.8 Pull the cover back and remove the filter element

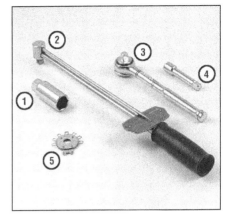

22.2a Tools required for changing spark plugs

1   *Spark plug socket* - This will have special padding inside to protect the spark plug's porcelain insulator
2   *Torque wrench* - Although not mandatory, using this tool is the best way to ensure the plugs are tightened properly
3   *Ratchet* - Standard hand tool to fit the spark plug socket
4   *Extension* - Depending on model and accessories, you may need special extensions and universal joints to reach one or more of the plugs
5   *Spark plug gap gauge* - This gauge for checking the gap comes in a variety of styles. Make sure the gap for your engine is included

Later models

Refer to illustrations 21.7 and 21.8

7   Detach the clips and pull the cover off the filter element **(see illustration)**.
8   Lift the filter element out and wipe out the inside of the air cleaner housing with a clean rag **(see illustration)**.
9   Seat the new filter in the housing, install the cover and secure the clips.

22   Spark plug check and replacement (every 30,000 miles or 24 months)

Refer to illustrations 22.2a, 22.2b, 22.5a, 22.5b, 22.6, 22.8, 22.9 and 22.10

1   The spark plug wires should be checked whenever new spark plugs are installed (see the next Section).
2   In most cases, the tools necessary for spark plug replacement include a spark plug socket which fits onto a ratchet (spark plug sockets are padded inside to prevent damage to the porcelain insulators on the new plugs), various extensions and a gap gauge to check and adjust the gap on the new plugs **(see illustration)**. On later models it is neces-

sary to remove the coil packs for access to the spark plugs **(see illustration)**. A torque wrench should be used to tighten the new plugs. It is a good idea to allow the engine to cool before removing or installing the spark plugs.
3   The best approach when replacing the spark plugs is to purchase the new ones in advance, adjust them to the proper gap and replace the plugs one at a time. When buying the new spark plugs, be sure to obtain the correct plug type for your particular engine. The plug type can be found in the Specifications at the front of this Chapter and on the Emission Control Information label located under the hood. If these two sources list different plug types, consider the emission control label correct.
4   Allow the engine to cool completely before attempting to remove any of the plugs. While you are waiting for the engine to cool, check the new plugs for defects and adjust the gap.
5   Check the gap by inserting the proper thickness gauge between the electrodes at the tip of the plug **(see illustration)**. The gap between the electrodes should be the same as the one specified on the Emissions Con-

trol Information label or in Chapter 5. The wire should slide between the electrodes with a slight amount of drag. If the gap is incorrect, use the adjuster on the gauge body to bend the curved side electrode slightly until the proper gap is obtained **(see illustration)**. If the side electrode is not exactly over the center electrode, bend it with the adjuster until it is. Check for cracks in the porcelain insulator (if any are found, the plug should not be used).

22.2b On some later models it will be necessary to unplug the electrical connectors and remove the bolts (arrows) and the coil pack for access to the spark plugs

22.5a Spark plug manufacturers recommend using a wire-type gauge when checking the gap - if the wire does not slide between the electrodes with a slight drag, adjustment is required

22.5b To change the gap, bend the side electrode only, as indicated by the arrows, and be very careful not to crack or chip the porcelain insulator surrounding the center electrode

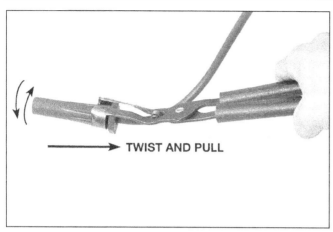

22.6 When removing the spark plug wires, pull only on the boot and twist it back-and-forth

22.8 Use the special socket and a ratchet to remove the spark plugs

6   With the engine cool, remove the spark plug wire as described in the next Section from one spark plug. Pull only on the boot at the end of the wire - do not pull on the wire. A plug wire removal tool should be used if available **(see illustration)**.

7   If compressed air is available, use it to blow any dirt or foreign material away from the spark plug hole. A common bicycle pump will also work. The idea here is to eliminate the possibility of debris falling into the cylinder as the spark plug is removed.

8   Place the spark plug socket over the plug and remove it from the engine by turning it in a counterclockwise direction **(see illustration)**.

9   Compare the spark plug with the chart shown on the inside back cover of this manual to get an indication of the general running condition of the engine. Before installing the new plugs, it is a good idea to apply a thin coat of anti-seize compound to the threads **(see illustration)**.

10   Thread one of the new plugs into the hole until you can no longer turn it with your fingers, then tighten it with a torque wrench (if available) or the ratchet. It's a good idea to slip a short length of rubber hose over the

end of the plug to use as a tool to thread it into place **(see illustration)**. The hose will grip the plug well enough to turn it, but will start to slip if the plug begins to cross-thread in the hole - this will prevent damaged threads and the accompanying repair costs.

11   Before pushing the spark plug wire onto the end of the plug, inspect the wire following the procedures outlined in the next Section.

12   Attach the plug wire to the new spark plug, again using a twisting motion on the boot until it's seated on the spark plug.

13   Repeat the procedure for the remaining spark plugs, replacing them one at a time to prevent mixing up the spark plug wires.

23   Spark plug wire check and replacement (30,000 miles or 24 months)

1   The spark plug wires should be checked whenever new spark plugs are installed.

2   Begin this procedure by making a visual check of the spark plug wires while the engine is running. In a darkened garage (make sure there is adequate ventilation) start the engine and observe each plug wire. Be

careful not to come into contact with any moving engine parts. If there is a break in the wire, you will see arcing or a small spark at the damaged area. If arcing is noticed, make a note to obtain new wires, then allow the engine to cool and check the distributor cap and rotor.

3   The spark plug wires should be inspected one at a time to prevent mixing up the order, which is essential for proper engine operation. Each original plug wire should be numbered to help identify its location. If the number is illegible, a piece of tape can be marked with the correct number and wrapped around the plug wire.

4   Disconnect the plug wire from the spark plug. Grasp the rubber boot, twist the boot half a turn and pull the boot free. Do not pull on the wire itself **(see illustration 22.6)**.

5   Check inside the boot for corrosion, which will look like a white crusty powder. Light corrosion can be removed with a small wire brush, but replace the wires if corrosion is heavy.

6   Push the wire and boot back onto the end of the spark plug. It should fit tightly onto the end of the plug. If it doesn't, remove the wire and use pliers to carefully crimp the metal connector inside the wire boot until the fit is snug.

7   Using a clean rag, wipe the entire length of the wire to remove built-up dirt and grease. Once the wire is clean, check for burns, cracks and other damage. Do not bend the wire sharply, because the conductor might break.

8   Disconnect the wire from the distributor or ignition coil pack. Pull only on the rubber boot. Check for corrosion and a tight fit. Replace the wire in the coil pack.

9   Inspect the remaining spark plug wires, making sure that each one is securely fastened at the coil pack, distributor and spark plug when the check is complete.

10   If new spark plug wires are required, purchase a set for your specific engine model. Remove and replace the wires one at a time to avoid mix-ups in the firing order.

11   Clean the coil pack or distributor with a

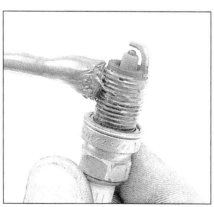

22.9 A light coat of anti-seize compound applied to the threads of the spark plugs will keep the threads in the cylinder head from being damaged the next time the plugs are removed

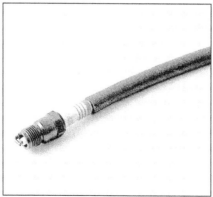

22.10 A piece of rubber hose will aid in getting the spark plug started in the hole

24.2 After detaching the clips the distributor cap can be lifted off

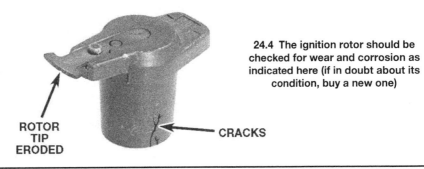

24.4 The ignition rotor should be checked for wear and corrosion as indicated here (if in doubt about its condition, buy a new one)

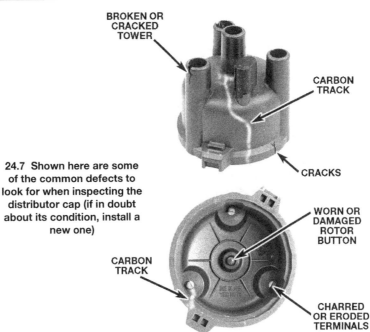

24.7 Shown here are some of the common defects to look for when inspecting the distributor cap (if in doubt about its condition, install a new one)

dampened cloth and dry them with a dampened cloth thoroughly.

12   Inspect the coil pack (1998 and later four-cylinder models) for cracks, damage and carbon tracking. Carbon tracks can usually be removed. If damage exists, refer to Chapter 5 for the replacement procedure.

24   Distributor cap and rotor check and replacement (models so equipped)

Refer to illustrations 24.2, 24.4 and 24.7

Note 1: *It's common practice to install a new distributor cap and rotor whenever new spark plug wires are installed.*

Note 2: *1998 and later four-cylinder models are not equipped with a distributor, so this procedure does not apply.*

1   Although the breakerless distributor used on these vehicles requires much less maintenance than a conventional distributor, periodic inspections should be performed at the intervals specified in the routine maintenance schedule and whenever any work is performed on the distributor.

2   Disconnect the ignition coil wire from the coil, then unsnap the spring clips that hold the cap to the distributor body **(see illustration)**. Detach the distributor cap and wires.

3   Place the cap, with the spark plug and coil wires still attached, out of the way. Use a length of wire or rope to secure it, if necessary.

4   The rotor is now visible on the end of the distributor shaft. Check it carefully for cracks and carbon tracks. Make sure the center terminal spring tension is adequate (not all models) and look for corrosion and wear on the rotor tip **(see illustration)**. If in doubt about its condition, replace it with a new one.

5   If replacement is required, detach the rotor from the shaft and install a new one.

6   While the distributor cap is off, check the air gap as described in Chapter 5.

7   Check the distributor cap for carbon tracks, cracks and other damage. Closely examine the terminals on the inside of the cap for excessive corrosion and damage **(see illustration)**. Slight deposits are normal. Again, if in doubt about the condition of the cap, replace it with a new one.

8   When replacing the cap, simply transfer the spark plug and coil wires, one at a time, from the old cap to the new cap. Be very careful not to mix up the wires!

9   Reattach the cap to the distributor, then tighten the screws or reposition the spring clips to hold it in place.

25   Carburetor choke check

Refer to illustration 25.3

1   The choke operates only when the engine is cold, so this check should be performed before the engine has been started for the day.

2   Open the hood and remove the air cleaner from the carburetor. It's held in place by a nut at the center. If any vacuum hoses must be disconnected, tag them to ensure reinstallation in their original positions.

3   Look at the center of the carburetor. You'll notice a flat plate at the carburetor opening **(see illustration)**.

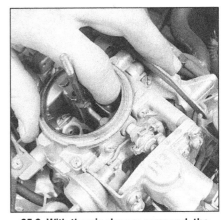

25.3 With the air cleaner removed, the choke plate is visible in the carburetor throat

4   Have an assistant press the throttle pedal to the floor. The plate should close completely. Start the engine while you watch the plate at the carburetor. Don't position your face near the carburetor, as the engine could backfire, causing serious burns! When the engine starts, the choke plate should open slightly.

5   Allow the engine to continue running at an idle speed. As the engine warms up to operating temperature, the plate should slowly open, allowing more air to enter through the top of the carburetor.

6   After a few minutes, the choke plate should be completely open to the vertical position. Blip the throttle to make sure the fast idle cam disengages.

7   You'll notice that engine speed corresponds to the plate opening. With the plate closed, the engine should run at a fast idle speed. As the plate opens and the throttle is moved to disengage the fast idle cam, the engine speed will decrease.

8   With the engine off and the throttle held half-way open, open and close the choke several times. Check the linkage to see if it's hooked up correctly and make sure it doesn't bind.

9   If the choke or linkage binds, sticks or works sluggishly, clean it with choke cleaner (an aerosol spray available at auto parts stores). If the condition persists after cleaning, replace the troublesome parts.

10   Visually inspect all vacuum hoses to be sure they're securely connected and look for cracks and deterioration. Replace as necessary.

11   Refer to Chapter 4 for more information on the choke.

26   Thermostatically controlled air cleaner check (carburetor equipped models only) - every 30,000 miles or 24 months

Refer to illustration 26.3

1   Carburetor-equipped engines have a thermostatically controlled air cleaner which draws air from different locations, depending on engine temperature.

2   This is a visual check, requiring the use of a small mirror.

3   When the engine is cold, locate the air control valve inside the air cleaner housing **(see illustration)**.

4   Disconnect The flexible air duct attached to the end of the snorkel. This will enable you to see the air control valve inside.

5   Start the engine and look through the end of the snorkel, which should move up and down to block off air. With the valve closed, air cannot enter through the snorkel, but instead enters the air cleaner through the flexible duct attached to the exhaust manifold and the heat stove passage.

6   As the engine warms up to operating temperature, the valve should move down to allow the air to enter through the snorkel end. Depending on outside temperature, this may take 10-to-15 minutes. To speed up this check you can reconnect the snorkel air duct, drive the vehicle, then check to see if the valve is completely open.

7   If the thermostatically controlled air cleaner isn't operating properly, see Chapter 6 for more information.

26.3 The air control valve is located in the air cleaner snorkel (arrow)

27   Idle speed check and adjustment (1994 and earlier models only) (every 30,000 miles or 24 months)

Primary idle speed

Refer to illustrations 27.5, 27.9a and 27.9b

Note 1: *On later models the idle speed is controlled by the ECM and is not adjustable. If the information in this Section differs from the Vehicle Emission Control Information label in the engine compartment of your vehicle, the label should be considered correct.*

Note 2: *1994 models with upgraded emissions systems also do not require adjustment. Check your VECI label.*

1   Engine idle speed is the speed at which the engine operates when no throttle pedal pressure is applied. The idle speed is critical to the performance of the engine itself, as well as many engine sub-systems.

2   A hand-held tachometer must be used when adjusting idle speed to get an accurate reading. The exact hook-up for these meters varies with the manufacturer, so follow the particular directions included.

3   Set the parking brake and block the wheels. Be sure the transaxle is in Neutral (manual transaxle) or Park (automatic

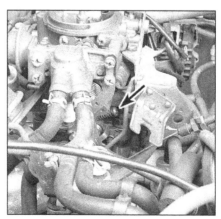

27.9a On carburetor-equipped models, the idle speed adjusting screw (arrow) is located at the rear of the carburetor

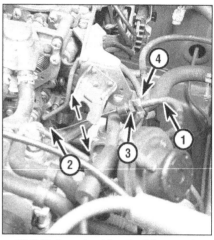

27.5 Before checking the idle speed, check the accelerator cable and make sure it isn't too tight (it should have about 1/8-inch play)

1   *Accelerator cable*
2   *Throttle lever*
3   *Cable adjusting nut*
4   *Locknut*

transaxle).

4   Turn off the air conditioner (if equipped), the headlights and any other accessories during this procedure.

5   Make sure the accelerator cable is not too tight. If necessary, turn the adjusting nut so there is approximately 1/8-inch play in the cable where it connects to the throttle lever **(see illustration)**.

6   On Metro models remove the spare fuse from the fuse block and install it in the diagnosis terminal.

7   Start the engine and allow it to reach normal operating temperature.

8   Check the engine idle speed with the tachometer and compare it to the VECI label on the hood.

9   If the idle speed is not correct, turn the idle speed adjusting screw (clockwise for faster, counterclockwise for slower) until the idle speed is correct **(see illustrations)**.

27.9b The idle speed adjusting screw (arrow) on fuel-injected models

27.12 The VSV (arrow) is located on the firewall in the engine compartment (Metro 1.0L three-cylinder engine)

28.4 Place your finger over the PCV valve and make sure suction can be felt - if there is no suction, the valve is faulty (earlier model)

28.5 Blow through the PCV valve (connect a piece of hose as shown here) - air should pass through the valve with some resistance

Idle speed/air conditioning VSV adjustment (1994 and earlier Metro 1.0L models)

Refer to illustration 27.12

10 With the engine operating at normal temperature, connect a tachometer and switch the air conditioner On.

11 Install the spare fuse into the diagnosis switch terminal in the passenger compartment fuse box.

12 Check and adjust the idle speed by adjusting the A/C vacuum switch valve (VSV) idle screw to the proper specification **(see illustration)**.

    Automatic transaxle   850 ± 50 rpm
    Manual transaxle   900 ± 50 rpm

13 Remove the spare fuse from the diagnosis switch terminal.

14 Adjust the idle speed to the rpm listed in the Chapter 1 Specifications (refer to idle speed adjustment in Chapter 1).

28 Positive Crankcase Ventilation (PCV) valve check and replacement (every 30,000 miles or 24 months)

Refer to illustration 28.4 and 28.5

1 The Positive Crankcase Ventilation (PCV) system directs blowby gases from the crankcase through the PCV valve and hose back into the intake manifold so they can be burned in the engine. The system consists of a hose leading from the valve cover to the intake manifold and a fresh air hose between the air cleaner assembly and the rocker arm cover.

2 The PCV valve and hose is located in the intake manifold on earlier models and in a hose between the valve cover and the intake manifold on later models.

3 With the engine idling at normal operating temperature, on earlier models detach the hose from the valve and on later models pull the valve (with hose attached) from the valve cover.

4 Place your finger over the valve opening

or hose **(see illustration)**. If there's no vacuum, check for a plugged hose, manifold port, or the valve itself. Replace any plugged or deteriorated hoses.

5 Turn off the engine. Connect a piece of hose to the PCV valve. Blow through the valve from the valve cover end **(see illustration)**. If the valve is operating properly, air should pass through the valve with some resistance. If no air passes through the valve or it passes too easily, install a new one.

6 To replace the valve, remove it from the manifold, noting its installed position.

7 When purchasing a replacement PCV valve, remove the screws and detach it from the manifold. Make sure it's for your particular vehicle and engine size. Compare the old valve with the new one to make sure they're the same.

8 Install the PCV valve and connect the hose. For further information on the PCV system refer to Chapter 6.

29 Cooling system servicing (draining, flushing and refilling) (every 30,000 miles or 24 months)

Warning : *Do not allow antifreeze to come in contact with your skin or painted surfaces of the vehicle. Rinse off spills immediately with plenty of water. Antifreeze is highly toxic if ingested. Never leave antifreeze lying around in an open container or in puddles on the floor; children and pets are attracted by it's sweet smell and may drink it. Check with local authorities about disposing of used antifreeze. Many communities have collection centers which will see that antifreeze is disposed of safely.*

1 Periodically, the cooling system should be drained, flushed and refilled to replenish the antifreeze mixture and prevent formation of rust and corrosion, which can impair the performance of the cooling system and cause engine damage.

2 At the same time the cooling system is serviced, all hoses and the radiator cap

should be inspected and replaced if defective (see Section 13).

3 Since antifreeze is a corrosive and poisonous solution, be careful not to spill any of the coolant mixture on the vehicle's paint or your skin. If this happens, rinse it off immediately with plenty of clean water. Consult local authorities about where to recycle or dispose of antifreeze before draining the cooling system. In many areas, reclamation centers have been set up to collect automobile oil and drained antifreeze/water mixtures, rather than allowing them to be added to the sewage system.

Draining

Refer to illustration 29.6

4 Apply the parking brake and block the wheels. If the vehicle has just been driven, wait several hours to allow the engine to cool down before beginning this procedure.

5 Once the engine is completely cool, remove the radiator cap. Also remove the coolant reservoir cap.

6 Drain the radiator by opening the drain plug at the bottom of the radiator **(see illustration)**. If the drain plug is corroded and

29.6 The radiator drain plug (arrow) is located at the bottom corner of the radiator

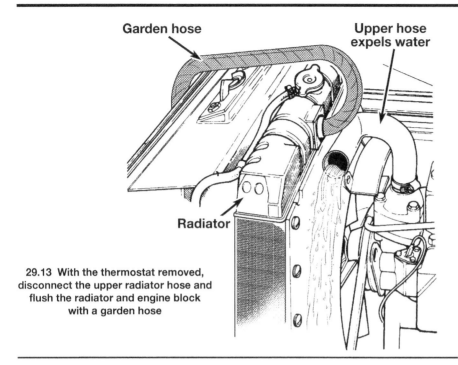

29.13 With the thermostat removed, disconnect the upper radiator hose and flush the radiator and engine block with a garden hose

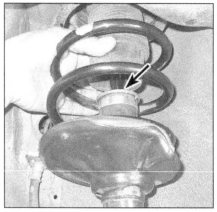

30.6 Check the front and rear shock absorbers for leakage where the rod enters the tube (arrow)

can't be turned easily, or if the radiator isn't equipped with a plug, disconnect the lower radiator hose to allow the coolant to drain. Be careful not to get antifreeze on your skin or in your eyes.

7   After the coolant stops flowing out of the radiator, disconnect the lower radiator hose from the radiator and allow the remaining coolant in the engine block to drain.

8   While the coolant is draining from the engine block, disconnect the hose from the coolant reservoir and remove the reservoir (see Chapter 3 if necessary). Flush the reservoir out with water until it's clean, and, if necessary, wash the inside with soapy water and a brush to make reading the fluid level easier.

9   While the coolant is draining, check the condition of the radiator hoses, heater hoses and clamps (refer to Section 15, if necessary).

10   Replace any damaged clamps or hoses (refer to Chapter 3 for detailed replacement procedures).

Flushing

Refer to illustration 29.13

11   Once the system is completely drained, remove the thermostat from the engine (see Chapter 3). Then reinstall the thermostat housing without the thermostat. This will allow the system to be thoroughly flushed.

12   Reconnect the lower radiator hose and tighten the radiator drain plug. Turn your heating system controls to Hot, so that the heater core will be flushed at the same time as the rest of the cooling system.

13   Disconnect the upper radiator hose, then place a garden hose in the upper radiator inlet and flush the system until the water runs clear at the upper radiator hose **(see illustration)**.

14   In severe cases of contamination or clogging of the radiator, remove the radiator

(see Chapter 3) and have a radiator repair facility clean and repair it if necessary.

15   Many deposits can be removed by the chemical action of a cleaner available at auto parts stores. Follow the procedure outlined in the manufacturer's instructions. **Note:** *When the coolant is regularly drained and the system refilled with the correct antifreeze/water mixture, there should be no need to use chemical cleaners or descalers.*

Refilling

16   To refill the system, install the thermostat, reconnect any radiator hoses and install the reservoir and the overflow hose.

17   Place the heater temperature control in the maximum heat position.

18   Make sure to use the proper coolant listed in this Chapter's Specifications. Slowly fill the radiator with the recommended mixture of antifreeze and water to the base of the filler neck. Then add coolant to the reservoir until it reaches the Full mark. Wait five minutes and recheck the coolant level in the radiator, adding if necessary.

19   Leave the radiator cap and run the engine in a well-ventilated area until the thermostat opens (coolant will begin flowing through the radiator and the upper radiator hose will become hot).

20   Turn the engine off and let it cool. Add more coolant mixture to bring the level back up to the base of the filler neck.

21   Squeeze the upper radiator hose to expel air, then add more coolant mixture if necessary. Replace the radiator cap.

22   Place the heater temperature control and the blower motor speed control to their maximum setting.

23   Start the engine, allow it to reach normal operating temperature and check for leaks.

24   If the coolant temperature rises above

normal, there is air trapped in the cooling system. Shut off the engine and allow it to cool completely; the system will automatically vent the trapped air. Repeat the procedure until the engine temperature stays at the normal position on the gauge.

30   Steering and suspension check (every 30,000 or 24 months)

Refer to illustrations 30.6, 30.9a, 30.9b, 30.9c, 30.11and 30.14
Note: *The steering linkage and suspension components should be checked periodically. Worn or damaged suspension and steering linkage components can result in excessive and abnormal tire wear, poor ride quality and vehicle handling and reduced fuel economy. For detailed illustrations of the steering and suspension components, refer to Chapter 10.*

Shock absorber check

1   Park the vehicle on level ground, turn the engine off and set the parking brake. Check the tire pressures.

2   Push down at one corner of the vehicle, then release it while noting the movement of the body. It should stop moving and come to rest in a level position within one or two bounces.

3   If the vehicle continues to move up-and-down or if it fails to return to its original position, a worn or weak shock absorber (which is part of the strut assembly) is probably the reason.

4   Repeat the above check at each of the three remaining corners of the vehicle.

5   Raise the vehicle and support it securely on jackstands.

6   Check the shock absorbers for evidence of fluid leakage **(see illustration)**. A light film of fluid is no cause for concern. Make sure that any fluid noted is from the shocks and not from some other source. If leakage is noted, replace the shocks as a set.

7   Check the shocks to be sure that they are securely mounted and undamaged.

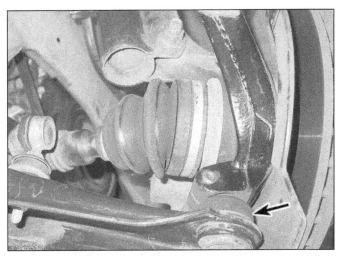

30.9a Inspect the balljoints (arrow) and the tie rod ends for torn grease seals

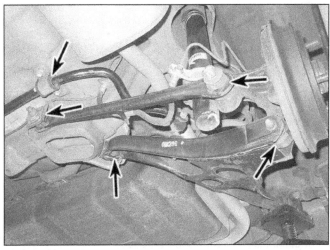

30.9b Examine the rear suspension components and bushings (arrows) for damage

Check the upper mounts for damage and wear. If damage or wear is noted, replace the shocks as a set (front or rear).

8   If the shocks must be replaced, refer to Chapter 10 for the procedure.

9   Visually inspect the steering and suspension components (front and rear) for damage and distortion. Look for damaged seals, boots and bushings and leaks of any kind. Examine the bushings where the lower control arm meets the chassis and on the stabilizer bar connections.

10   Clean the lower end of the steering knuckle. Have an assistant grasp the lower edge of the tire and move the wheel in-and-out while you look for movement at the steering knuckle-to-control arm balljoint. If there is any movement, the suspension balljoint(s) must be replaced.

11   Grasp each front tire at the front and rear edges, push in at the front, pull out at the rear and feel for play in the steering system components. If any freeplay is noted, check the steering gear and the tie-rod ends for looseness **(see illustration)**.

12   Additional steering and suspension system information and illustrations can be found in Chapter 10.

Driveaxle boot check

13   The driveaxle boots are very important because they prevent dirt, water and foreign material from entering and damaging the constant velocity (CV) joints. Oil and grease can cause the boot material to deteriorate prematurely, so it's a good idea to wash the boots with soap and water. Because it constantly pivots back and forth following the steering action of the front hub, the outer CV boot wears out sooner and should be inspected regularly.

14   Inspect the boots for tears and cracks as well as loose clamps **(see illustration)**. If there is any evidence of cracks or leaking lubricant, they must be replaced as described in Chapter 8.

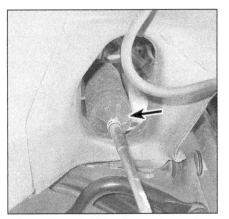

30.9c Flex each steering gear boot for cracks and leaking steering fluid

31   Automatic transaxle fluid change (every 30,000 miles or 24 months)

Refer to illustrations 31.7, 31.12 and 31.14

1   At the specified time intervals, the automatic transaxle fluid should be drained and replaced.

2   Before beginning work, purchase the specified transmission fluid (see *Recommended lubricants and fluids* at the front of this Chapter).

3   Other tools necessary for this job include jackstands to support the vehicle in a raised position, 3/8-inch drive ratchet, a drain pan capable of holding least six quarts, newspapers and clean rags.

4   The fluid should be drained immediately after the vehicle has been driven. Hot fluid is more effective than cold fluid at removing built-up sediment. **Warning:** *Fluid temperature can exceed 350-degrees F in a hot transaxle. Wear protective gloves.*

5   After the vehicle has been driven to warm up the fluid, raise it and place it on jackstands for access to the transaxle drain plug.

30.11 With the steering wheel in the lock position and the vehicle raised, grasp the front tire as shown and try to move it back-and-forth - if any play is noted, check the steering gear mounts and tie-rod ends for looseness

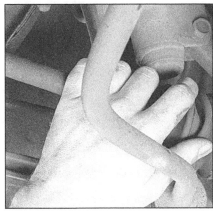

30.14 Flex the inner and outer driveaxle boots by hand to check for cracks and/or tears

6   Move the necessary equipment under the vehicle, being careful not to touch any of the hot exhaust components.

31.7 The automatic transaxle drain plug (arrow) is located in a recess in the bottom of the pan - use a socket wrench to avoid rounding off the hex

7   Place the drain pan under the transaxle and remove the drain plug **(see illustration)**. Be sure the drain pan is in position, as fluid will come out with some force. Once the fluid has drained, clean the drain plug and reinstall it securely.

8   Remove the transaxle pan bolts.
9   Carefully pry the transaxle pan loose with a screwdriver. Don't damage the pan or transaxle gasket surface or leaks could develop.
10   Remove the pan and gasket. Carefully clean the gasket surface of the transaxle to remove all traces of the old gasket and sealant.
11   Drain any remaining fluid from the transaxle pan, clean it with solvent and dry it thoroughly. Make sure to install magnets in their original positions (indentations) in the pan.
12   Remove the retaining bolts and detach the old filter from the transaxle **(see illustration)**.
13   Install the new filter and tighten the bolts.
14   Make sure the gasket surface on the transaxle pan is clean, then install a new gasket. Put the pan in place against the transaxle and install the bolts. Apply sealant to the two cross grooved headed bolts before installing them **(see illustration)**. Working around the pan, tighten each bolt a little at a time until the final torque figure listed in this Chapter's Specifications is reached. Don't overtighten

the bolts!
15   Lower the vehicle.
16   Pull out the dipstick and add new fluid to the transaxle through the dipstick tube (see *Recommended lubricants and fluids* for the recommended fluid type and capacity). Use a funnel to prevent spills. It is best to add a little fluid at a time, continually checking the level with the dipstick (see Section 6). Allow the fluid time to drain into the pan.
17   Install the dipstick.
18   Start the engine and shift the selector into all positions from P through 2, then shift into P and apply the parking brake.
19   Turn off the engine and check the fluid level. Add fluid to bring the level into the notched area on the dipstick.

32   Manual transaxle lubricant change - (every 30,000 or 24 months)

1   At the specified time intervals, the manual transaxle lubricant should be drained and replaced.
2   Before beginning work, purchase the

31.12 Remove the filter bolts (arrows) and detach the filter

31.14 Apply thread sealant to the two bolts (arrows) with cross grooved heads before installing them

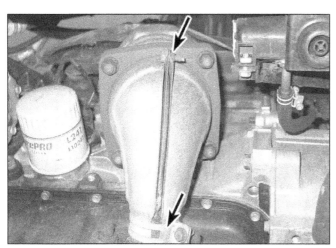

33.2a Check all of the flanged and slip-jointed exhaust connections (arrows) - look for stains that indicate exhaust leakage

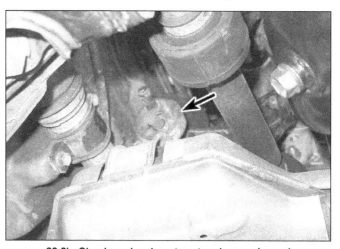

33.2b Check each exhaust system hanger (arrow) for damage and cracks

34.9 The charcoal canister (arrow) on earlier models is located in the engine compartment on the driver's side

specified transaxle lubricant (see *Recommended lubricants and fluids* and *Capacities* at the beginning of this Chapter).

3   Other tools necessary for this job include jackstands to support the vehicle in a raised position, 3/8-inch drive ratchet, a drain pan capable of holding at least four quarts, newspapers and clean rags.

4   After the vehicle has been driven to warm up the fluid, raise it and place it on jackstands for access to the transaxle drain plug. Place the drain pan under the transaxle, remove the drain plug and allow the old oil to drain into the pan (see Section 18).

5   Reinstall the drain plug securely.

6   Add new fluid through the filler hole until it begins to run out of the filler hole (see Section 18). Install the check/fill plug and tighten it securely.

33   Exhaust system check (every 30,000 miles or 24 months)

Refer to illustrations 33.2a and 33.2b

1   With the engine cold (at least three hours after the vehicle has been driven), check the complete exhaust system from the engine to the end of the tailpipe. Ideally, the inspection should be done with the vehicle on a hoist to permit unrestricted access. If a hoist isn't available, raise the vehicle and support it securely on jackstands.

2   Check the exhaust pipes and connections for evidence of leaks, severe corrosion and damage. Make sure that all brackets and hangers are in good condition and tight **(see illustrations)**.

3   At the same time, inspect the underside of the body for holes, corrosion, open seams, etc. which may allow exhaust gases to enter the passenger compartment. Seal all body openings with silicone or body putty.

4   Rattles and other noises can often be traced to the exhaust system, especially the mounts and hangers. Try to move the pipes, muffler and catalytic converter. If the components can come in contact with the body or suspension parts, secure the exhaust system with new mounts.

5   Check the running condition of the

engine by inspecting inside the end of the tailpipe. The exhaust deposits here are an indication of engine state-of-tune. If the pipe is black and sooty or coated with white deposits, the engine may need a tune-up, including a thorough fuel system inspection and adjustment.

34   Fuel system check (every 30,000 miles or 24 months)

Warning: *Gasoline is extremely flammable, so take extra precautions when you work on any part of the fuel system. Don't smoke or allow open flames or bare light bulbs near the work area, and don't work in a garage where a natural gas-type appliance (such as a water heater or clothes dryer) with a pilot light is present. Since gasoline is carcinogenic, wear latex gloves when there's a possibility of being exposed to fuel, and, if you spill any fuel on your skin, rinse it off immediately with soap and water. Mop up any spills immediately and do not store fuel-soaked rags where they could ignite. When you perform any kind of work on the fuel system, wear safety glasses and have a Class B type fire extinguisher on hand. The system on fuel injected models is under constant pressure, so, before any lines are disconnected, the fuel system pressure must be relieved (see Chapter 4).*

Refer to illustration 34.9

1   If you smell gasoline while driving or after the vehicle has been sitting in the sun, inspect the fuel system immediately.

2   Remove the gas filler cap and inspect it for damage and corrosion. The gasket should have an unbroken sealing imprint. If the gasket is damaged or corroded, install a new cap.

3   Inspect the fuel feed and return lines for cracks. Make sure that the connections between the fuel lines and the fuel injection system and between the fuel lines and the in-line fuel filter are tight. **Warning:** *If your vehicle is fuel injected, you must relieve the fuel system pressure before servicing fuel system components. The fuel system pressure-relief procedure is outlined in Chapter 4.*

4   If the fuel injectors are visible, look for

signs of fuel leakage (wet spots) around any of the injectors, they may need new O-rings (see Chapter 4).

5   Since some components of the fuel system - the fuel tank and part of the fuel feed and return lines, for example - are underneath the vehicle, they can be inspected more easily with the vehicle raised on a hoist. If that's not possible, raise the vehicle and support it on jackstands.

6   With the vehicle raised and safely supported, inspect the gas tank and filler neck for punctures, cracks and other damage. The connection between the filler neck and the tank is particularly critical. Sometimes a rubber filler neck will leak because of loose clamps or deteriorated rubber. Inspect all fuel tank mounting brackets and straps to be sure that the tank is securely attached to the vehicle. **Warning:** *Do not, under any circumstances, try to repair a fuel tank (except rubber components). A welding torch or any open flame can easily cause fuel vapors inside the tank to explode.*

7   Carefully check all rubber hoses and metal lines leading away from the fuel tank. Check for loose connections, deteriorated hoses, crimped lines and other damage. Repair or replace damaged sections as necessary (see Chapter 4).

8   The evaporative emissions control system can also be a source of fuel odors. The function of the system is to store fuel vapors from the fuel tank in a charcoal canister until they can be routed to the intake manifold where they mix with incoming air before being burned in the combustion chambers.

9   The most common symptom of a faulty evaporative emissions system is a strong odor of fuel in the engine compartment. If a fuel odor has been detected, and you have already checked the areas described above, check the charcoal canister, located next to the brake fluid reservoir or at the left front corner of the vehicle behind the bumper on earlier models and the hoses connected to it **(see illustration)**. On later models the canister is located on top of the fuel tank, refer to Chapter 6 for more information.

35   Fuel filter replacement (every 30,000 miles or 24 months)

1   Replace the fuel filter at the specified intervals. The fuel filter on 1999 and later models is located inside the fuel tank and doesn't require periodic replacement.

Carbureted engines

Refer to illustration 35.5

2   Release the residual pressure from the fuel system by removing, then reinstalling the fuel filler cap.

3   The fuel filter is located on the firewall, below the brake master cylinder.

4   Release the hose clamps at the filter fitting and slide them back up the hoses.

5   Detach the filter from the bracket and

disconnect the hoses **(see illustration)**. Now would be a good time to replace the hoses if they are deteriorated.

6   Push the hoses onto the new filter and position the clamps approximately 1/4-inch back from the ends.

7   Push the filter back into the bracket. Check to make sure it's held securely and the hoses aren't kinked.

8   Start the engine and check for fuel leaks.

Fuel injected engines

Refer to illustrations 35.14 and 35.15

9   Refer to Chapter 4 and relieve the fuel system pressure.

10   The fuel filter is located under the rear of the vehicle at the left front corner of the fuel tank.

11   Disconnect the negative battery cable, raise the vehicle and support it securely on jackstands **(see illustration)**.

12   Place a metal drain pan under the filter.

13   Slide back the hose clamps, then disconnect the fuel filter inlet and outlet hoses.

14   Remove the bolts and detach the filter from the vehicle **(see illustration)**. Remove the bracket and transfer it to the new filter.

15   Attach the bracket to the new filter. Be sure to line up the match marks **(see illustration)**.

16   Install the fuel filter and bracket assembly. Tighten the bolts securely.

17   Connect the inlet and outlet hoses.

18   Start the engine and check for fuel leaks at the procedure.

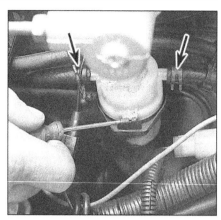

35.5 Loosen the clamps (arrows), detach the hoses, then use a screwdriver to disconnect the filter from the bracket

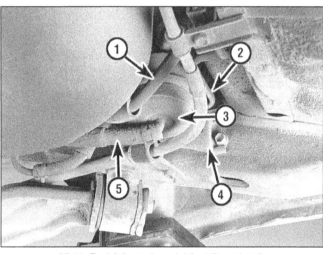

35.14 Fuel-injected model fuel filter details

1	Outlet hose	4	Mounting bracket
2	Fuel feed line	5	Inlet hose
3	Fuel filter		

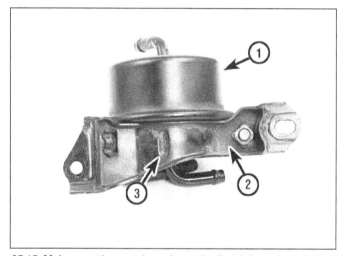

35.15 Make sure the match marks on the fuel-injected model fuel filter and bracket line up

1	Fuel filter	3	Match marks
2	Bracket		

36.1 The fuel cutoff solenoid valve should make a clicking sound when the ignition switch is turned on

1	Carburetor	2	Fuel cutoff solenoid valve

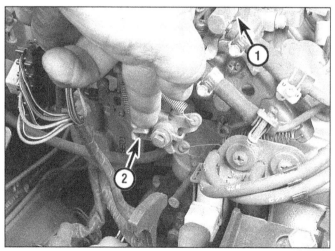

36.4 Push down on the idle switch lever

1	Carburetor	2	Idle switch lever

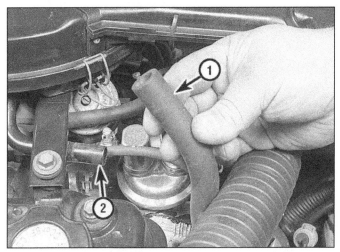

37.2 With the engine cold and the second air hose (1) disconnected from the air cleaner housing (2), a bubbling sound should be heard as air is drawn in

37.4a On 1985 and 1986 models, unplug the connector (arrow) and short the harness side of the connector with a jumper wire

36 Fuel cutoff system check (carburetor-equipped models only)

Refer to illustrations 36.1 and 36.4
1   Turn the ignition switch on and off several times and listen for a clicking sound from the fuel cutoff solenoid valve (located on the carburetor) indicating that it is operating (see illustration).
2   Start the engine and allow it to idle and warm up to normal operating temperature.
3   On manual transaxle is engaged (transaxle in Neutral).
4   Have an assistant push down on the accelerator and hold the engine at a fast idle speed (between 3000 and 4000 rpm), then push down on the idle switch lever on the carburetor and make sure the engine speed changes (see illustration).
5   On manual transaxle models, repeat this check with the clutch pedal depressed (clutch disengaged). The engine speed should not change when the idle lever is pushed.

37 Pulse Air System check (carburetor-equipped models only)

Refer to illustrations 37.2, 37.4a and 35.4b
1   Inspect the Pulse Air System hoses, pipe and electrical connector connected to the air cleaner housing for damage and loose connections.
2   When the engine is cold, disconnect the second air hose from the air cleaner housing, start the engine and allow it to idle, then make sure that air can be heard being drawn into the hose (listen for a bubbling sound) (see illustration).
3   Allow the engine to warm up to normal operating temperature. Air should no longer be heard being drawn into the hose. Increase the engine speed to around 3000 rpm, then release the accelerator. Air should now be drawn into the hose as the engine speed drops.
4   Unplug the coolant temperature sensor electrical connector. Install a jumper wire in the wiring harness side of the connector

(1985 and 1986 models) or on 1987 and 1988 models, a jumper wire incorporating a 10K ohm resistance. Shorting the connector in this manner will cause air again to be drawn into the hose (see illustration).
5   Refer to Chapter 6 for more information on the Pulse Air System.

38 Oxygen sensor system light check and resetting (1985 and 1986 models only)

1   At the specified intervals the "Sensor" light in the instrument cluster will flash when the engine reaches normal operating temperature, indicating the oxygen sensor and feedback carburetor system is in good condition. If this is the case, remove the fuse block cover and turn the Cancel switch Off to reset the light.
2   If the light doesn't flash at the specified intervals, remove the instrument cluster (Chapter 12), make sure the wiring is in good condition and the bulb isn't burned out.
3   If the wiring and bulb are in good condition, push the Cancel switch to the On position. Turn the ignition switch on (don't start the engine) and make sure the "Sensor" light goes on but doesn't flash.
4   Start the engine and allow it to warm up to normal operating temperature.
5   Increase engine speed to between 1500 and 2000 rpm. The Sensor light should now flash.
6   If it does, turn off the engine and reset the Cancel switch.
7   If it doesn't, there is a fault in the carburetor feedback and/or oxygen sensor systems. Refer to Chapters 4 and 6.

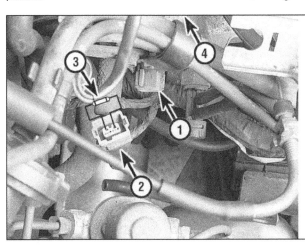

37.4b 1987 and 1988 model Pulse Air system check details

1   Coolant temperature sensor
2   Disconnected plug
3   Jumper wire incorporating a 10,000 ohm resistor
4   Intake manifold

Notes

Chapter 2 Part A Engines

Contents

Specifications

Three-cylinder engine

General
Firing order	1-3-2
Cylinder numbers (drivebelt end-to-transaxle end)	1-2-3

FRONT

Firing order 1-3-2

2A-00 Specs HAYNES

Cylinder location and distributor rotation (three-cylinder engine).
The blackened terminal shown on the distributor cap indicates the Number One spark plug wire position

Camshaft
Cam lobe height
Sprint (intake and exhaust)
Standard	1.5012 in (38.132 mm)
Wear limit	1.4973 in (38.032 mm)

Metro (intake and exhaust)
1991 and earlier
Standard	1.5601 to 1.5664 in (39.638 to 39.788 mm)
Wear limit	1.5562 in (39.528 mm)

1992 through 1998
Base and LSi models, and all 1995 through 1998 models
Standard	1.5911 to 1.5974 in (40.415 to 40.575 mm)
Wear limit	1.5872 in (40.315 mm)

XFi models
Standard	1.5602 to 1.5665 in (39.628 to 39.778 mm)
Wear limit	1.5562 in (39.528 mm)

1999 on
Intake
Standard	1.4241 to 1.4303 in (36.171 to 36.331 mm)
Limit	1.4202 in (36.071 mm)

Exhaust
Standard	1.4314 to 1.4376 in (36.356 to 36.516 mm)
Limit	1.4275 inch (36.256 mm)

Camshaft bearing oil clearance
Sprint
Standard	0.002 to 0.0035 in ((0.05 to 0.09 mm)
Service limit	0.006 in (0.15 mm)

Metro
1997 and earlier
Standard	0.008 to 0.0024 in (0.020 to 0.062 mm)
Service limit	0.0047 in (0.12 mm)

1998 on
Standard	0.0016 to 0.0032 in (0.040 to 0.082 mm)
Service limit	0.0047 in (0.12 mm)

Three-cylinder engine (continued)

Camshaft (continued)

Camshaft journal diameter (front-to-rear)
  Sprint
    1 .... 1.7372 to 1.7381 in (44.125 to 44.150 mm)
    2 .... 1.7451 to 1.7460 in (44.325 to 44.350 mm)
    3 .... 1.7530 to 1.7539 in (44.525 to 44.550 mm)
    4 .... 1.7609 to 1.7618 in (44.725 to 44.750 mm)
  Metro (through 1997)*
    1 .... 1.0220 to 1.0228 in (25.959 to 25.980 mm)
    2 and 3 .... 1.1795 to 1.1803 in (29.959 to 29.980 mm)
Camshaft journal bore (inside) diameter
  Sprint
    1 .... 1.7402 to 1.7407 in (44.200 to 44.216 mm)
    2 .... 1.7480 to 1.7486 in (44.400 to 44.416 mm)
    3 .... 1.7560 to 1.7565 in (44.600 to 44.616 mm)
    4 .... 1.7638 to 1.7644 in (44.800 to 44.816 mm)
  Metro (through 1997)*
    1 .... 1.0236 to 1.0244 in (26.000 to 26.021 mm)
    2 and 3 .... 1.1811 to 1.1819 in (30.000 to 30.021 mm)

Specification not available for 1998 and later models.

Valves and related components

Sprint
  Rocker-arm shaft diameter .... 0.628 to 0.629 in (15.973 to 15.988 mm)
  Rocker-arm inside diameter .... 0.629 to 0.630 in (16.000 to 16.018 mm)
  Shaft-to-rocker arm clearance
    Standard .... 0.0005 to 0.0017 in (0.012 to 0.045 mm)
    Service limit.. 0.0035 in (0.09 mm)
Metro
  Valve lifter outside diameter .... 1.2188 to 1.2194 in (30.959 to 30.975 mm)
  Valve lifter bore inside diameter .... 1.2205 to 1.2214 in (31.000 to 31.025 mm)
  Valve lifter-to-bore clearance
    Standard .... 0.0010 to 0.0025 in (0.025 to 0.066 mm)
    Wear limit .... 0.0059 in (0.15 mm)

Oil pump clearances

Outer gear-to-oil pump housing (service limit) .... 0.0122 in (0.310 mm)
Gear endplay limit... 0.0059 in (0.15 mm)

Four-cylinder engine

General

Firing order 1-3-4-2
Cylinder numbers (drivebelt end-to-transaxle end) 1-2-3-4

Camshaft

Camshaft lobe height
  1995 through 1998
    Standard .... 1.5024 in (38.136 mm)
    Wear limit .... 1.4975 in (38.036 mm)

Cylinder location and distributor rotation (four-cylinder engine).
The blackened terminal shown on the distributor cap indicates the Number One spark plug wire position

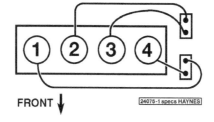

FRONT ↓

Cylinder location and spark-plug wire routing (1998 four-cylinder models)

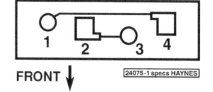

FRONT ↓

Cylinder location and spark-plug wire routing (1999 and later four-cylinder models)

1999 on	
Intake	
Standard	1.4241 to 1.4303 in (36.171 to 36.331 mm)
Wear limit	1.4202 in (36.071 mm)
Exhaust	
Standard	1.4314 to 1.4376 in (35.356 to 35.516 mm)
Wear limit	1.4275 in (36.256 mm)
Camshaft journal clearance	
1995 through 1998	
Standard	0.0020 to 0.0036 in (0.050 to 0.091 mm)
Wear limit	0.0059 in (0.15 mm)
1999 on	
Standard	0.0016 to 0.0032 in (0.040 to 0.082 mm)
Wear limit	0.0047 in (0.12 mm)
Camshaft journal bore diameter	
1995 through 1998	
Journal bore diameter A	1.7402 to 1.7407 in (44.200 to 44.216 mm)
Journal bore diameter B	1.7480 to 1.7486 in (44.400 to 44.416 mm)
Journal bore diameter C	1.7560 to 1.7565 in (44.600 to 44.616 mm)
Journal bore diameter D	1.7638 to 1.7644 in (44.800 to 44.816 mm)
Journal bore diameter E	1.7716 to 1.7723 in (45.000 to 45.016 mm)
1999 on	1.1024 to 1.1031 in (28.000 to 28.021 mm)
Camshaft journal outside diameter	
1995 through 1998	
Journal outside diameter A	1.7372 to 1.7381 in (44.125 to 44.140 mm)
Journal outside diameter B	1.7451 to 1.7460 in (44.325 to 44.350 mm)
Journal outside diameter C	1.7530 to 1.7539 in (44.525 to 44.550 mm)
Journal outside diameter D	1.7609 to 1.7618 in (44.725 to 44.750 mm)
Journal outside diameter E	1.7687 to 1.7697 in (44.925 to 44.950 mm)
1999 on	1.1000 to 1.1008 in (27.939 to 27.960 mm)

Oil pump clearances

Outer gear-to-oil pump housing (service limit)	0.0122 in (0.310 mm)
Gear endplay limit	0.0059 in (0.15 mm)

Torque specifications

	Ft lbs (unless otherwise indicated)
Camshaft bearing cap bolts	
Three-cylinder engine Metro	96 in-lbs
1998 and later four-cylinder engine Metro	96 in-lbs
Camshaft timing belt sprocket bolt	
Three-cylinder engine	44
Four-cylinder engine	43
Cylinder head bolts	
Sprint	48
Metro	
1998 and earlier	54
1999 on	
Three-cylinder engine	54
Four-cylinder engine	49
Crankshaft pulley bolts	
1994 and earlier	96 in-lbs
1995 on	
Three-cylinder engine	144 in-lbs
Four-cylinder engine	
1995 through 1998	97 in-lbs
1999 on	144 in-lbs
Crankshaft timing belt sprocket bolt	
1994 and earlier	
Sprint	52
Metro	81
1995 on	
Three-cylinder engine	
1995 and 1996	81
1997 on	96
Four-cylinder engine	
1995 and 1996	79
1997 on	96

Four-cylinder engine (continued)

Torque specifications (continued)   Ft lbs (unless otherwise indicated)

Distributor gear case	96 in-lbs
Exhaust manifold	17
Intake manifold	17
Flywheel/driveplate bolts	
1994 and earlier	45
1995 on	
Three-cylinder engine	
1995 and 1996	52
1997 and 1998	45
1999 on	55
Four-cylinder engine	
1995 and 1996	45
1997 and 1998	
Automatic	45
Manual	56.8
1999 on	
Automatic	69
Manual	57
Oil pump	
Oil pump pick-up tube bolt	97 in-lbs
Oil pump pick-up tube bracket bolt	97 in-lbs
Oil pump-to-block bolts	97 in-lbs
Oil pan bolts/nuts	97 in-lbs
Rear main seal retainer bolts	
1994 and earlier	96 in-lbs
1995 through 1998	106 in-lbs
1999 on	
Three-cylinder engine	106 in-lbs
Four-cylinder engine	97 in-lbs
Rocker-arm adjustment screw locknuts	156 in-lbs
Rocker-arm shaft retaining bolts/screws	97 in-lbs
Timing belt cover bolts/nuts	97 in-lbs
Timing belt tensioner	
Timing belt tensioner bolt	
1999 and earlier	20
2000 on	
Three-cylinder engine	20
Four-cylinder engine	18
Timing belt tensioner nut	97 in-lbs
Valve cover bolts/nuts	
1994 and earlier	96 in-lbs
1995 on	
Three-cylinder engine	44 in-lbs
Four-cylinder engine	
1995 through 1998	44 in-lbs
1999 on	97 in-lbs

1   General information

This Part of Chapter 2 is devoted to in-vehicle repair procedures for the engine. All information concerning engine removal and installation and engine block and cylinder head overhaul can be found in Part B of this Chapter.

The following repair procedures are based on the assumption the engine is installed in the vehicle. If the engine has been removed from the vehicle and mounted on a stand, many of the steps outlined in this Part of Chapter 2 will not apply.

2   Repair operations possible with the engine in the vehicle

Many major repair operations can be accomplished without removing the engine from the vehicle.

Clean the engine compartment and the exterior of the engine with some type of degreaser before any work is done. It'll make the job easier and help keep dirt out of the internal areas of the engine.

Depending on the components involved, it may be helpful to remove the hood to improve access to the engine as repairs are performed (see Chapter 11, if necessary). Cover the fenders to prevent damage to the paint. Special pads are available, but an old bedspread or blanket will also work.

If vacuum, exhaust, oil or coolant leaks develop, indicating a need for gasket or seal replacement, the repairs can generally be made with the engine in the vehicle. The intake and exhaust manifold gaskets, oil pan gasket, crankshaft oil seals and cylinder head gasket are all accessible with the engine in place.

Exterior engine components, such as the intake and exhaust manifolds, the oil pan, the oil pump, the water pump, the starter motor, the alternator, the distributor and the fuel system components can be removed for repair with the engine in place.

Since the cylinder head can be removed without pulling the engine, valve component servicing can also be accomplished with the engine in the vehicle. Replacement of the camshaft, timing belt and sprockets is also possible with the engine in the vehicle.

In extreme cases caused by a lack of necessary equipment, repair or replacement of piston rings, pistons, connecting rods and rod bearings is possible with the engine in the vehicle. However, this practice is not recommended because of the cleaning and preparation work that must be done to the components involved.

3   Top Dead Center (TDC) for number one piston - locating

Note: *The following procedure is based on the assumption that the spark plug wires and the distributor (if equipped), are correctly installed. If you're trying to locate TDC to install the distributor correctly, piston position must be determined by feeling for compression at the number one spark plug hole, then aligning the ignition timing marks as described in Step 8.*

1   Top Dead Center (TDC) is the highest point in the cylinder that each piston reaches as it travels up-and-down when the crankshaft turns. Each piston reaches TDC on the compression stroke and again on the exhaust stroke, but TDC generally refers to piston position on the compression stroke. The timing marks on the vibration damper/crankshaft pulley installed on the front of the crankshaft are referenced to the number one piston at TDC.

2   Positioning the piston(s) at TDC is an essential part of many procedures such as rocker arm removal, camshaft and timing belt/ sprocket replacement and distributor removal.

3   Before beginning this procedure, be sure to place the transaxle in Neutral (or Park on automatics), apply the parking brake and block the rear wheels.

4   On vehicles with a distributor-type ignition system, disable the ignition system by detaching the coil wire from the center terminal of the distributor cap and grounding it on the block with a jumper wire. On vehicles with a distributorless ignition system, disable the ignition system by disconnecting the primary electrical connectors at the ignition coil pack/modules (see Chapter 5). On all vehicles, remove the spark plugs (see Chapter 1).

5   In order to bring any piston to TDC, the crankshaft must be turned using one of the methods outlined below. When looking at the front of the engine, normal crankshaft rotation is clockwise.

a) *The preferred method is to turn the crankshaft with a socket and breaker bar attached to the bolt threaded into the front of the crankshaft.*

b) *A remote starter switch, which may save some time, can also be used. Follow the instructions included with the switch. Once the piston is close to TDC, use a socket and breaker bar as described in the previous paragraph.*

c) *If an assistant is available to turn the ignition switch to the Start position in short bursts, you can get the piston close to TDC without a remote starter switch. Make sure your assistant is out of the vehicle, away from the ignition switch, then use a socket and breaker bar as described in Paragraph a) to complete the procedure.*

Vehicles with a distributor

Refer to illustrations 3.6, 3.8 and 3.9

6   Note the position of the terminal for the number one spark plug wire on the distributor cap. If the wire isn't marked, follow the plug wire from the number one cylinder spark plug to the cap. Use a felt-tip pen or chalk to make a mark on the distributor body directly under the terminal **(see illustration)**.

7   Detach the cap from the distributor and

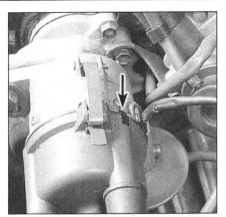

3.6 Mark the distributor housing adjacent to the number one terminal on the cap

set it aside (see Chapter 1 if necessary).

8   Turn the crankshaft clockwise (see Step 3 above) until the notch in the crankshaft pulley is aligned with the 0 on the timing plate (located at the front of the engine) **(see illustration)**.

9   Look at the distributor rotor - it should be pointing directly at the mark you made on the distributor body **(see illustration)**. If it is, go to Step 12.

10   If the rotor is 180-degrees off, the number one piston is at TDC on the exhaust stroke. Go to Step 11.

11   If the rotor is 180-degrees off, turn the crankshaft one complete turn (360-degrees) clockwise. The rotor should now be pointing at the mark on the distributor. When the rotor is pointing at the number one spark plug wire terminal in the distributor cap and the ignition timing marks are aligned, the number one piston is at TDC on the compression stroke.

12   After the number one piston has been positioned at TDC on the compression stroke, TDC for any of the remaining pistons can be located by turning the crankshaft 240-degrees and following the firing order. Mark the remaining spark plug wire terminal locations on the distributor body just like you did for the number one terminal, then number the

3.8 Turn the crankshaft until the notch in the pulley is aligned with the "0" (zero) on the timing belt cover (arrows)

3.9 When the number one piston is at Top Dead Center on the compression stroke, the rotor should point toward the mark you made on the distributor

marks to correspond with the cylinder numbers. As you turn the crankshaft, the rotor will also turn. When it's pointing directly at one of the marks on the distributor, the piston for that particular cylinder is at TDC on the compression stroke.

Vehicles without a distributor

Refer to illustration 3.13

13   Remove the spark plugs (see Chapter 1) and install a compression gauge in the number one cylinder **(see illustration)**. Turn the crankshaft clockwise with a socket and breaker bar as described above.

14   When the piston approaches TDC, compression will be noted on the compression gauge. Continue turning the crankshaft until the notch in the crankshaft damper is aligned with the TDC mark on the front cover **(see illustration 3.8)**. At this point the number one cylinder is at TDC on the compression stroke. If the marks aligned but there was no compression, the piston was on the exhaust stroke. Continue rotating the crankshaft 360-degrees (1-turn) and then line-up the marks. **Note:** *If a compression gauge is not available, put the No. 1 piston at TDC by simultaneously*

aligning the timing mark on the camshaft timing belt sprocket with the index mark on the timing belt cover **(see illustration 7.19b)***, and then verify that the timing mark on the crankshaft damper is aligned with the TDC mark on the front cover.*

15   After the number one piston has been positioned at TDC on the compression stroke, TDC for any of the remaining cylinders can be located by turning the crankshaft 180 degrees and following the firing order (refer to the Specifications). Rotating the engine 180 degrees past TDC #1 will put the engine at TDC compression for cylinder #3.

---

4   Valve cover - removal and installation

Three-cylinder engines

Refer to illustrations 4.4, 4.5a, 4.5b and 4.5c

1   Disconnect the negative cable from the battery.

2   Remove the air cleaner assembly (see Chapter 4).

3   Detach the spark plug wires from the

3.13 On vehicles without a distributor, use a compression gauge in the number one spark plug hole to help find TDC

plugs, unclip the wire loom from the cover, then set the wires aside, leaving them attached to the loom.

4   Disconnect the breather hose from the cover **(see illustration)**.

5   Remove the valve cover nuts/bolts **(see illustrations)** and then lift off the cover. If the

4.4 Disconnect the breather hose (arrow) from the valve cover (Sprint shown, Metro similar)

4.5a Sprint engine valve cover bolt locations (arrows)

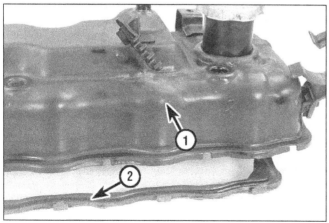

4.5b Engine valve cover gasket installation details (Sprint)

1	Valve cover	2	Gasket

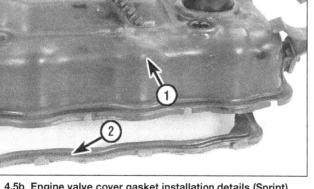

4.5c Engine valve cover details (typical three-cylinder Metro)

1	Nut sealing washer locations	2	Spark plug wire holders
		3	Valve cover

4.11 Disconnect the breather hose (left arrow) and the PCV valve and hose (right arrow) from the valve cover (four-cylinder Metro)

4.13 To detach the valve cover from a four-cylinder engine, remove these six bolts (arrows)

cover sticks to the cylinder head, tap on it with a soft-face hammer or place a block of wood against the cover and tap on the wood with a hammer.

6   Thoroughly clean the valve cover and remove all traces of old gasket material with a scraper.

7   Install a new gasket on the cover, using RTV sealant to hold it in place. Place new grommets in the holes in the cover, position the cover, then install the nuts/bolts.

8   Working from the center out, tighten the nuts/bolts to the torque listed in this Chapter's Specifications.

9   The remaining steps are the reverse of removal. When finished, run the engine and check for oil leaks.

Four-cylinder engines

Refer to illustrations 4.11, 4.13, 4.14a and 4.14b

10   Disconnect the negative battery cable.

11   Disconnect the breather hose and PCV valve and hose **(see illustration)** from the valve cover.

12   Remove the spark plugs wires and the ignition coils from the valve cover (see Chapter 5).

13   Remove the six valve cover bolts **(see illustration)**.

14   Remove and discard the old valve cover gasket and the spark plug hole seals **(see illustrations)**.

15   Installation is the reverse of removal. Be sure to use a new valve cover gasket and new spark plug hole seals. Tighten the valve cover bolts to the torque listed in this Chapter's Specifications.

---

5   Intake manifold - removal and installation

Removal

Note: *It's not necessary to remove the carburetor or throttle body from the intake manifold when removing the manifold.*

1   On fuel-injected models, relieve the fuel

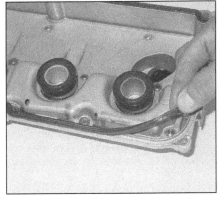

4.14a Whenever the valve cover is removed, be sure to remove and discard the old valve cover gasket

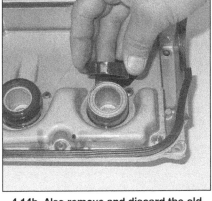

4.14b Also remove and discard the old spark plug hole seals

pressure (see Chapter 4) and then disconnect the negative cable from the battery.

2   Drain the cooling system (see Chapter 1).

3   Remove the air cleaner assembly (see Chapter 4).

Carbureted models and models with throttle body injection

Refer to illustrations 5.8 and 5.9

4   Clearly label, then disconnect all hoses, wires, brackets and emission lines that run to the carburetor or throttle body and to the intake manifold.

5   Disconnect the fuel lines from the carburetor or the throttle body and then cap the fittings to prevent leakage (see Chapter 4).

6   Disconnect the throttle cable from the carburetor or throttle body (see Chapter 4).

7   On models with an automatic transaxle, disconnect the throttle valve cable from the carburetor or throttle body (see Chapter 7B).

8   Unscrew the bolts and nuts and remove the intake manifold from the engine **(see illustration)**. If it sticks, tap the manifold with a soft-face hammer. **Caution:** *Do not pry between the gasket sealing surfaces or tap on the carburetor/throttle body.*

5.8 To detach the manifold on carbureted models and Metro models with throttle body injection, remove these nuts (arrows) and the nuts (not visible in this photo) along the lower edge of the intake manifold (Sprint shown, Metro models similar)

5.9 Remove all traces of old
gasket material

9   Thoroughly clean the manifold and
cylinder head mating surfaces, removing all
traces of gasket material (see illustration).

Models with multiport fuel injection

Refer to illustrations 5.10, 5.11 and 5.15

Note: *It's not necessary to remove the throttle body from the intake manifold in order to remove the manifold. So the following procedure assumes that you are going to remove the intake manifold and the throttle body as a single assembly. However, if you are going to replace the manifold, you will need to remove the throttle body (see Chapter 4).*

10   Disconnect the EVAP canister purge hose and the brake booster hose (see illustration) from the intake manifold:
11   Remove the intake manifold support bracket (see illustration) and then remove the breather hose and the PCV valve and hose (see illustration 4.11).
12   Disconnect the following electrical connectors, and remove the following components, from the intake manifold:

a) *All ground wires attached to the intake manifold.*
b) *The fuel injector electrical connectors, the fuel rail and the fuel injectors (see Chapter 4).*
c) *The Throttle Position (TP) sensor electrical connector (see Chapter 6).*

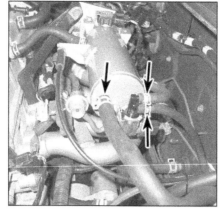

5.10 Loosen the hose clamps (arrows) and then disconnect the brake booster hose from the air intake plenum and the two smaller hoses from the EVAP canister purge valve; if you're going to replace the intake manifold, remove the purge valve too (four-cylinder Metro models)

d) *The Idle Air Control (IAC) valve electrical connector (see Chapter 6).*
e) *The EVAP canister purge valve electrical connector and, if you're going to replace the intake manifold, the EVAP canister purge valve (see Chapter 6).*
f) *The MAP sensor electrical connector and, if you're going to replace the intake manifold, the MAP sensor (see Chapter 6)*

13   Disconnect the accelerator cable from the throttle body (see Chapter 4).
14   Disconnect the coolant hoses from the throttle body (see *Throttle body - removal and installation* in Chapter 4).
15   Remove the intake manifold mounting nuts and bolts (see illustration) and then remove the manifold. If the manifold is stuck, tap it loose with a soft-face hammer. **Caution:** *Do not try to loosen the manifold by prying between the gasket sealing surfaces or by tapping on the throttle body.*
16   Thoroughly clean the manifold and cylinder head mating surfaces. Remove all traces of old gasket material (see illustration 5.9).

5.11 To remove the air intake manifold support bracket on four-cylinder Metro models, remove these three bolts (arrows)

Installation

17   Install the manifold and a new gasket and tighten the bolts and nuts in several steps, working from the center out, until you reach the torque listed this Chapter's Specifications.
18   Reinstall the remaining parts in the reverse order of removal.
19   Add coolant (see Chapter 1).
20   Run the engine and then check for exhaust and/or coolant leaks.

---

6   Exhaust manifold - removal and installation

Removal

Refer to illustrations 6.3a, 6.3b, 6.5, 6.6a, 6.6b, 6.7 and 6.8

Warning: *Allow the engine to cool completely before beginning this procedure.*

1   Disconnect the negative cable from the battery.
2   Set the parking brake and block the rear wheels. Raise the front of the vehicle and support it securely on jackstands.
3   Working from under the vehicle, remove

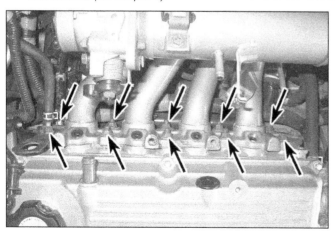

5.15 To detach the intake manifold on 1998 and later four-cylinder Metro models, remove these nuts and bolts (arrows)

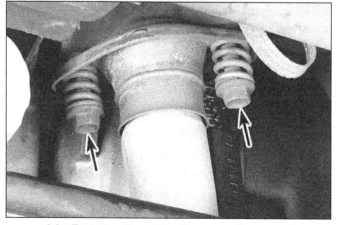

6.3a Typical early-model exhaust pipe-to-exhaust manifold flange bolts (arrows)

the nuts that secure the exhaust system to the bottom of the exhaust manifold (see illustrations). Apply penetrating oil to the threads to make removal easier.

4 On some air conditioned models, the lower adjusting brace for the compressor is in the way. Loosen and remove the compressor drivebelt and then remove the lower adjusting brace, if necessary.

5 Trace the oxygen sensor electrical lead to the connector and then unplug it (see illustration).

6 Remove the fasteners that secure the heat shield to the exhaust manifold (see illustrations), then detach the heat shield.

7 Disconnect the air injection tube, if equipped (see illustration).

8 Apply penetrating oil to the threads, then remove the exhaust manifold mounting bolts or nuts (see illustration).

9 Separate the exhaust manifold from the cylinder head.

10 Clean and inspect all threaded fasteners.

11 Remove all traces of old gasket material from the mating surfaces and inspect them for cracks and other damage.

6.3b Typical late-model exhaust pipe/catalytic converter-to-exhaust manifold flange bolts

Installation

12 Position a new gasket on the cylinder head, install the manifold and tighten the fasteners in several steps, working from the center out, to the torque listed in this Chapter's Specifications.

6.5 Trace the oxygen sensor lead from the sensor to the electrical connector and then unplug the connector (arrow) (typical early-model shown)

13 Reinstall the remaining parts in the reverse order of removal. Be sure to use a new exhaust pipe seal and gasket.

14 Run the engine and check for exhaust leaks.

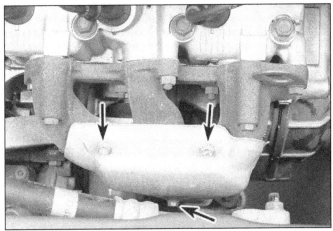

6.6a Remove the heat shield retaining bolts (arrows) (typical early-model heat shield shown, some early shields have four bolts)

6.6b Typical late-model heat shield with oxygen sensor in manifold; to remove it from the exhaust manifold, remove these bolts (arrows); if you're going to replace the manifold, remove the oxygen sensor too

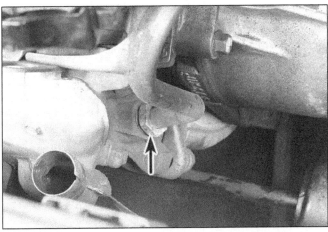

6.7 On earlier models with air injection, disconnect the air injection tube by unscrewing this nut (arrow)

6.8 To detach the exhaust manifold, remove these bolts and nuts (arrows) (late-model four-cylinder Metro shown, other models similar)

7.4 On Sprint and on 1998 and later Metro four-cylinder models, loosen the locknuts and back off the valve adjustment screws

1   Rocker arm
2   Locknut
3   Adjustment screw

7   Timing belt and sprockets - removal, inspection and installation

Removal

Refer to illustrations 7.4, 7.8a, 7.8b, 7.9, 7.10, 7.11, 7.14, 7.15a, 7.15b and 7.15c

Warning: *The air conditioning system is under high pressure. Do not loosen any fittings or disconnect any components until after the system has been discharged by an air conditioning technician. Always wear eye protection when disconnecting refrigerant fittings.*

Caution: *Do not try to turn the crankshaft with the camshaft sprocket bolt and do not rotate the crankshaft counterclockwise.*

1   Disconnect the negative cable from the battery.
2   Position the number one piston at Top Dead Center (see Section 3).
3   Remove the valve cover (see Section 4).
4   On Sprint and on 1998 and later Geo Metro models, loosen the locknuts and back off the valve adjustment screws until they're no longer in contact with the valves **(see illustration)**.
5   Set the parking brake and block the rear wheels. Raise the front of the vehicle and support it securely on jackstands.
6   Loosen the four water pump pulley nuts, then remove the drivebelts (see Chapter 1).
7   Remove the rubber plug or plastic splash-shield from the right inner fender **(see illustration)**.
8   Remove the crankshaft and water pump drivebelt pulleys **(see illustration)**. **Note:** *If you're only replacing the timing belt, it's not necessary to remove the crankshaft center bolt; however, if you will be removing the crankshaft sprocket to replace the oil pump or oil seal, you must remove the center bolt. Do this before you remove the pulley bolts.*

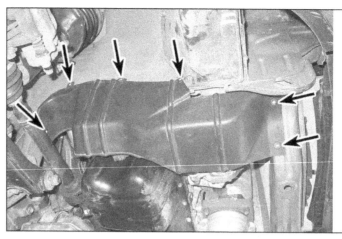

7.7 Remove the fasteners (arrows) and pull off the plastic splash shield (if equipped)

7.8a Typical crankshaft pulley (early three-cylinder engine shown, other models similar)

1   Crankshaft pulley bolts
2   Indexing notch
3   Center bolt

The center bolt is very tight; to break it loose, remove the splash pan from beneath the front of the engine, wrap a rag or duct tape around the pulley and attach a chain wrench to immobilize the pulley **(see illustration)**. *Use a breaker bar and socket to loosen the bolt.*

9   Remove the bolts that secure the timing

7.8b Wrap duct tape or a rag around the pulley and grip it with a chain wrench

belt cover and detach the cover **(see illustration)**.

10   If you plan to reuse the timing belt, and it doesn't already have arrows painted on it, paint one on to indicate the direction of rotation (clockwise) **(see illustration)**.
11   Loosen the adjusting nut and pulley bolt. Move the tensioner pulley towards the water pump as far as possible **(see illustration)**.
12   Temporarily secure the tensioner pulley by tightening the adjusting nut.

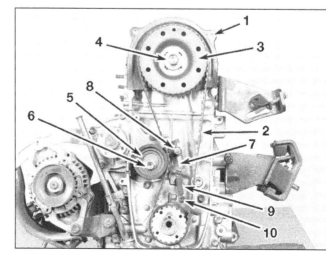

7.9 Typical engine timing belt and related components (1.0L engine shown, 1.3L engine similar)

1   Inside cover
2   Timing belt
3   Camshaft sprocket
4   Bolt
5   Tensioner
6   Tensioner bolt
7   Tensioner plate
8   Tensioner stud
9   Tensioner spring
10   Spring screw

7.10 If the timing belt doesn't have arrows like these to indicate direction of rotation, make one with chalk

7.11 Loosen the adjusting nut and pulley bolt and move the tensioner pulley as far as possible towards the water pump (three-cylinder engine shown, four-cylinder engine similar)

    A   Adjusting nut
    B   Pulley bolt
    C   Tensioner pulley

7.14 The belt guide has a notch which allows it to fit over the crankshaft key (arrows) (three-cylinder engine shown, four-cylinder engine similar)

7.15a Keep the camshaft sprocket from turning with a large screwdriver

13   Slip the timing belt off the sprockets and set it aside.

14   If you intend to replace the oil pump or crankshaft front oil seal, slide off the crankshaft sprocket and the belt guide located behind it **(see illustration)**. When

removing the guide, note how it's installed (the chamfered side faces out).

15   If you intend to replace the camshaft or oil seal, unscrew the camshaft sprocket bolt and slide the sprocket off - a large screwdriver inserted through a hole in the sprocket will keep it from turning while you remove the bolt **(see illustration)**. **Note:** *On Metro models, the camshaft can be kept from turning by inserting a (10 mm) metal rod through the hole* **(see illustration)**. *Unbolt the cover* **(see illustration)** *to access the seal.*

Inspection

Refer to illustrations 7.16 and 7.17

16   Rotate the tensioner pulley by hand and move it from side-to-side to detect roughness and excess play **(see illustration)**. Visually inspect the sprockets for damage and wear. Replace parts as necessary.

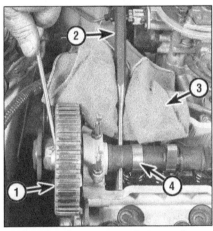

7.15b On Metro three-cylinder engines, you can hold the camshaft with a metal rod

    1   Camshaft sprocket
    2   Rod
    3   Shop cloth
    4   Camshaft

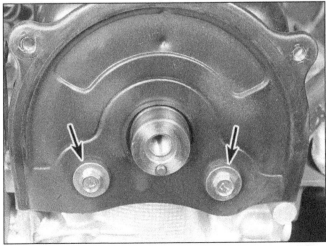

7.15c To get to the camshaft seal on Sprint models and on 1995 through 1997 Metro four-cylinder models, remove these two mounting bolts (arrows) and detach the cover (Sprint model shown, 1995 through 1997 Metro four-cylinder models similar)

7.16 Check the tensioner pulley for roughness and excess play

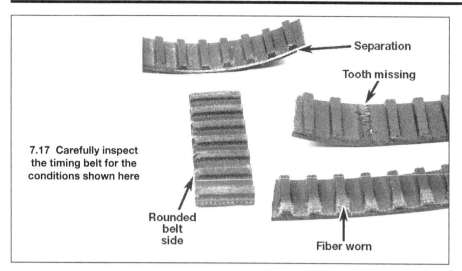

7.17 Carefully inspect the timing belt for the conditions shown here

Separation

Tooth missing

Rounded belt side

Fiber worn

7.18a The camshaft sprocket is indexed by a dowel (arrow) (Sprint engine shown, Metro models similar)

17   Check the timing belt for cracks, ply separation, wear, missing teeth and oil contamination. Replace the belt if it's worn or damaged **(see illustration)**. **Note:** *Unless the engine has very low mileage, it's common practice to replace the timing belt with a new one every time it's removed. Don't reinstall the original belt unless it's in like-new condition. Never reuse a belt in questionable condition.*

Installation

Refer to illustrations 7.18a, 7.18b, 7.19a and 7.19b

18   Reinstall the camshaft seal cover and timing belt sprockets, if they were removed. Note that the camshaft sprocket is indexed by a dowel pin **(see illustration)**. Slip the belt guide onto the crankshaft before installing the crankshaft sprocket - the chamfered side of the guide faces away from the belt. The crankshaft sprocket has a keyway which matches the key in the crankshaft **(see illustration)**.
19   Align the valve timing marks located on the crankshaft and camshaft sprockets **(see illustrations)**.
20   Slip the timing belt onto the crankshaft

sprocket. While maintaining tension on the side of the belt opposite the tensioner, slip the belt onto the camshaft sprocket.
21   Release the tensioner adjusting nut to allow spring tension to apply pressure against the belt. Rotate the crankshaft clockwise two complete revolutions (720-degrees). Retighten the nut.
22   Temporarily install the crankshaft pulley, taking care to align the notch in the pulley with the raised area on the sprocket. Install the crankshaft pulley bolts and the center bolt, if removed. Tighten the center bolt to the torque listed in this Chapter's Specifications. When tightening the bolts, immobilize the crankshaft as discussed in Step 8. Remove the pulley, leaving the center bolt in place.
23   Using the bolt in the center of the crankshaft sprocket, turn the crankshaft clockwise through two complete revolutions (720-degrees). Recheck the alignment of the valve timing marks. If the marks do not align properly, loosen the tensioner, slip the belt off the camshaft sprocket, align the marks, reinstall the belt, and check the alignment again.

24   Tighten the tensioner bolt and nut to the torque listed in this Chapter's Specifications. Start with the nut, then tighten the bolt.
25   Reinstall the remaining parts in the reverse order of removal.
26   Start the engine, set the ignition timing (see Chapter 1) and road test the vehicle.

8   Crankshaft front oil seal - replacement

Refer to illustrations 8.2 and 8.4

1   Remove the timing belt, crankshaft sprocket and inner belt guide (see Section 7).
2   Wrap the tip of a small screwdriver with tape. Working from below, use the screwdriver to pry the seal out of the bore **(see illustration)**. Be careful not to damage the crankshaft or the seal bore.
3   Clean and inspect the seal bore and the seal contact surface on the crankshaft. Minor imperfections can be removed with emery cloth. If there's a groove worn in the crankshaft seal surface (from contact with the seal), installing a new seal will probably not

7.18b The crankshaft sprocket has a slot (keyway) which must align with the key in the crankshaft (three-cylinder engine shown, four-cylinder engines similar)

7.19a The mark on the lower timing belt sprocket must align with the mark on the engine front cover (arrows) (three-cylinder engine shown, four-cylinder engines similar)

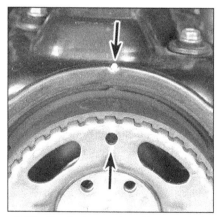

7.19b The timing mark on the camshaft sprocket must align with the V mark on the camshaft seal cover (arrows) - on some later models, the mark takes the form of an "E" stamped into the sprocket

8.2 Carefully pry the seal out with a small screwdriver - wrap the tip with tape to prevent damage to the seal bore and crankshaft sealing surface

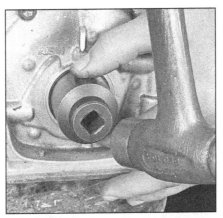

8.4 Install the new seal by gently tapping it into place with a socket and hammer

9.2 Carefully pry the camshaft oil seal out with a small screwdriver - wrap the tip with tape to prevent damage to the seal bore and camshaft sealing surface

stop the leak. Try installing a repair sleeve which fits over the crankshaft sealing surface. These are normally available at larger auto parts stores.

4   Lubricate the outer edge of the new seal with multi-purpose grease or engine oil. Also lubricate the seal lip. Drive the seal into place with a hammer and socket **(see illustration)**.

5   Reinstall the timing belt and related components as described in Section 7.

6   Run the engine and check for oil leaks.

9   Camshaft oil seal - replacement

Refer to illustrations 9.2 and 9.4

1   Remove the timing belt, camshaft sprocket and camshaft seal cover (see Section 7).

2   Note how far the seal is seated in the bore, then carefully pry it out with a small screwdriver **(see illustration)**. Wrap the tip of the screwdriver with tape so you don't scratch the bore or damage the camshaft in the process (if the camshaft is damaged, the new seal will end up leaking).

3   Clean the bore and coat the outer edge

of the new seal with engine oil or multi-purpose grease. Also lubricate the seal lip.

4   Using a socket with an outside diameter slightly smaller than the outside diameter of the seal, carefully drive the new seal into place with a hammer **(see illustration)**. Make sure it's installed squarely and driven in to the same depth as the original. If a socket isn't available, a short section of pipe will also work.

5   Reinstall the seal cover, if applicable, camshaft sprocket and timing belt (see Section 7).

6   Run the engine and check for oil leaks at the camshaft seal.

10   Camshaft, rocker arms and shafts (Sprint models and Metro four-cylinder models) - removal, inspection and installation

1   Set the number one piston at top dead center (see Section 3) and then remove the valve cover (see Section 4).

2   Remove the timing belt and the camshaft timing belt sprocket (see Section 7).

Sprint models and 1995 through 1997 Metro four-cylinder models

Removal

Rocker arms and shafts

Refer to illustrations 10.4, 10.6 and 10.7

3   On Sprint models, loosen the locknuts and back off the valve adjustment screws until they're no longer in contact with the valves **(see illustration 7.4)**. (1995 through 1997 1.3L Metro models have hydraulic lash adjusters, so the rocker arms don't use adjustment screws.)

4   Remove the rocker-arm shaft retaining screws **(see illustration)**. They can be very tight and might require an impact driver.

5   Number the rocker arms with a scribe. Start with number one at the timing belt end and work your way toward the rear of the engine in a criss-cross pattern until all of them are marked. When you're done, the rocker arms on the intake manifold side should be numbered 1, 3 and 5, and the ones on the exhaust manifold side should be marked 2, 4, and 6.

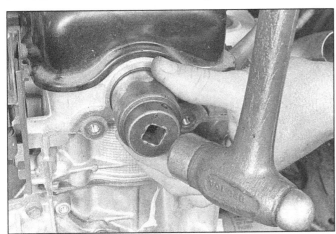

9.4 Gently tap the new oil seal into place with a socket and hammer

10.4 Remove the eight rocker-arm shaft retaining screws (arrows) (Sprint model shown, 1995 through 1997 Metro four-cylinder models similar)

10.6 Push on the rear of the rocker-arm shaft until it protrudes enough at the front to grip it - if necessary, use an Allen-head driver to turn the shaft as shown here (Sprint model shown, 1995 through 1997 Metro four-cylinder models similar)

10.7 Pull the shaft out the front of the cylinder head - wrap it with a rag and twist it out with locking pliers

6   Push on the rear of one rocker-arm shaft until it protrudes enough at the front to grasp it with pliers **(see illustration)**.
7   Slowly pull each rocker-arm shaft out the front of the engine **(see illustration)**. Lift the rocker arms and springs out of the head as they are released from the shaft.

Camshaft

Refer to illustrations 10.10 and 10.11
8   Remove the timing belt, cam sprocket and inner timing belt cover (see Section 7).
9   Remove the distributor and fuel pump (see Chapters 4 and 5).
10   Remove the air injection tube and unbolt the distributor case **(see illustration)**.
11   Carefully guide the camshaft out of the cylinder head **(see illustration)**. Don't nick the bearing surfaces in the head with the cam lobes.

Inspection

Refer to illustrations 10.13, 10.14a, 10.14b, 10.15 and 10.16
12   Thoroughly clean the parts in solvent and wipe them off with a lint-free cloth.
13   Measure the inside diameter of each rocker arm and the outside diameter of the

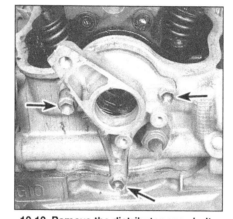

10.10 Remove the distributor case bolts (arrows) (Sprint model shown, 1995 through 1997 Metro four-cylinder models similar)

shaft where the rocker arm rides **(see illustration)**. Compare the measurements to this Chapter's Specifications. Subtract the rocker-arm shaft diameter from the rocker arm inside diameter to calculate the shaft-to-arm clearance.
14   Visually inspect the camshaft, rocker

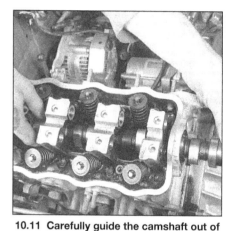

10.11 Carefully guide the camshaft out of the cylinder head to avoid nicking the bearing surfaces (Sprint model shown, 1995 through 1997 Metro four-cylinder models similar)

arms and springs for wear and damage **(see illustrations)**. Replace any components that are damaged or excessively worn.

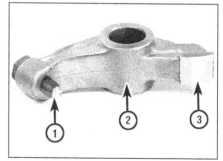

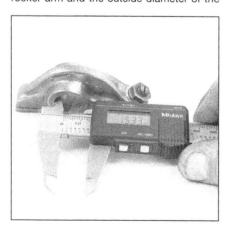

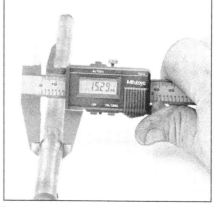

10.13 Measure the inside diameter of the rocker arms and the outside diameter of the rocker-arm shafts

10.14a Inspect the tips of the valve adjusting screws and the cam-riding faces (Sprint rocker arm shown, 1995 through 1997 Metro four-cylinder rocker arms are "roller" type rockers)

1   Valve adjusting screw
2   Rocker arm
3   Cam-riding face

10.14b Check the cam lobes for pitting, wear and score marks - if scoring is excessive, as is the case here, install a new camshaft

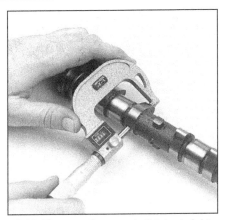

10.15 Measure the camshaft lobe heights

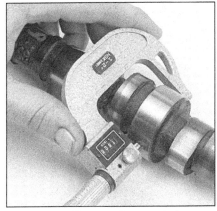

10.16 Measure the camshaft journal diameters

15   Using a micrometer or accurate caliper, measure the cam lobe height **(see illustration)** and compare it to this Chapter's Specifications. If the lobe height is less than the minimum allowable, the camshaft is worn and must be replaced.

16   Using a micrometer or accurate caliper, measure the diameter of each bearing journal **(see illustration)** and compare it to this Chapter's specifications. If the journals are worn or damaged, replace the camshaft.

17   Using a bore gauge, measure the inside diameter of the camshaft bearing bores and record the results. Subtract the camshaft bearing journal outside diameter readings from the bore diameters to determine the bearing clearance. Do this for each bearing and compare the results to the clearances in this Chapter's Specifications. If the oil clearance is excessive, a new cylinder head may be needed (especially if the bearing journals aren't worn excessively).

Installation

Refer to illustration 10.19

18   Lubricate the camshaft bearing journals and lobes with engine assembly lube or moly-base grease, then reinstall the camshaft in the head.

19   Apply the same lubricant to the rocker-arm bores, then slowly push a rocker-arm shaft into the cylinder head while guiding it into the rocker arms and springs. Note the rocker arm numbers and be sure to install them in the same positions they were in originally. Also note that the shafts are different and must be installed with the stepped sides facing the correct direction **(see illustration)**.

20   When all the rocker-arm components are positioned correctly, install the retaining screws and tighten them to the torque listed in this Chapter's Specifications. You may have to rotate the shafts to get the bolt holes to line up.

21   Install the remaining components in the reverse order of removal.

22   On Sprint models, adjust the valves (see Chapter 1).

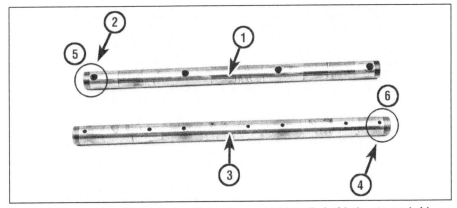

10.19 The shaft on the intake manifold side must be installed with the stepped side toward the camshaft sprocket - the shaft on the exhaust side must be installed with the stepped side toward the distributor (Sprint rocker-arm shaft shown, 1995 through 1997 Metro four-cylinder rocker arms shafts have same stepped ends with same dimensions)

1   Intake rocker-arm shaft
2   0.55-inch (14 mm)
3   Exhaust rocker-arm shaft
4   0.59-inch (15 mm)
5   Camshaft sprocket side
6   Distributor side

1998 and later Metro four-cylinder models

Removal

Refer to illustration 10.26

23   Remove the timing belt and the camshaft timing belt sprocket (see Section 7).

24   Remove the Camshaft Position (CMP) sensor from the camshaft sensor housing (see Chapter 6).

25   Loosen all valve adjusting screw locknuts **(see illustration 7.4)**. Be sure to back off the adjusting screws far enough to allow the rocker arms to move freely.

26   Gradually and evenly loosen the 12 camshaft bearing cap bolts in the indicated sequence **(see illustration)** and then remove the camshaft bearing caps. Keep the caps in order.

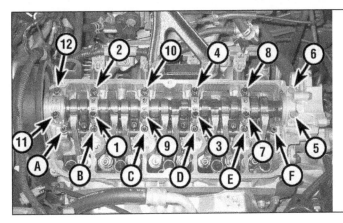

10.26 Camshaft bearing cap bolt loosening sequence (four-cylinder models); rocker arm shaft bolts are designated A, B, C, etc.

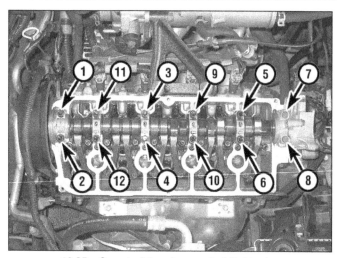

10.35a Camshaft bearing cap bolt tightening sequence (four-cylinder models)

10.35b Make sure that the cam bearing caps are installed in ascending numerical order, starting with the No. 1 cap at the timing belt end of the head

27   Remove the camshaft from the cylinder head and then slide the old camshaft oil seal off the front end of the cam. Discard the old seal.
28   Unclip and then remove the intake rocker arms from the rocker-arm shaft.
29   Remove the six rocker-arm shaft bolts **(see illustration 10.26)**.
30   Carefully slide the rocker-arm shaft out of the cylinder head and remove the exhaust rocker arms and the springs from the rocker-arm shaft.

Inspection

31   Refer to Steps 12 through 16.
32   Inspect the camshaft bearing surfaces in the cylinder head and in the bearing caps for scoring. If the bearing surfaces are in good shape, wipe them off, lay the cam in position in the cylinder head, lay a strip of Plastigage across the number one journal, install the bearing caps and tighten the cap bolts to the torque listed in this Chapter's Specifications. Then remove the bearing cap bolts, remove the bearing caps and the compare the width of the crushed Plastigage strip with the scale printed on the edge of the Plastigage container. When you have the determined the width of the Plastigage, compare your measurement to the camshaft journal bearing oil clearance listed in this Chapter's Specifications. If the cam bearing journals you measured in Step 16 were okay, but the cam journal bearing oil clearance is excessive, replace the cylinder head.

Installation

Refer to illustrations 10.35a and 10.35b
33   Lubricate the rocker-arm shaft and the rocker arms with clean engine oil and then carefully slide the rocker-arm shaft into the cylinder head with the holes facing up and install the rocker arms and springs in the correct sequence.
34   Install the rocker-arm shaft bolts and then tighten them to the torque listed in this Chapter's Specifications.

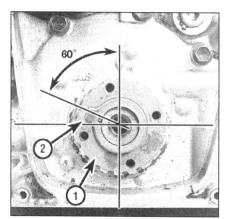

11.2 Position the crankshaft as shown here

1   Crankshaft sprocket
2   Key

35   Lubricate the bearing surfaces of the cam journals and the cam lobes and then place the cam in position. Install the cam bearing caps, coat the threads of the cam bearing cap bolts with clean engine oil, install the cam bearing cap bolts and then tighten the bolts in the indicated sequence **(see illustration)** to the torque listed in this Chapter's Specifications. Note that the caps are numbered, and must be installed in ascending numerical order, starting with cap No. 1 at the timing belt end of the head **(see illustration)**.
36   Coat the lip of a new camshaft oil seal with clean engine oil and then carefully slide the seal onto the nose of the cam. Make sure that the seal is square to the bore and then drive it into place with a socket slightly smaller in diameter than the outside diameter of the seal **(see illustration 9.4)**. Drive the seal onto the cam until it's flush with the front surface of the cylinder head and the number one camshaft bearing cap.
37   The remainder of installation is the reverse of removal.

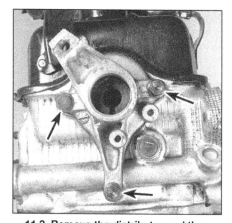

11.3 Remove the distributor and then remove the three distributor case attaching bolts (arrows)

11   Camshaft and lifters - removal, inspection and installation (Metro three-cylinder models)

Note: *If the valvetrain is making noise, and you suspect the lifters are causing it, before removing them for inspection, make sure the engine oil level is correct and allow some time after a cold start for them to quiet down. Hydraulic lifters are not adjustable or repairable - if they're defective, they must be discarded and new ones must be installed.*

Removal

Refer to illustrations 11.2, 11.3, 11.4 and 11.7
1   Remove the valve cover (see Section 4).
2   Remove the timing belt cover, timing belt and camshaft sprocket (see Section 7). Temporarily turn the crankshaft 60-degrees counterclockwise **(see illustration)**.
3   Remove the distributor (see Chapter 5) and case **(see illustration)** from the cylinder head.
4   The camshaft rides in three bearings. Each bearing cap is held by two bolts. The

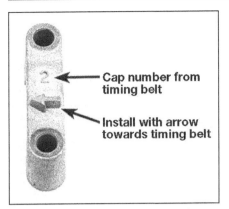

11.4 Camshaft bearing cap marks

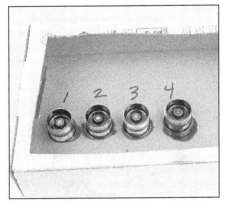

11.7 Store the lifters in a pan partially filled with oil

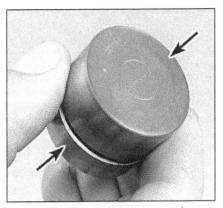

11.9a Check the lobe contact surfaces and the bore surfaces (arrows) of the lifters for wear

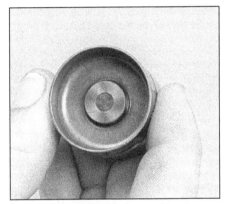

11.9b Check the valve side of the lifters too

caps are marked with numbers that indicate their position and arrows that point to the timing belt end of the engine (see illustration).

5 Loosen each of the cap bolts 1/4-turn at a time to relieve valve spring tension evenly until the caps are loose. If any of the caps stick, gently tap them with a soft-face hammer. Caution: *Failure to follow this procedure could tilt the camshaft in the bearings, which could damage the bearings or the camshaft.*

6 Lift out the camshaft, wipe it off with a clean shop towel and set it aside.

7 Remove the lifters, wipe each one off with a clean shop towel and number it with a felt-tip pen. Set them aside in a pan of oil to keep the oil from running out of them (see illustration).

Inspection

Camshaft

8 Refer to Section 10, Steps 12 through 16 and Step 32.

Lifters

Refer to illustrations 11.9a, 11.9b, 11.10, 11.11 and 11.13

9 Inspect the lifters for wear, galling and signs of seizure (see illustrations). If aluminum from the cylinder head is adhering to them, replace them. If any of the lifter bores are rough, scored or worn, replace the cylinder head.

10 Using a telescoping gauge and outside micrometer or a dial bore gauge, measure each lifter bore inside diameter and record the results (see illustration).

11 Measure each lifter outside diameter and record the results (see illustration).

12 Subtract the lifter outside diameter from the corresponding bore inside diameter to determine the clearance. Compare the

results to this Chapter's Specifications and replace parts as necessary.

13 Check the oil clearance for each camshaft journal as follows:

a) *Clean the bearing caps and the camshaft journals with lacquer thinner or acetone.*

b) *Carefully lay the camshaft in the head. Don't install the lifters and don't use any lubrication.*

c) *Lay a strip of Plastigage on each journal.*

d) *Install the bearing caps in their original locations.*

e) *Tighten the bolts to the torque listed in this Chapter's specifications in 1/4-turn increments.* Caution: *Don't turn the camshaft while the Plastigage is in place.*

f) *Remove the bolts and detach the caps.*

g) *Compare the width of the crushed Plastigage (at its widest point) to the scale on the Plastigage envelope* (see illustration).

h) *If the clearance is greater than specified, replace the camshaft and/or cylinder head.*

i) *Scrape off the Plastigage with your fingernail or the edge of a credit card - don't scratch or nick the journals or bearing caps.*

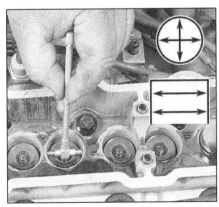

11.10 Measure each lifter bore inside diameter with a bore gauge . . .

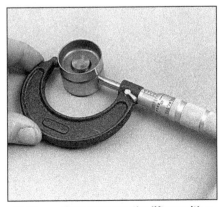

11.11 . . . then measure the lifters with a micrometer - subtract each lifter diameter from the corresponding bore diameter to obtain the lifter-to-bore clearances

11.13 Compare the width of the crushed Plastigage to the scale on the envelope

11.18 Position the dowel as shown

11.19 Apply sealant to the areas indicated (arrows)

14   If any of the conditions described above are noted, the cylinder head is probably getting insufficient lubrication or dirty oil, so make sure you track down the cause of this problem (low oil level, low oil pump capacity, clogged oil passage, etc.) before installing a new head, camshaft or lifters.

Installation

Refer to illustrations 11.18, 11.19, 11.20 and 11.22

Note: *The engine must be allowed to sit without starting it for at least 30 minutes after reassembly so the lifters can seat.*

15   Clean the camshaft, the bearing surfaces in the head and the lifters. Remove all sludge and dirt. Wipe off all components with a clean, lint-free cloth.

16   Lightly lubricate the lifters and bores with assembly lube or moly-base grease. Refer to the numbers marked on them and install the lifters in the head.

17   Lubricate the camshaft bearing surfaces in the head and the bearing journals and lobes on the camshaft with assembly lube or moly-base grease.

18   Slide a new oil seal onto the front of the camshaft, then carefully lower the camshaft into position with the dowel at the five o'clock position **(see illustration)**. **Caution:** *Failure to adequately lubricate the camshaft and related components can cause serious damage to bearing and friction surfaces during the first few seconds after engine start-up.*

19   Apply a thin coat of sealant to the mating surfaces of the number one and three bearing caps **(see illustration)**.

20   Apply a thin coat of assembly lube or moly-base grease to the bearing surfaces of the camshaft bearing caps and install the caps in their original locations **(see illustration)**.

21   Install the bolts for the bearing caps. Gradually tighten all fasteners - 1/4-turn at a time - until the camshaft is drawn down and seated in the bearing saddles. Don't tighten the fasteners completely at this time.

22   Following the recommended sequence **(see illustration)**, tighten the fasteners for the bearing caps to the torque listed in this Chapter's Specifications.

23   Install the camshaft sprocket, timing belt, timing belt cover and related components (see Section 7).

24   Remove the spark plugs and carefully rotate the crankshaft clockwise with a socket and breaker bar to make sure the valve timing is correct. After two revolutions, the timing marks on the sprockets should still be aligned. If they're not, re-index the timing belt to the sprockets (see Section 7). **Note:** *If you feel resistance while rotating the crankshaft, stop immediately and check the valve timing.*

12 Valve spring, retainer and seal - replacement

Refer to illustrations 12.4, 12.8 and 12.16

Note: *On Sprint models, and on Metro 1.3L models, broken valve springs and defective valve stem seals can be replaced without removing the cylinder head. Two special tools*

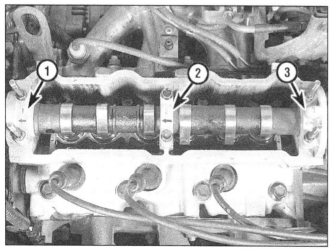

11.20 Bearing cap positions

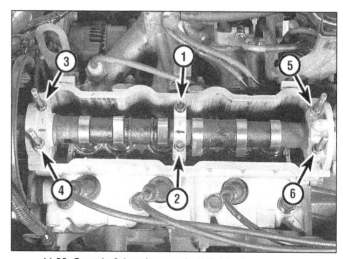

11.22 Camshaft bearing cap bolt tightening sequence

1	No. 1 bearing	3	No. 3 bearing
2	No. 2 bearing		

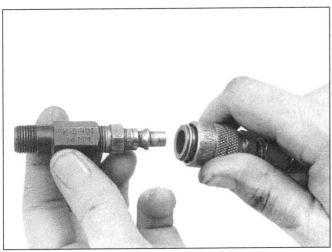

12.4 This is what the air hose adapter that threads into the spark plug hole looks like - they're commonly available from auto parts stores

12.8 Compress the spring enough to release the keepers (arrows)

and a compressed air source are normally required to perform this operation, so read through this Section carefully and rent or buy the tools before beginning the job. If compressed air isn't available, a length of nylon rope can be used to keep the valves from falling into the cylinder during this procedure.

1   If you're working on a Sprint, refer to Section 10 and remove the rocker arms and shafts from the cylinder head. If you're working on a Metro, refer to Section 11 and remove the camshaft and lifters.

2   Remove the spark plug from the cylinder which has the defective component. If all of the valve stem seals are being replaced, all of the spark plugs should be removed.

3   If you're replacing all of the valve stem seals, begin with cylinder number one and work on the valves for one cylinder at a time.

4   Thread an adapter into the spark plug hole (see illustration) and connect an air hose from a compressed air source to it. Most auto parts stores can supply the air hose adapter. Note: Many cylinder compression gauges utilize a screw-in fitting that may work with your air hose quick-disconnect fitting.

5   Apply compressed air to the cylinder. Warning: The piston may be forced down by compressed air, causing the crankshaft to turn suddenly. If a wrench is still attached to the bolt in the crankshaft nose, it could cause damage or injury when the crankshaft moves. Keep your hands clear of the drivebelts and rotating engine components.

6   The valves should be held in place by the air pressure. If the valve faces or seats are in poor condition, leaks may prevent air pressure from retaining the valves - refer to the alternative procedure below.

7   If you don't have access to compressed air, an alternative method can be used. Position the piston at a point just before Top Dead Center, then feed a long piece of nylon rope through the spark plug hole until it fills the combustion chamber. Be sure to leave

the end of the rope hanging out of the engine so it can be removed easily. Use a large ratchet and socket to rotate the crankshaft in the normal direction of rotation (clockwise) until slight resistance is felt.

8   Stuff shop rags into the cylinder head holes above and below the valves to prevent parts and tools from falling into the engine, then use a valve spring compressor to compress the spring (see illustration). Remove the keepers with a small needle-nose pliers or a magnet. Note: A couple of different types of tools are available for compressing the valve springs with the head in place. One type, shown here, grips the lower spring coils and presses on the retainer as the handle is turned, while another type utilizes the rocker-arm shaft for leverage.

9   Remove the spring retainer and valve spring, then remove the guide seal. Note: If air pressure fails to hold the valve in the closed position during this operation, the valve face and/or seat is probably damaged. If so, the cylinder head will have to be removed for additional repair operations.

10   Wrap a rubber band or tape around the top of the valve stem so the valve won't fall into the combustion chamber, then release the air pressure. Note: If a rope was used instead of air pressure, turn the crankshaft slightly in the direction opposite normal rotation.

11   Inspect the valve stem for damage. Rotate the valve in the guide and check the end for eccentric movement, which would indicate the valve is bent.

12   Move the valve up-and-down in the guide and make sure it doesn't bind. If the valve stem binds, either the valve is bent or the guide is damaged. In either case, the head will have to be removed for repair.

13   Reapply air pressure to the cylinder to retain the valve in the closed position, then remove the tape or rubber band from the valve stem. If a rope was used instead of air pressure, rotate the crankshaft clockwise

until slight resistance is felt.

14   Lubricate the valve stem with engine oil and install a new guide seal.

15   Install the spring in position over the valve.

16   Install the valve spring retainer. Compress the valve spring and carefully position the keepers in the groove. Apply a small dab of grease to the inside of each keeper to hold it in place if necessary (see illustration).

17   Remove the pressure from the spring tool and make sure the keepers are seated.

18   Disconnect the air hose and remove the adapter from the spark plug hole. If a rope was used in place of air pressure, pull it out of the cylinder.

19   Refer to Section 10 and install the rocker arms and shafts or Section 11 and install the lifters and camshaft.

20   Install the spark plugs(s) and connect the wire(s).

21   Refer to Section 4 and install the valve cover.

22   Start and run the engine, then check for oil leaks and unusual sounds coming from the valve cover area.

12.16 Apply a small dab of grease to each keeper before installation to hold them in place on the valve stem until the spring is released

13.10a Cylinder head bolt loosening sequence (Sprint engine shown, Metro three-cylinder engine uses the same sequence)

A   Timing belt end
B   Distributor end

---

13 Cylinder head - removal and installation

Removal

Refer to illustrations 13.10a, 13.10b, 13.10c and 13.11

Caution: *Allow the engine to cool completely before beginning this procedure.*

1   Position the number one piston at Top Dead Center (see Section 3).
2   Disconnect the negative cable from the battery.
3   Drain the cooling system and remove the spark plugs (see Chapter 1).
4   Remove the intake manifold (see Section 5).
5   Remove the exhaust manifold (see Section 6).
6   On models with a distributor, remove the distributor (see Chapter 5) and unbolt the distributor case from the cylinder head.
7   On carbureted models, remove the fuel pump and pushrod (see Chapter 4).
8   Remove the valve cover (see Section 4).

9   Remove the timing belt (see Section 7).
10   Loosen the cylinder head bolts, 1/4-turn at a time, in the specified sequence **(see illustrations)** until they can be removed by hand.
11   Carefully detach the cylinder head from the block and place it on wooden blocks to prevent damage to the sealing surfaces. If the head sticks to the engine block, dislodge it by prying against a protrusion on the head casting **(see illustration)**. **Note:** *Cylinder head disassembly and inspection procedures are covered in Sections 10 and 11 and in Chapter 2, Part B. It's a good idea to inspect the camshaft and have the head checked for warpage and cracks, even if you're just replacing the gasket.*

Installation

Refer to illustrations 13.12, 13.14, 13.15, 13.16a, 13.16b and 13.16c

12   The mating surfaces of the cylinder head and block must be perfectly clean when the head is installed. Use a gasket scraper to remove all traces of carbon and old gasket material **(see illustration)**, then clean the mating surfaces with lacquer thinner or acetone. If there's oil on the mating surfaces when the head is installed, the gasket may not seal correctly and leaks may develop. When working on the block, stuff the cylin-

---

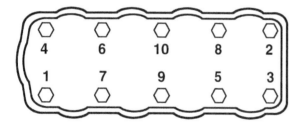

| 4 | 6 | 10 | 8 | 2 |
| 1 | 7 | 9 | 5 | 3 |

24075-2a-13.10b HAYNES

13.10b Cylinder head bolt loosening sequence (1995 through 1997 Metro four-cylinder engine)

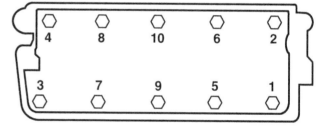

| 4 | 8 | 10 | 6 | 2 |
| 3 | 7 | 9 | 5 | 1 |

24075-2a-13.10c HAYNES

13.10c Cylinder head bolt loosening sequence (1998 and later Metro four-cylinder engine)

13.11 Use casting protrusions to pry against - don't pry between the gasket surfaces

13.12 Remove all traces of old gasket material

13.14 All bolt holes in the block - particularly the main bearing cap and head bolt holes - should be cleaned and restored with a tap (be sure to remove debris from the holes after this is done)

ders with clean shop rags to keep out debris. Use a vacuum cleaner to remove material that falls into the cylinders. Since the head is made of aluminum, aggressive scraping can cause damage. Be extra careful not to nick or gouge the mating surfaces with the scraper.

13 Check the block and head mating surfaces for nicks, deep scratches and other damage. If damage is slight, it can be removed with a file; if it's excessive, machining may be the only alternative.

14 Use a tap of the correct size to chase the threads in the head bolt holes **(see illustration)**. Mount each bolt in a vise and run a die down the threads to remove corrosion and restore the threads. Dirt, corrosion, sealant and damaged threads will affect torque readings.

15 Place a new gasket on the block with the side marked TOP facing up **(see illustration)**. Make sure the openings in the gasket and block are lined up, then set the cylinder head in position. **Note:** *Before installing the cylinder head, make sure the oil check valve is in place and the opening is clean and free of obstructions.*

16 Tighten the cylinder head bolts in three equal steps following the proper sequence **(see illustrations)** until they're all at the torque listed in this Chapter's Specifications.

17 Reinstall the timing belt (see Section 7).

18 Reinstall the remaining parts in the reverse order of removal.

19 Be sure to refill the cooling system and check all fluid levels. Rotate the crankshaft clockwise slowly by hand through two complete revolutions. Recheck the camshaft timing marks (see Section 7).

20 Start the engine and set the ignition timing (see Chapter 1). Run the engine until normal operating temperature is reached. Check for leaks and proper operation. Shut off the engine, remove the valve cover and retorque the cylinder head bolts, unless the gasket manufacturer states otherwise. Recheck the valve adjustment.

14 Oil pan - removal and installation

Refer to illustrations 14.6a, 14.6b, 14.7, 14.10 and 14.11

1 Disconnect the negative cable from the battery. Raise the front of the vehicle and support it securely on jackstands.

2 Drain the engine oil (see Chapter 1).

3 Remove the splash shield from under the engine, if equipped.

4 Disconnect the front exhaust pipe from the exhaust manifold (see Chapter 4) and then lower it for access to the oil pan.

5 If it's in the way, remove the bellhousing lower plate.

13.15 Position a new gasket with the "top" mark facing up (A) - location of oil check valve (B)

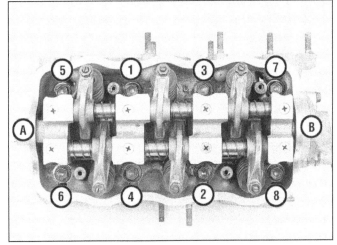

13.16a Cylinder head bolt tightening sequence (Sprint engine shown, Metro three-cylinder engine uses the same sequence)

A *Timing belt end*   B *Distributor end*

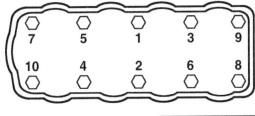

24075-2a-13.10b HAYNES

13.16b Cylinder head bolt tightening sequence (1995 through 1997 Metro four-cylinder engine)

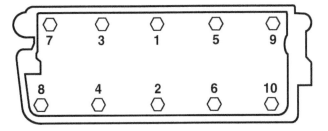

24075-2a-13.16c HAYNES

13.16c Cylinder head bolt tightening sequence (1998 and later Metro four-cylinder engine)

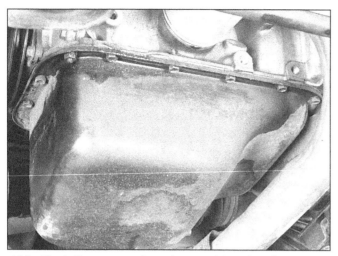

14.6a The bolts are spaced evenly around the edge of the oil pan

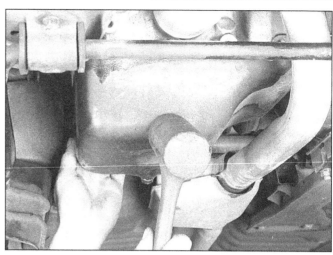

14.6b Use a soft-face hammer to dislodge the oil pan

14.7 Remove the oil pickup tube bolts (arrows)

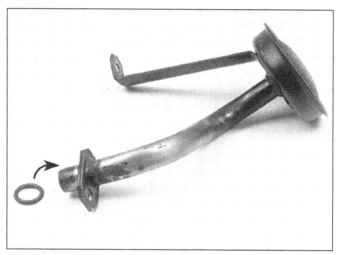

14.10 Place the O-ring onto the oil pickup tube (arrow)

6   Remove the bolts/nuts and detach the oil pan **(see illustration)**. Don't pry between the block and pan or damage to the sealing surfaces may result and oil leaks could develop. Use a soft-face hammer to dislodge

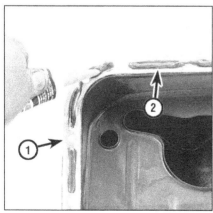

14.11 Apply a bead of RTV sealant to the oil pan flange as shown

1   Oil pan
2   Sealant

the pan if it's stuck **(see illustration)**.
7   The oil pickup tube can now be removed if necessary **(see illustration)**.
8   Use a scraper to remove all traces of old gasket material and sealant from the block and oil pan. Clean the gasket sealing surfaces with lacquer thinner or acetone and make sure the bolt holes in the block are clean.
9   Check the oil pan flange for distortion, particularly around the bolt holes. If necessary, place the pan on a block of wood and use a hammer to flatten and restore the gasket surface.
10   If it was removed, install the oil pickup tube with a new seal **(see illustration)**.
11   Before installing the oil pan, apply a thin coat of RTV sealant to the flange **(see illustration)**. Attach the new gasket to the pan (make sure the bolt holes are aligned).
12   Position the oil pan against the engine block and install the mounting bolts/nuts. Tighten them in a criss-cross pattern to the torque listed in this Chapter's Specifications.
13   Wait at least 30 minutes before filling the engine with oil, then start the engine and check the pan for leaks.

15 Oil pump - removal, inspection and installation

Removal

Refer to illustrations 15.3, 15.4a, 15.4b, 15.5 and 15.7

1   Remove the timing belt, tensioner, crankshaft sprocket and belt guide (see Section 7).
2   Remove the oil pan and oil pickup tube (see Section 14).
3   Remove the alternator (see Chapter 5) and bracket **(see illustration)**. Unbolt the dipstick tube and pull the tube out of the engine front case.
4   Remove the front cover-to-block bolts and carefully separate the cover from the engine **(see illustrations)**.
5   Remove the screws, detach the cover from the rear of the case and remove the inner and outer oil pump gears **(see illustration)**.
6   Remove the crankshaft oil seal from the front of the case (see Section 8).

15.3 Oil pump mounting details

1 Alternator bracket nut
2 Dipstick tube mounting bolt

15.4a Remove the front cover-to-block bolts - note that they come in different lengths

1 Short bolts
2 Long bolts

15.4b Carefully separate the front cover from the engine with a prybar (arrow)

7 Remove the pressure relief valve retainer, spring and plunger **(see illustration)**.

Inspection

Refer to illustrations 15.9a, 15.9b and 15.9c

8 Clean all parts thoroughly and remove all traces of old gasket material from the sealing surfaces. Visually inspect all parts for wear, cracks and other damage. Replace parts as necessary.

9 Install the oil pump outer and inner gears and measure the clearances **(see illustrations)**. Compare the clearances to this Chapter's Specifications. Replace parts as necessary. Pack the pump cavity with petroleum jelly and install the cover. Tighten the cover screws to the torque listed in this Chapter's Specifications.

Installation

Refer to illustrations 15.12a and 15.12b

10 Install a new crankshaft front oil seal (see Section 8).

11 Install the pressure relief valve components.

12 Using a new gasket, position the front

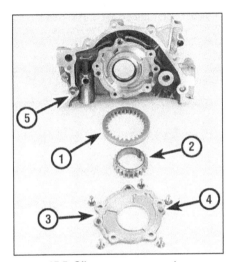

15.5 Oil pump components - exploded view

1 Oil pump outer gear
2 Oil pump inner gear
3 Cover
4 Pin
5 Dowel pin

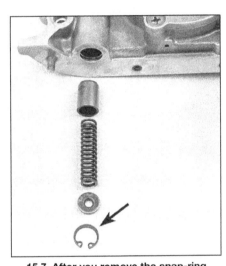

15.7 After you remove the snap-ring (arrow) with snap-ring pliers, the retainer, spring and plunger can be removed - if the plunger sticks in the bore, tap the front cover on a block of wood to dislodge it

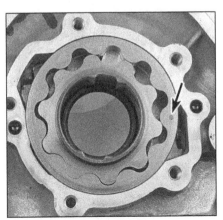

15.9a Install the outer gear with the dot (arrow) visible - the inner gear will only fit one way: with the raised center collar facing out

15.9b Measure the outer gear-to-oil pump housing clearance with a feeler gauge

15.9c Measure the gear end play with a precision straightedge and feeler gauge

15.12a The oil pump has two tangs (arrows) . . .

cover on the engine **(see illustrations)**. Install the bolts and tighten them to the torque listed in this Chapter's Specifications.

13   Reinstall the remaining parts in the reverse order of removal.

14   Add oil, start the engine and check for oil pressure and leaks.

16   Flywheel/driveplate - removal and installation

Refer to illustrations 16.4, 16.8a and 16.8b

1   Raise the vehicle and support it securely on jackstands, then refer to Chapter 7 and remove the transaxle. If it's leaking, now would be a very good time to replace the front pump seal/O-ring (automatic transaxle only).

2   Remove the pressure plate and clutch disc (see Chapter 8) (manual transaxle equipped models). Now is a good time to check/replace the clutch components and pilot bearing.

3   On automatic transaxle models, mark the relationship between the driveplate and crankshaft to ensure correct alignment during reinstallation.

4   Remove the bolts that secure the flywheel/driveplate to the crankshaft. If the crankshaft turns, immobilize it by jamming a heavy screwdriver into the ring gear teeth **(see illustration)**.

5   Remove the flywheel/driveplate from the crankshaft. Since the flywheel is fairly heavy,

be sure to support it while removing the last bolt.

6   Clean the flywheel to remove grease and oil. Inspect the surface for cracks, rivet grooves, burned areas and score marks. Light scoring can be removed with emery cloth. Check for cracked and broken ring gear teeth. Lay the flywheel on a flat surface and use a straightedge to check for warpage.

7   Clean and inspect the mating surfaces of the flywheel/driveplate and the crankshaft. If the crankshaft oil seal is leaking, replace it before reinstalling the flywheel/driveplate (see Section 17).

8   Position the flywheel/driveplate against the crankshaft. On manual transaxle models, be sure to align the hole in the flywheel with the alignment dowel in the crankshaft **(see illustrations)**. On automatic transaxle models, align the marks you made during removal. Before installing the bolts, apply thread locking compound to the threads.

9   Tighten the bolts to the torque listed in this Chapter's Specifications. Keep the crankshaft from turning as described above.

10   The remainder of installation is the reverse of the removal procedure.

15.12b . . . which must be aligned with the flats on the crankshaft (arrows)

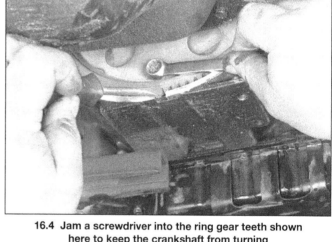

16.8a The alignment dowel (arrow) . . .

16.4 Jam a screwdriver into the ring gear teeth shown here to keep the crankshaft from turning

16.8b . . . must be aligned with the dowel hole in the flywheel (arrow) (they ensure the flywheel can only be installed one way)

17.2a Gently pry out the old seal with a screwdriver - wrap the tip with tape to prevent damage to the seal bore and crankshaft sealing surfaces

17.2b Gently tap the new seal into place with a large socket and hammer

17   Crankshaft rear oil seal - replacement

Refer to illustrations 17.2a, 17.2b, 17.5 and 17.6

1   The transaxle must be removed from the vehicle for this procedure (see Chapter 7).

2   The seal can be replaced without removing the oil pan or the seal housing. However, this method is not recommended because the lip of the seal is quite stiff and it's possible to cock the seal in the housing bore or damage it during installation. If you want to take the chance, pry out the old seal **(see illustration)**. Apply multi-purpose grease to the crankshaft seal journal and the lip of the new seal and carefully tap the new seal into place **(see illustration)**. The lip is stiff so carefully work it onto the seal journal of the crankshaft with a smooth object like the end of an extension as you tap the seal into place. Don't rush it or you may damage the seal.

3   The following method is recommended

but requires removal of the oil pan (see Section 14) and the seal housing.

4   After the oil pan has been removed, remove the bolts, detach the seal housing and peel off all the old gasket material.

5   Position the seal and housing assembly on a couple of wood blocks on a workbench and drive the old seal out from the back side with a punch and hammer **(see illustration)**.

6   Drive the new seal into the housing with a block of wood **(see illustration)** or a section of pipe slightly smaller in diameter than the outside diameter of the seal.

7   Lubricate the crankshaft seal journal and the lip of the new seal with multi-purpose grease. Position a new gasket on the engine block.

8   Slowly and carefully push the seal onto the crankshaft. The seal lip is stiff, so work it onto the crankshaft with a smooth object such as the end of an extension as you push the housing against the block.

9   Install and tighten the housing bolts to the torque listed in this Chapter's Specifications.

10   The remaining steps are the reverse of removal.

11   Run the engine and check for oil leaks.

18   Engine mounts - check and replacement

1   Engine mounts seldom require attention, but broken or deteriorated mounts should be replaced immediately or the added strain placed on the driveline components may cause damage or wear.

Check

2   During the check, the engine must be raised slightly to remove the weight from the mounts.

3   Raise the vehicle and support it securely on jackstands, then position a jack under the engine oil pan. Place a large block of wood between the jack head and the oil pan, then carefully raise the engine just enough to take the weight off the mounts. **Warning:** *DO NOT place any part of your body under the engine when it's supported only by a jack!*

4   Check the mounts to see if the rubber is cracked, hardened or separated from the metal plates. Sometimes the rubber will split right down the center.

5   Check for relative movement between the mount plates and the engine or frame (use a large screwdriver or pry bar to attempt to move the mounts). If play is noted, lower the engine and tighten the mount fasteners.

6   Rubber preservative should be applied to the mounts to slow deterioration.

Replacement

Refer to illustrations 18.8a, 18.8b, 18.8c, 18.8d, 18.8e and 18.8f

7   Disconnect the negative battery cable from the battery, then raise the vehicle and support it securely on jackstands (if not already done).

8   Raise the engine slightly with a jack or

17.5 After removing the retainer from the block, support it on two wood blocks and drive out the old seal with a punch and hammer

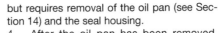

17.6 Drive the new seal into the retainer with a wood block or a section of pipe, if you have one large enough - make sure you don't cock the seal in the retainer bore

18.8a Typical late-model left (automatic) transaxle mount (left manual transaxle mount similar); to remove the mount *insulator* (the big rubber part), remove this through-bolt (arrow)

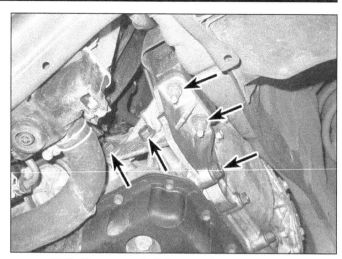

18.8b To remove the left transaxle mount *bracket*, remove these bolts (arrows)

18.8c Typical late-model right engine mount; to remove the mount *insulator*, remove the through-bolt (lower arrow) and then unbolt the insulator from the bracket by removing these two bolts (upper arrows); to remove the right mount *bracket*, remove the two bracket mounting bolts (not shown) from underneath the vehicle

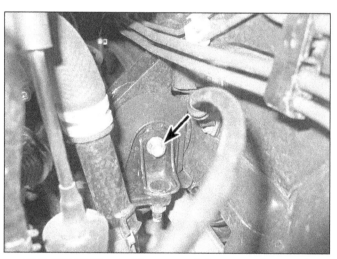

18.d Typical late-model rear mount; to remove the mount *insulator*, remove this nut (arrow) . . .

hoist. Remove the mount-to-bracket fasteners **(see illustrations)** and detach the mount.
9   Installation is the reverse of removal. Use thread locking compound on the mount bolts and nuts and be sure to tighten them securely.

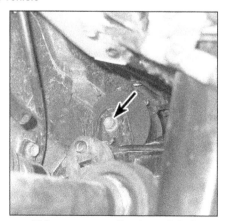

18.8e . . . remove the through-bolt (arrow) . . .

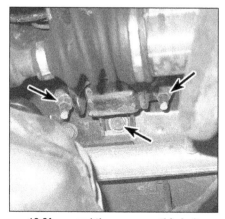

18.8f . . . and then remove this bolt (middle arrow); to remove the rear mount *bracket*, remove these two nuts (outer arrows) and then pull the bracket straight up

Chapter 2 Part B
General engine overhaul procedures

Contents

Specifications

General

Displacement	
Three-cylinder engine...	61 cubic inches (1.0 liter)
Four-cylinder engine	79 cubic inches (1.3 liters)
Bore and Stroke	
Three-cylinder engine...	2.91 x 3.03 inches (74 x 77 mm)
Four-cylinder engine	2.91 x 3.54 inches (74 x 90 mm)
Cylinder compression pressure (all models)	
Standard..	199 psi
Minimum..	156.4 psi
Difference between cylinders	14.2 psi or less
Oil pressure (engine warmed up)	
Sprint..	43 to 54 psi at 3000 rpm
Metro	
1989..	43 to 59 psi at 3000 rpm
1990 through 1992...	48 to 56 psi at 4000 rpm
1993..	43 to 54 psi at 3000 rpm
1994..	46 to 61 psi at 3000 rpm
1995 through 2001...	47 to 61 psi at 4000 rpm

Torque specifications

Connecting rod bearing cap nuts (all models)..	26
Main bearing cap bolts (all models)...	40

2.2 The oil pressure sending unit (arrow) is located on the front of the block , right above the oil filter (sending unit shown is on a four-cylinder engine; sending unit on three-cylinder engines is in same location)

2.3 To check the oil pressure, unscrew the sending unit and connect an oil pressure gauge in its place

1   General information - engine overhaul

Included in this portion of Chapter 2 are general information and diagnostic testing procedures for determining the overall mechanical condition of your engine.

The information ranges from advice concerning preparation for an overhaul and the purchase of replacement parts and/or components to detailed, step-by-step procedures covering removal and installation.

The following Sections have been written to help you determine whether your engine needs to be overhauled and how to remove and install it once you've determined it needs to be rebuilt. For information concerning in-vehicle engine repair, see Chapter 2A.

The Specifications included in this Part are general in nature and include only those necessary for testing the oil pressure and checking the engine compression. Refer to Chapter 2A for additional engine Specifications.

It's not always easy to determine when, or if, an engine should be completely overhauled, because a number of factors must be considered.

High mileage is not necessarily an indication that an overhaul is needed, while low mileage doesn't preclude the need for an overhaul. Frequency of servicing is probably the most important consideration. An engine that's had regular and frequent oil and filter changes, as well as other required maintenance, will most likely give many thousands of miles of reliable service. Conversely, a neglected engine may require an overhaul very early in its service life.

Excessive oil consumption is an indication that piston rings, valve seals and/or valve guides are in need of attention. Make sure that oil leaks aren't responsible before decid-

ing that the rings and/or guides are bad. Perform a cylinder compression check to determine the extent of the work required (see Section 3). Also check the vacuum readings under various conditions (see Section 4).

Check the oil pressure with a gauge installed in place of the oil pressure sending unit and compare it to this Chapter's Specifications (see Section 2). If it's extremely low, the bearings and/or oil pump are probably worn out.

Loss of power, rough running, knocking or metallic engine noises, excessive valve train noise and high fuel consumption rates may also point to the need for an overhaul, especially if they're all present at the same time. If a complete tune-up doesn't remedy the situation, major mechanical work is the only solution.

An engine overhaul involves restoring the internal parts to the specifications of a new engine. During an overhaul, the piston rings are replaced and the cylinder walls are reconditioned (rebored and/or honed). If a rebore is done by an automotive machine shop, new oversize pistons will also be installed. The main bearings, connecting rod bearings and camshaft bearings are generally replaced with new ones and, if necessary, the crankshaft may be reground to restore the journals. Generally, the valves are serviced as well, since they're usually in less-than-perfect condition at this point. While the engine is being overhauled, other components, such as the distributor, starter and alternator, can be rebuilt as well. The end result should be a like new engine that will give many trouble free miles. **Note:** *Critical cooling system components such as the hoses, drivebelts, thermostat and water pump should be replaced with new parts when an engine is overhauled. The radiator should be checked carefully to ensure that it isn't clogged or leaking (see Chapter 3). If you purchase a rebuilt engine or short block, some rebuilders will not warranty their engines unless the radiator has been*

professionally flushed. Also, we don't recommend overhauling the oil pump - always install a new one when an engine is rebuilt.

Overhauling the internal components on today's engines is a difficult and time-consuming task which requires a significant amount of specialty tools and is best left to a professional engine rebuilder. A competent engine rebuilder will handle the inspection of your old parts and offer advice concerning the reconditioning or replacement of the original engine, never purchase parts or have machine work done on other components until the block has been thoroughly inspected by a professional machine shop. As a general rule, time is the primary cost of an overhaul, especially since the vehicle may be tied up for a minimum of two weeks or more. Be aware that some engine builders only have the capability to rebuild the engine you bring them while other rebuilders have a large inventory of rebuilt exchange engines in stock. Also be aware that many machine shops could take as much as two weeks time to completely rebuild your engine depending on shop workload. Sometimes it makes more sense to simply exchange your engine for another engine that's already rebuilt to save time.

2   Oil pressure check

Refer to illustrations 2.2 and 2.3

1   Low engine oil pressure can be a sign of an engine in need of rebuilding. A "low oil pressure" indicator (often called an "idiot light") is not a test of the oiling system. Such indicators only come on when the oil pressure is dangerously low. Even a factory oil pressure gauge in the instrument panel is only a relative indication, although much better for driver information than a warning light. A better test is with a mechanical (not electrical) oil pressure gauge.

3.6 Screw a compression gauge with a threaded fitting into the spark plug hole (we don't recommend the type that requires hand pressure to maintain the seal); be sure to keep the throttle plate open as far as possible during the compression check

4.4 A vacuum gauge is handy for diagnosing engine condition and performance; make sure that you hook up the vacuum gauge hose to a good *manifold* vacuum source, downstream from the throttle plate

2   Locate the oil pressure indicator sending unit on the front of the engine block **(see illustration)**.

3   Unscrew and remove the oil pressure sending unit and then screw in the hose for your oil pressure gauge **(see illustration)**. If necessary, install an adapter fitting. Use Teflon tape or thread sealant on the threads of the adapter and/or the fitting on the end of your gauge's hose.

4   Connect an accurate tachometer to the engine, according to the tachometer manufacturer's instructions.

5   Check the oil pressure with the engine running (normal operating temperature) at the specified engine speed, and compare it to this Chapter's Specifications. If it's extremely low, the bearings and/or oil pump are probably worn out.

3   Cylinder compression check

Refer to illustration 3.6

1   A compression check will tell you what mechanical condition the upper end of your engine (pistons, rings, valves, head gaskets) is in. Specifically, it can tell you if the compression is down due to leakage caused by worn piston rings, defective valves and seats or a blown head gasket. **Note:** *The engine must be at normal operating temperature and the battery must be fully charged for this check.*

2   Begin by cleaning the area around the spark plugs before you remove them (compressed air should be used, if available). The idea is to prevent dirt from getting into the cylinders as the compression check is being done.

3   Remove all of the spark plugs from the engine (see Chapter 1).

4   Block the throttle wide open.

5   Disable the ignition system by discon-

necting the primary (low voltage) electrical connectors from the coil packs (see Chapter 5). The fuel pump circuit should also be disabled (see Chapter 4).

6   Install a compression gauge in the spark plug hole **(see illustration)**.

7   Crank the engine over at least seven compression strokes and watch the gauge. The compression should build up quickly in a healthy engine. Low compression on the first stroke, followed by gradually increasing pressure on successive strokes, indicates worn piston rings. A low compression reading on the first stroke, which doesn't build up during successive strokes, indicates leaking valves or a blown head gasket (a cracked head could also be the cause). Deposits on the undersides of the valve heads can also cause low compression. Record the highest gauge reading obtained.

8   Repeat the procedure for the remaining cylinders and compare the results to this Chapter's Specifications.

9   Add some engine oil (about three squirts from a plunger-type oil can) to each cylinder, through the spark plug hole, and repeat the test.

10   If the compression increases after the oil is added, the piston rings are definitely worn. If the compression doesn't increase significantly, the leakage is occurring at the valves or head gasket. Leakage past the valves may be caused by burned valve seats and/or faces or warped, cracked or bent valves.

11   If two adjacent cylinders have equally low compression, there's a strong possibility that the head gasket between them is blown. The appearance of coolant in the combustion chambers or the crankcase would verify this condition.

12   If one cylinder is slightly lower than the others, and the engine has a slightly rough idle, a worn lobe on the camshaft could be the cause.

13   If the compression is unusually high,

the combustion chambers are probably coated with carbon deposits. If that's the case, the cylinder head(s) should be removed and decarbonized.

14   If compression is way down or varies greatly between cylinders, it would be a good idea to have a leak-down test performed by an automotive repair shop. This test will pinpoint exactly where the leakage is occurring and how severe it is.

4   Vacuum gauge diagnostic checks

Refer to illustrations 4.4 and 4.6

A vacuum gauge provides valuable information about what is going on in the engine at a low-cost. You can check for worn rings or cylinder walls, leaking head or intake manifold gaskets, incorrect carburetor adjustments, restricted exhaust, stuck or burned valves, weak valve springs, improper ignition or valve timing and ignition problems.

Unfortunately, vacuum gauge readings are easy to misinterpret, so they should be used in conjunction with other tests to confirm the diagnosis.

Both the absolute readings and the rate of needle movement are important for accurate interpretation. Most gauges measure vacuum in inches of mercury (in-Hg). The following references to vacuum assume the diagnosis is being performed at sea level. As elevation increases (or atmospheric pressure decreases), the reading will decrease. For every 1,000 foot increase in elevation above approximately 2000 feet, the gauge readings will decrease about one inch of mercury.

Connect the vacuum gauge directly to intake manifold vacuum, not to ported (throttle body) vacuum **(see illustration)**. Be sure no hoses are left disconnected during the test or false readings will result.

Before you begin the test, allow the engine to warm up completely. Block the wheels and set the parking brake. With the transmission in Park, start the engine and allow it to run at normal idle speed. **Warning:** *Keep your hands and the vacuum gauge clear of the fans.*

Read the vacuum gauge; an average, healthy engine should normally produce about 17 to 22 in-Hg with a fairly steady needle **(see illustration)**. Refer to the following vacuum gauge readings and what they indicate about the engine's condition:

1   A low steady reading usually indicates a leaking gasket between the intake manifold and cylinder head(s) or throttle body, a leaky vacuum hose, late ignition timing or incorrect camshaft timing. Check ignition timing with a timing light and eliminate all other possible causes, utilizing the tests provided in this Chapter before you remove the timing chain cover to check the timing marks.

2   If the reading is three to eight inches below normal and it fluctuates at that low reading, suspect an intake manifold gasket leak at an intake port or a faulty fuel injector.

3   If the needle has regular drops of about two-to-four inches at a steady rate, the valves are probably leaking. Perform a compression check or leak-down test to confirm this.

4   An irregular drop or down-flick of the needle can be caused by a sticking valve or an ignition misfire. Perform a compression check or leak-down test and read the spark plugs.

5   A rapid vibration of about four in-Hg vibration at idle combined with exhaust smoke indicates worn valve guides. Perform a leak-down test to confirm this. If the rapid vibration occurs with an increase in engine speed, check for a leaking intake manifold gasket or head gasket, weak valve springs, burned valves or ignition misfire.

6   A slight fluctuation, say one inch up and down, may mean ignition problems. Check all the usual tune-up items and, if necessary, run the engine on an ignition analyzer.

7   If there is a large fluctuation, perform a compression or leak-down test to look for a weak or dead cylinder or a blown head gasket.

8   If the needle moves slowly through a wide range, check for a clogged PCV system, incorrect idle fuel mixture, carburetor/throttle body or intake manifold gasket leaks.

9   Check for a slow return after revving the engine by quickly snapping the throttle open until the engine reaches about 2,500 rpm and let it shut. Normally the reading should drop to near zero, rise above normal idle reading (about 5 in-Hg over) and then return to the previous idle reading. If the vacuum returns slowly and doesn't peak when the throttle is snapped shut, the rings may be worn. If there is a long delay, look for a restricted exhaust system (often the muffler or catalytic converter). An easy way to check this is to temporarily disconnect the exhaust ahead of the suspected part and redo the test.

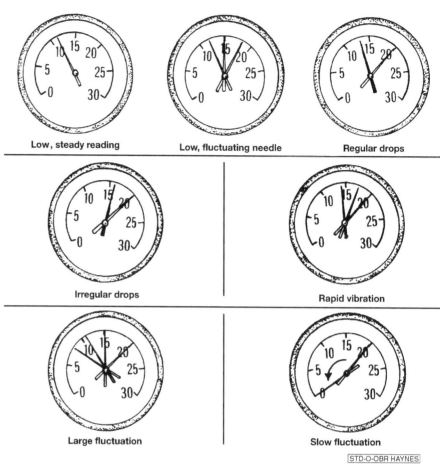

Low, steady reading   Low, fluctuating needle   Regular drops

Irregular drops   Rapid vibration

Large fluctuation   Slow fluctuation

STD-O-OBR HAYNES

4.6 Typical vacuum gauge readings

5   Engine rebuilding alternatives

The do-it-yourselfer is faced with a number of options when purchasing a rebuilt engine. The major considerations are cost, warranty, parts availability and the time required for the rebuilder to complete the project. The decision to replace the engine block, piston/connecting rod assemblies and crankshaft depends on the final inspection results of your engine. Only then can you make a cost effective decision whether to have your engine overhauled or simply purchase an exchange engine for your vehicle.

Some of the rebuilding alternatives include:

Individual parts - If the inspection procedures reveal that the engine block and most engine components are in reusable condition, purchasing individual parts and having a rebuilder rebuild your engine may be the most economical alternative. The block, crankshaft and piston/connecting rod assemblies should all be inspected carefully by a machine shop first.

Short block - A short block consists of an engine block with a crankshaft and piston/connecting rod assemblies already installed. All new bearings are incorporated and all clearances will be correct. The existing camshafts, valve train components, cylinder head and external parts can be bolted to the short block with little or no machine shop work necessary.

Long block - A long block consists of a short block plus an oil pump, oil pan, cylinder head, valve cover, camshaft and valve train components, timing sprockets and chain or gears and timing cover. All components are installed with new bearings, seals and gaskets incorporated throughout. The installation of manifolds and external parts is all that's necessary.

Low mileage used engines - Some companies now offer low mileage used engines which is a very cost effective way to get your vehicle up and running again. These engines often come from vehicles which have been totaled in accidents or come from other countries which have a higher vehicle turn over rate. A low mileage used engine also usually has a similar warranty like the newly remanufactured engines.

Give careful thought to which alternative is best for you and discuss the situation with local automotive machine shops, auto parts dealers and experienced rebuilders before ordering or purchasing replacement parts.

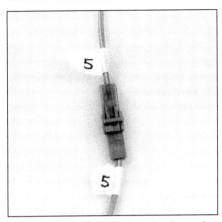

7.5 Label both ends of each wire and hose before disconnecting it

7.18a Typical front left lifting bracket (arrow)

6 Engine removal - methods and precautions

If you've decided that an engine must be removed for overhaul or major repair work, several preliminary steps should be taken.

Locating a suitable place to work is extremely important. Adequate work space, along with storage space for the vehicle, will be needed. If a shop or garage isn't available, at the very least a flat, level, clean work surface made of concrete or asphalt is required.

Cleaning the engine compartment and engine before beginning the removal procedure will help keep tools clean and organized.

An engine hoist or A-frame will also be necessary. Make sure the equipment is rated in excess of the combined weight of the engine and transaxle. Safety is of primary importance, considering the potential hazards involved in lifting the engine out of the vehicle.

If the engine is being removed by a novice, a helper should be available. Advice and aid from someone more experienced would also be helpful. There are many instances when one person cannot simultaneously perform all of the operations required when lifting the engine out of the vehicle.

Plan the operation ahead of time. Arrange for or obtain all of the tools and equipment you'll need prior to beginning the job. Some of the equipment necessary to perform engine removal and installation safely and with relative ease are (in addition to an engine hoist) a heavy duty floor jack, complete sets of wrenches and sockets as described in the front of this manual, wooden blocks and plenty of rags and cleaning solvent for mopping up spilled oil, coolant and gasoline. If the hoist must be rented, make sure that you arrange for it in advance and perform all of the operations possible without it beforehand. This will save you money and time.

Plan for the vehicle to be out of use for quite a while. A machine shop will be required to perform all of the work which is beyond the scope of the home mechanic. These shops often have a busy schedule, so it would be a good idea to consult them before removing the engine in order to accurately estimate the amount of time required to rebuild or repair components that may need work.

Always be extremely careful when removing and installing the engine. Serious injury can result from careless actions. Plan ahead, take your time and a job of this nature, although major, can be accomplished successfully.

7 Engine - removal and installation

Removal

Refer to illustrations 7.5, 7.18a, 7.18b, 7.23a and 7.23b

Warning 1: *The models covered by this manual are equipped with Supplemental Restraint systems (SRS), more commonly known as airbags. Always disable the airbag system before working in the vicinity of the impact sensors, steering column or instrument panel to avoid the possibility of accidental deployment of the airbag, which could cause personal injury (see Chapter 12).*

Warning 2: *Gasoline is extremely flammable, so take extra precautions when you work on any part of the fuel system. Don't smoke or allow open flames or bare light bulbs near the work area, and don't work in a garage where a natural gas-type appliance (such as a water heater or a clothes dryer) with a pilot light is present. Since gasoline is carcinogenic, wear latex gloves when there's a possibility of being exposed to fuel, and, if you spill any fuel on your skin, rinse it off immediately with soap and water. Mop up any spills immediately and do not store fuel-soaked rags where they could ignite. The fuel system is under constant pressure, so, if any fuel lines are to be disconnected, the fuel pressure in the system must be relieved first (see Chapter 4 for more information). When you perform any kind of work on the fuel system, wear safety glasses and have a Class B type fire extin-*

guisher on hand.

Warning 3: *The air conditioning system is under high pressure! If you have to disconnect air conditioning hoses for engine removal, first have a dealer service department or service station discharge the system. Carefully note the routing of air conditioning system refrigerant lines before beginning engine removal to see if line disconnection, and therefore professional discharging, is necessary.*

1   On fuel-injected vehicles, relieve the fuel system pressure (see Chapter 4), and then disconnect the negative cable from the battery.

2   Remove the hood (see Chapter 11) and then cover the fenders and cowl. Special pads are available to protect the fenders, but an old bedspread or blanket will also work.

3   Remove the air cleaner assembly.

4   Drain the cooling system (see Chapter 1).

5   Label the vacuum lines, emissions system hoses, wiring connectors, ground straps and fuel lines, to ensure correct reinstallation, then detach them. Pieces of masking tape with numbers or letters written on them work well **(see illustration)**. If there's any possibility of confusion, make a sketch of the engine compartment and clearly label the lines, hoses and wires.

6   Label and detach all coolant hoses from the engine.

7   Remove the cooling fan, shroud and radiator (see Chapter 3).

8   Remove the drivebelts (see Chapter 1).

9   Disconnect the fuel lines running from the engine to the chassis (see Chapter 4). Plug or cap all open fittings/lines. **Warning:** *Gasoline is extremely flammable, so extra precautions must be taken when working on any part of the fuel system. DO NOT smoke or allow open flames or bare light bulbs near the vehicle. Also, don't work in a garage if a natural gas appliance with a pilot light is present.*

10   Disconnect the throttle linkage (and TV linkage cable, if equipped) from the engine (see Chapter 4).

11   On air-conditioned models, unbolt the compressor (see Chapter 3) and set it aside. Do not disconnect the hoses unless absolutely necessary (see Warning 3 above).

12   Drain the engine oil and then remove the filter (see Chapter 1).

13   Remove the starter motor (see Chapter 5).

14   Remove the alternator (see Chapter 5).

15   Unbolt the exhaust system from the engine (see Chapter 4).

16   If you're working on a vehicle with an automatic transaxle, remove the torque converter-to-driveplate fasteners (see Chapter 7B).

17   Support the transaxle with a jack. Position a block of wood between them to prevent damage to the transaxle. Special transaxle jacks with safety chains are available - use one if possible.

18   Attach an engine sling or a length of chain to the lifting brackets on the engine **(see illustrations)**.

7.18b Typical rear right lifting bracket (arrow)

19   Roll the hoist into position and connect the sling to it. Take up the slack in the sling or chain, but don't lift the engine. **Warning:** *DO NOT place any part of your body under the engine when it's supported only by a hoist or other lifting device.*

20   Remove the transaxle-to-engine block bolts.

21   Remove the engine mount-to-frame bolts.

22   Recheck to be sure nothing is still connecting the engine to the transaxle or vehicle. Disconnect anything still remaining.

23   Raise the engine slightly. Carefully work it sideways to separate it from the transaxle **(see illustration)**. If you're working on a vehicle with an automatic transaxle, be sure the torque converter stays in the transaxle (clamp a pair of vise-grips to the housing to keep the converter from sliding out). If you're working on a vehicle with a manual transaxle, the input shaft must be completely disengaged from the clutch. Slowly raise the engine out of the engine compartment **(see illustration)**. Check carefully to make sure nothing is hanging up.

24   Remove the flywheel/driveplate and mount the engine on an engine stand.

Installation

25   Check the engine and transaxle mounts. If they're worn or damaged, replace them.

26   If you're working on a manual transaxle equipped vehicle, install the clutch and pressure plate (see Chapter 8). Now is a good time to install a new clutch.

27   Carefully lower the engine into the engine compartment - make sure the engine mounts line up.

28   If you're working on an automatic transaxle equipped vehicle, guide the torque converter into the crankshaft following the procedure outlined in Chapter 7.

29   If you're working on a manual transaxle equipped vehicle, apply a dab of high-temperature grease to the input shaft and guide it into the crankshaft pilot bearing until the bell-housing is flush with the engine block.

30   Install the transaxle-to-engine bolts and tighten them securely. **Caution:** *DO NOT use the bolts to force the transaxle and engine together!*

31   Reinstall the remaining components in the reverse order of removal.

32   Add coolant, oil and transaxle fluid as needed.

33   Run the engine and check for leaks and proper operation of all accessories, then install the hood and test drive the vehicle.

34   Have the air conditioning system recharged and leak tested, if it was discharged.

---

8   Engine overhaul - disassembly sequence

---

1   It's much easier to remove the external components if it's mounted on a portable engine stand. A stand can often be rented quite cheaply from an equipment rental yard. Before the engine is mounted on a stand, the flywheel/driveplate should be removed from the engine.

2   If a stand isn't available, it's possible to remove the external engine components with

it blocked up on the floor. Be extra careful not to tip or drop the engine when working without a stand.

3   If you're going to obtain a rebuilt engine, all external components must come off first, to be transferred to the replacement engine. These components include:

> *Emissions-related components*
> *Distributor (if equipped)*
> *Spark plug wires and spark plugs*
> *Ignition coils*
> *Thermostat and housing assembly*
> *Water pump*
> *Fuel injection components*
> *Intake/exhaust manifolds*
> *Oil filter*
> *Engine mounts and mount brackets*
> *Clutch and flywheel (models with manual transaxle)*
> *Driveplate (models with automatic transaxle)*

Note: *When removing the external components from the engine, pay close attention to details that may be helpful or important during installation. Note the installed position of gaskets, seals, spacers, pins, brackets, washers, bolts and other small items.*

4   If you're going to obtain a short block (assembled engine block, crankshaft, pistons and connecting rods), then remove the timing belt, cylinder head, oil pan, oil pump pick-up tube, oil pump and water pump from your engine so that you can turn in your old short block to the rebuilder as a core (see Chapter 2A). See Engine rebuilding alternatives for additional information regarding the different possibilities to be considered.

---

9   Engine overhaul - reassembly sequence

---

1   Before beginning engine reassembly, make sure you have all the necessary new parts, gaskets and seals as well as the following items on hand:

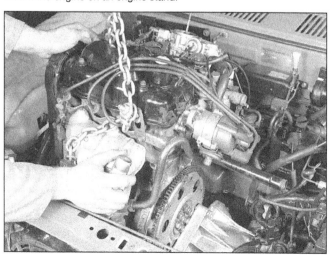

7.23a Pull the engine sideways to separate it from the transaxle . . .

7.23b and then slowly raise the engine until it clears the body

Common hand tools
A 1/2-inch drive torque wrench
New engine oil
Gasket sealant
Thread locking compound

2   If you obtained a short block it will be necessary to install the cylinder head, the oil pump and pick-up tube, the oil pan, the water pump, the timing belt and timing cover, and the valve cover (see Chapter 2A). In order to save time and avoid problems, the external components must be installed in the following general order:

Thermostat and housing cover
Water pump
Intake and exhaust manifolds
Carburetor or fuel injection components
Emission control components
Distributor (if equipped)
Spark plug wires and spark plugs
Ignition coils
Oil filter
Engine mounts and mount brackets
Clutch and flywheel (manual transaxle)
Driveplate (automatic transaxle)

10 Initial start-up and break-in after overhaul

Warning: *Have a fire extinguisher handy when starting the engine for the first time.*

1   Once the engine has been installed in the vehicle, double-check the engine oil and coolant levels.

2   With the spark plugs out of the engine and the ignition system and fuel pump disabled (see Section 4), crank the engine until oil pressure registers on the gauge or the light goes out.

3   Install the spark plugs, hook up the plug wires and restore the ignition system and fuel pump functions.

4   Start the engine. It may take a few moments for the fuel system to build up pressure, but the engine should start without a great deal of effort.

5   After the engine starts, it should be allowed to warm up to normal operating temperature. While the engine is warming up, make a thorough check for fuel, oil and coolant leaks.

6   Shut the engine off and recheck the engine oil and coolant levels.

7   Drive the vehicle to an area with minimum traffic, accelerate from 30 to 50 mph, then allow the vehicle to slow to 30 mph with the throttle closed. Repeat the procedure 10 or 12 times. This will load the piston rings and cause them to seat properly against the cylinder walls. Check again for oil and coolant leaks.

8   Drive the vehicle gently for the first 500 miles (no sustained high speeds) and keep a constant check on the oil level. It is not unusual for an engine to use oil during the break-in period.

9   At approximately 500 to 600 miles, change the oil and filter.

10   For the next few hundred miles, drive the vehicle normally. Do not pamper it or abuse it.

11   After 2000 miles, change the oil and filter again and consider the engine broken in.

Notes

Chapter 3
Cooling, heating and air conditioning systems

Contents

Specifications

General

Radiator cap pressure rating	13 psi
Thermostat rating (opening temperature)	
1994 and earlier	180 degrees F (82 degrees C)
1995 on	190 to 197 degrees F (82 degrees C)
Cooling system capacity	See Chapter 1
Refrigerant type	
1993 and earlier	R-12*
1994 on	R-134a
Refrigerant capacity	
1994 and earlier	1.3 to 1.7 lbs
1995 through 1998	1.21 to 1.32 lbs
1999 on	1.10 to 1.32 lbs
Refrigerant oil type	
1993 and earlier	525 viscosity refrigerant oil
1994 on	Polyalkylene glycol (PAG)
Refrigerant oil capacity	
Condenser	0.7 to 1 fluid ounce
Receiver/drier	0.3 fluid ounce
System	3.4 fluid ounces

Torque specifications

	Ft-lbs
Thermostat housing cap (upper/outer half of thermostat housing) bolts	15
Thermostat housing-to-cylinder head bolts	20
Water pump-to-engine block bolts	
Sprint	132 in-lbs
Metro	
1991 and earlier	108 in-lbs
1992 on	115 in-lbs
Water pump pulley bolts	17 to 18
Coolant inlet pipe nuts	14

Check underhood labels to be sure the vehicle has not been converted to R-134a.

1   General information

Engine cooling system

All vehicles covered by this manual employ a pressurized engine cooling system with thermostatically-controlled coolant circulation. An impeller-type water pump mounted on the drivebelt end of the block pumps coolant through the engine. The coolant flows around each cylinder and toward the transaxle end of the engine. Cast-in coolant passages direct coolant around the intake and exhaust ports, near the spark plug areas and in close proximity to the exhaust valve guides.

A wax pellet-type thermostat is located in a housing near the transaxle end of the engine. During warm up, the closed thermostat prevents coolant from circulating through the radiator. As the engine nears normal operating temperature, the thermostat opens and allows hot coolant to travel through the radiator, where it's cooled before returning to the engine.

The cooling system is sealed by a pressure type radiator cap, which raises the boiling point of the coolant and increases the cooling efficiency of the radiator. If the system pressure exceeds the cap pressure relief value, the excess pressure in the system forces the spring-loaded valve inside the cap off its seat and allows the coolant to escape through the overflow tube into a coolant reservoir. When the system cools, the excess coolant is automatically drawn from the reservoir back into the radiator.

The coolant reservoir does double duty as both the point at which fresh coolant is added to the cooling system to maintain the proper fluid level and as a holding tank for overheated coolant.

This type of cooling system is known as a closed design because coolant that escapes past the pressure cap is saved and reused.

Heater

The heater consists of a blower fan and heater core located in the heater box, the hoses connecting the heater core to the engine cooling system and the heater/air conditioning control head on the dashboard. Hot engine coolant is circulated through the heater core. When the heater is activated, a flap door opens to expose the heater box to the passenger compartment. A fan switch on the control head activates the blower motor, which forces air through the core, heating the air.

Air conditioning system

The air conditioning system consists of a condenser mounted in front of the radiator, an evaporator mounted adjacent to the heater core, a compressor mounted on the engine, a receiver/drier which contains a high pressure relief valve and the plumbing connecting all of the above components.

A blower fan forces the warmer air of the passenger compartment through the evaporator core (sort of a radiator-in-reverse), transferring the heat from the air to the refrigerant. The liquid refrigerant boils off into low pressure vapor, taking the heat with it when it leaves the evaporator.

2   Antifreeze - general information

Warning: *Do not allow antifreeze to come in contact with your skin or painted surfaces of the vehicle. Rinse off spills immediately with plenty of water. Never leave antifreeze lying around in an open container or in a puddle in the driveway or on the garage floor. Children and pets are attracted by its sweet smell. Antifreeze is fatal if ingested in sufficient quantity. Check with local authorities before disposing of used antifreeze. Many communities have collection centers which will see that antifreeze is disposed of safely. Antifreeze is also combustible, so don't store or use it near open flames.*

The cooling system should be filled with a water/ethylene glycol based antifreeze solution, which will prevent freezing down to at least -20-degrees F, or lower if local climate requires it. It also provides protection against corrosion and increases the coolant boiling point.

The cooling system should be drained, flushed and refilled at the specified intervals (see Chapter 1). Old or contaminated antifreeze solutions are likely to cause damage and encourage the formation of corrosion and scale in the system. Use distilled water with the antifreeze.

Before adding antifreeze, check all hose connections, because antifreeze tends to search out and leak through very minute openings. Engines don't normally consume coolant, so if the level goes down, find the cause and correct it.

The exact mixture of antifreeze-to-water which you should use depends on the relative weather conditions. The mixture should contain at least 50-percent antifreeze, but should never contain more than 70-percent antifreeze. Consult the mixture ratio chart on the antifreeze container before adding coolant. Hydrometers are available at most auto parts stores to test the coolant. Use antifreeze which meets the vehicle manufacturer's specifications.

3   Thermostat - check and replacement

Warning: *Do not remove the radiator cap, drain the coolant or replace the thermostat until the engine has cooled completely.*

Check

1   Before assuming the thermostat is to blame for a cooling system problem, check the coolant level, drivebelt tension (see Chapter 1) and temperature gauge operation.

3.10 To get at the thermostat on a Sprint, unplug the thermo switch electrical connector (middle arrow) and then remove the thermostat housing cover bolts (outer arrows)

2   If the engine seems to be taking a long time to warm up (based on heater output or temperature gauge operation), the thermostat is probably stuck open - replace it with a new one.
3   If the engine runs hot, use your hand to check the temperature of the upper radiator hose. If the hose isn't hot, but the engine is, the thermostat is probably stuck closed, preventing the coolant inside the engine from escaping to the radiator. Replace the thermostat. **Caution:** *Don't drive the vehicle without a thermostat. The lack of a thermostat may cause the computer to stay in open loop, causing emissions and fuel economy to suffer.*
4   If the upper radiator hose is hot, it means the coolant is flowing and the thermostat is open. Consult the *Troubleshooting* section at the front of this manual for cooling system diagnosis.

Replacement

Refer to illustrations 3.10, 3.11a, 3.11b, 3.11c, 3.12, 3.13a and 3.13b
5   Disconnect the negative battery cable from the battery.
6   Drain the cooling system (see Chapter 1). If the coolant is relatively new or in good condition, save it and reuse it.
7   Remove the air cleaner.
8   Follow the upper radiator hose to the engine to locate the thermostat housing. Loosen the hose clamp, then detach the hose from the fitting. If it's stuck, grasp it near the end with a pair of adjustable pliers and twist it to break the seal, then pull it off. If the hose is old or deteriorated, cut it off and install a new one.
9   If the outer surface of the large fitting that mates with the hose is deteriorated (corroded, pitted, etc.) it may be damaged further by hose removal. If it is, the thermostat housing cover will have to be replaced.
10   On Sprint models, unplug the electrical connector for the engine cooling fan switch **(see illustration)**.

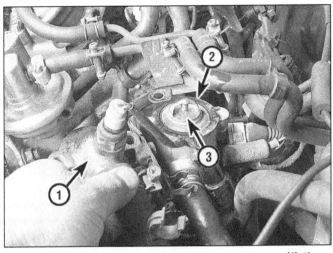

3.11a On Sprint models, remove the thermostat cover (1), the gasket (2) and the thermostat (3)

3.11b On Metro models, the thermostat and housing are located under the distributor (1997 and earlier models) . . .

3.11c . . . or under the Camshaft Position Sensor (CMP) housing (1998 and later models); remove the thermostat housing cover bolts (arrows) and then detach the cover

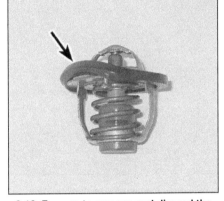

3.12 Be sure to remove and discard the old thermostat gasket (arrow)

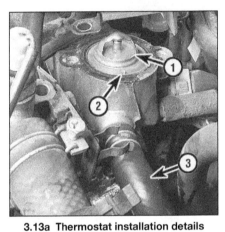

3.13a Thermostat installation details (Sprint models)

1   Thermostat
2   Air bleed valve
3   Heater inlet hose

11   Remove the bolts and detach the housing cover and bracket, if equipped (see illustrations). If the cover is stuck, tap it with a soft-face hammer to jar it loose. Be prepared for some coolant to spill as the gasket seal is broken.

12   Note how the thermostat is installed. If there's an air bleed hole or air bleed valve in the thermostat, note its orientation. Then remove the thermostat. Remove and discard the old gasket (see illustration). Thoroughly clean the sealing surfaces.

13   Install the new thermostat in the housing. Make sure that the air bleed hole or air bleed valve is correctly positioned and that the spring end of the thermostat is facing into the thermostat housing, i.e. toward the engine (see illustrations).

14   Install a new gasket over the thermostat.

15   Install the cover and bolts. Tighten the bolts to the torque listed in this Chapter's Specifications.

16   Reattach the hose to the fitting and tighten the hose clamp securely.

3.13b When installing the thermostat on Metro models, make sure that the bleed hole (arrow) is up

17   Refill the cooling system (see Chapter 1).

18   Start the engine and allow it to reach normal operating temperature, then check for leaks and proper thermostat operation (as described in Steps 2 through 4).

4   Engine cooling fan - check and replacement

Check

Refer to illustrations 4.5a and 4.5b

1   The engine cooling fan is controlled by a temperature switch mounted in the thermostat housing cover. When the coolant reaches a predetermined temperature, the switch closes the ground return for the fan motor relay, completing the circuit.

2   First, check the fuses (see Chapter 12).

3   To test the fan motor, unplug the motor connector and use fused jumper wires to connect the fan directly to the battery. If the fan still doesn't work, replace the motor.

4   If the motor tested okay, the fault lies in the coolant temperature switch, the fan relay or the wiring harness (see Chapter 12).

5   Test the temperature switch by unplugging the connector and bridging the terminals in the wire harness (see illustrations) with the ignition switch On.

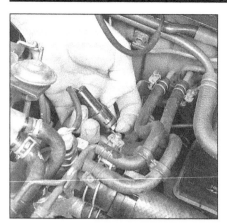

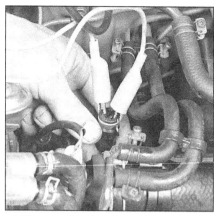

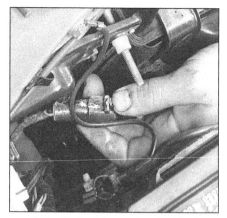

4.5a To check the engine cooling fan temperature switch, unplug the connector . . .

4.5b . . . and bridge the terminals with a jumper wire (Sprint model shown, Metro models similar)

4.11a Typical fan motor electrical connector (early Metro models)

4.13 To detach the fan shroud from the lower end of the radiator, remove these two retaining bolts (arrows) (1.3L Metro shown, other models similar)

4.11b Typical fan motor electrical connector (arrow) (late-model Metros)

6   If the fan does not operate, check the wiring and the relay (see Chapter 12).

Replacement

Refer to illustrations 4.11a, 4.11b, 4.13, 4.14a, 4.14b, 4.15 and 4.16

7   Disconnect the negative battery cable.
8   Drain the cooling system (see Chapter 1).
9   On 1991 and earlier models, remove the air cleaner intake duct (see Chapter 4).
10   Loosen the upper radiator hose clamp (see Section 5) and then disconnect the upper radiator hose from the radiator.
11   Trace the electrical lead from the fan motor to the electrical connector and then unplug the connector **(see illustrations)**.
12   Raise the front of the vehicle and place it securely on jackstands.
13   Remove the lower fan shroud retaining bolt **(see illustration)**.
14   Remove the two upper radiator/fan shroud mounting bolts **(see illustrations)** and then remove the fan/shroud assembly from the engine compartment. After removing the fan/shroud assembly, secure the radiator by reinstalling the two upper bolts.
15   Remove the nut and detach the fan blade assembly from the motor shaft **(see illustration)**.
16   Take out the screws holding the fan motor to the bracket and detach the motor **(see illustration)**.
17   Installation is the reverse of removal.

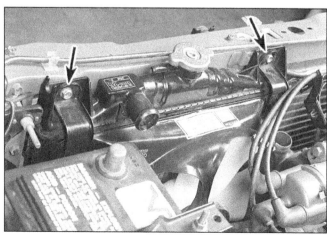

4.14a Upper radiator/fan shroud mounting bolts (arrows) (early models)

4.14b Upper fan shroud mounting bolts (arrows) (late-model 1.3L model shown, other models similar)

4.15 Remove the nut and detach the fan blade assembly from the motor shaft

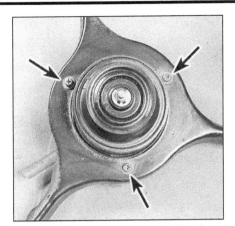

4.16 Take out the screws holding the fan motor to the bracket and detach the motor

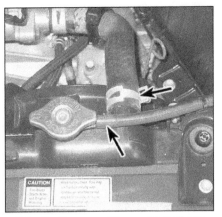

5.4a To disconnect the upper radiator hose, loosen this hose clamp (upper arrow); also disconnect the coolant reservoir overflow hose (lower arrow) from the filler neck

5.4b To disconnect the lower radiator hose, loosen this hose clamp (arrow)

5 Radiator - removal and installation

Warning 1: *Wait until the engine is completely cool before beginning this procedure.*
Warning 2: *Some models covered by this manual are equipped with an airbag. Always disable the airbag system before working in the vicinity of the steering column, impact*

sensors or dashboard to avoid the accidental deployment of the airbag, which could cause personal injury. See Chapter 12 for the airbag disarming procedure.

Removal

Refer to illustrations 5.4a, 5.4b and 5.7
1   Disconnect the negative battery cable from the battery.
2   Raise the front of the vehicle and support it securely on jackstands.
3   Drain the cooling system (see Chapter 1). If the coolant is relatively new or in good condition, save it and reuse it.
4   Loosen the hose clamps **(see illustrations)**, then detach the upper and lower radiator hoses from the radiator. If a hose is stuck, grasp it near the end with a pair of adjustable pliers, twist it to break the seal and then pull it off, but be careful not to damage the radiator fittings. If the hoses are old or deteriorated, cut them off and install new ones.
5   Disconnect the coolant reservoir hose from the radiator filler neck **(see illustration 5.4a)**.
6   Unplug the cooling fan electrical connector (see Section 4).

7   If the vehicle is equipped with an automatic transaxle, disconnect the cooler lines **(see illustration)** and plug the lines and fittings.
8   Carefully lift out the radiator and the cooling fan and shroud as a single assembly. Don't spill coolant on the vehicle or scratch the paint. To separate the cooling fan shroud assembly from the radiator, remove the lower shroud-to-radiator bolts (see Section 4).
9   While the radiator is removed, inspect it for leaks and damage. If it's leaking or damaged, have it repaired by a radiator shop or dealer service department.
10   Remove bugs and dirt from the radiator with a garden hose or a soft brush. Be careful not to bend the cooling fins. (If you do bent the cooling fins, straighten them with a "fin comb," which is available from tool suppliers and at some auto parts stores.)

Installation

Refer to illustration 5.11
11   Installation is the reverse of the removal procedure. Be sure the rubber cushions are seated properly **(see illustration)**.

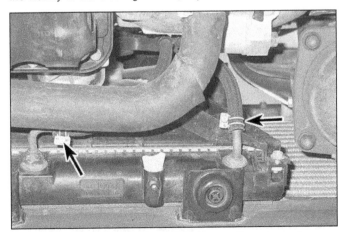

5.7 On automatic transaxle models, loosen these hose clamps (arrows) and then disconnect the cooling lines from the bottom of the radiator

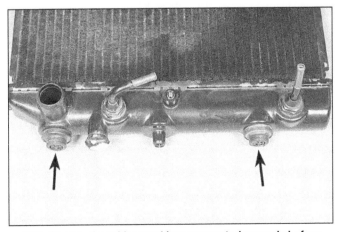

5.11 Be sure the rubber cushions are seated properly before reinstalling the radiator

12 After installation, fill the cooling system with the recommended mixture of antifreeze and water. Refer to Chapter 1 if necessary.

13 Start the engine and check for leaks. Allow the engine to reach normal operating temperature, indicated by the upper radiator hose becoming hot. Recheck the coolant level and add more if required.

14 If you're working on an automatic transaxle equipped vehicle, check and add fluid as needed.

6 Coolant reservoir - removal and installation

1994 and earlier models

Refer to illustrations 6.1 and 6.2

1 Lift the cap off the coolant reservoir **(see illustration)**.

2 Lift the coolant reservoir straight up to remove it **(see illustration)**.

3 Installation is the reverse of removal.

1995 and later models

Refer to illustration 6.5

4 Remove the left headlight (see Chapter 12).

5 Remove the two bolts from the left inner fender **(see illustration)**.

6 Disconnect the overflow hose and cap from the coolant reservoir (see Section 4 in Chapter 1).

7 Remove the coolant reservoir by lifting it out of the fender mount.

8 If you're planning to reuse the coolant reservoir, be sure to wash it thoroughly with soap and water.

9 Installation is the reverse of removal.

7 Water pump - check

Refer to illustrations 7.4 and 7.5

1 A failure in the water pump can cause serious engine damage due to overheating.

2 There are three ways to check the operation of the water pump while it's installed on the engine. If the pump is defective, it should be replaced with a new or rebuilt unit.

3 With the engine running at normal operating temperature, squeeze the upper radiator hose. If the water pump is working properly, a pressure surge should be felt as the hose is released.

4 Water pumps are equipped with weep or vent holes. If a failure occurs in the pump seal, coolant will leak from the hole. In most cases you'll need a flashlight to find the hole on the water pump to check for leaks **(see illustration)**.

5 If the water pump shaft bearings fail there may be a howling sound at the drivebelt end of the engine while it's running. Shaft wear can be felt if the water pump pulley is rocked up-and-down **(see illustration)**. Don't mistake drivebelt slippage, which causes a squealing sound, for water pump bearing failure.

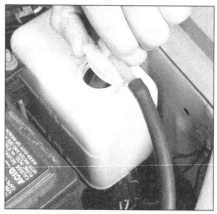

6.1 Lift the coolant reservoir cap off and remove the hose (1994 and earlier models)

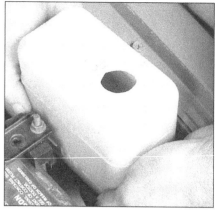

6.2 The coolant reservoir slips out of/into the bracket (1994 and earlier models)

8 Water pump - replacement

Refer to illustrations 8.1, 8.2, 8.6a, 8.6b, 8.9 and 8.12

Warning: *Wait until the engine is completely cool before beginning this procedure.*

1 Disconnect the negative battery cable and then loosen the water pump pulley bolts **(see illustration)** while the drivebelt is still installed (the drivebelt helps to hold the pulley while you're breaking loose the pulley bolts). If the pulley turns when you try to loosen the pulley bolts, this method won't work; use a chain or strap wrench to hold the pulley.

2 On late-model vehicles, disconnect the suction tube bracket (air conditioned models), remove the engine oil level dipstick tube and detach the alternator adjustment bracket from the water pump **(see illustration)**.

3 Drain the cooling system (see Chapter 1).

4 Remove the drivebelt(s) (see Chapter 1) and then remove the pulley from the end of the water pump shaft.

5 Remove the timing belt (see Chapter 2A).

6.5 To detach the coolant reservoir from the body on 1995 and later models, remove these two bolts (arrows) from underneath the front end

7.4 The water pump weep hole (arrow) will drip coolant when the seal on the pump shaft fails)

7.5 Rock the shaft up-and-down to check for play

8.1 Before removing the drivebelt, try loosening the water pump pulley bolts; if the drivebelt can't hold the pulley, remove the belt and then hold the pulley with a chain wrench or a strap wrench

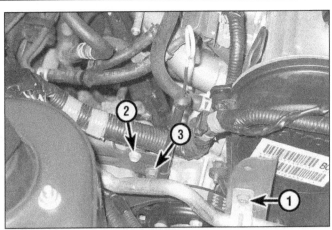

8.2 On late-model air-conditioned vehicles, remove the suction tube bracket bolt (1) and then pull the suction tube away from the timing belt cover; on all newer models, remove the dipstick tube bracket bolt (2) and remove the dipstick tube, then remove the alternator adjustment bracket bolt (3) and pivot the bracket up to clear the water pump

8.6a Water pump mounting bolts (arrows) on a typical 1.0L installation; when installing the pump, be sure to install new rubber seals in the ridge gaps at the top and bottom of the pump

8.6b Water pump mounting bolts (arrows) on a typical late-model 1.3L installation; note the locations of the two studs and make sure that they're installed in the same location when installing the pump

6   Remove the bolts (see illustrations) and then detach the water pump from the engine. Note the locations of the various brackets and the different lengths/diameters of the bolts as they're removed to ensure correct installation.

7   Clean the bolt threads and the threaded holes in the engine to remove corrosion and sealant.

8   Compare the new pump to the old one to make sure they're identical.

9   Remove all traces of old gasket material from the sealing surfaces (see illustration).

10   Use a bead of RTV sealant to hold the pump gasket in place. Carefully mate the pump to the engine. Slip a couple of bolts through the pump mounting holes to hold the pump in place.

11   Install the remaining bolts. Tighten them to the torque listed in this Chapter's Specifications in 1/4-turn increments. Don't overtighten them or the pump may be damaged.

12   Place new rubber seals between the water pump and oil pump and between the

8.9 Remove all traces of old gasket material from the engine block before installing the new pump

water pump and cylinder head (see illustration). Reinstall all parts removed for access to the pump.

8.12 After installing the water pump, be sure to install new rubber seals (arrows) in the gaps between the water pump and the cylinder head, and between the water pump and the oil pump

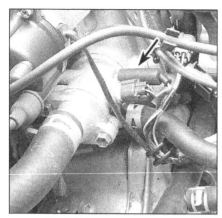

9.3a If you're working on a Sprint, the coolant temperature sending unit is located on the firewall side of the intake manifold below the thermostat

9.3b If you're working on a Metro, the coolant temperature sending unit is located in the intake manifold near the base of the throttle body (arrow)

10.2 Unplug the hose and electrical connector from the motor, remove the three screws, then maneuver the blower unit out from under the dash (Sprint models)

13 Refill the cooling system and check the drivebelt tension (see Chapter 1). Run the engine and check for leaks.

9 Coolant temperature sending unit (1996 and earlier models) - check and replacement

Note: The following procedure applies only to 1996 and earlier models, which use a coolant temperature sending unit for the coolant temperature gauge, and a separate Engine Coolant Temperature (ECT) sensor for the Powertrain Control Module (PCM). On 1997 and later models, the ECT sensor functions as both an information sensor for the PCM and as the coolant temperature sending unit for the coolant temperature gauge. For these models, refer to Chapter 6 for the replacement procedure for the ECT sensor.

Check

Refer to illustrations 9.3a and 9.3b
1 If the coolant temperature gauge is inoperative, check the fuses first (see Chapter 12).
2 If the temperature indicator shows excessive temperature after running a while, see the *Troubleshooting* section in the front of the manual.
3 If the temperature gauge indicates Hot shortly after the engine is started cold, disconnect the wire at the coolant temperature sending unit **(see illustrations)**. If the gauge reading drops, replace the sending unit. If the reading remains high, the wire to the gauge may be shorted to ground or the gauge is faulty.
4 If the coolant temperature gauge fails to indicate after the engine has been warmed up (approximately 10 minutes) and the fuses checked out okay, shut off the engine. Disconnect the wire at the sending unit and, using a jumper wire, connect it to a clean ground on the engine. Turn on the ignition without starting the engine. If the gauge now indicates Hot, replace the sending unit.

5 If the gauge still doesn't work, the circuit may be open or the gauge may be faulty. See Chapter 12 for additional information.

Replacement

Warning: *The engine must be completely cool before removing the sending unit.*
6 With the engine completely cool, remove the cap from the radiator to release any pressure, then reinstall the cap. This reduces coolant loss during sending unit replacement.
7 Disconnect the wire from the sending unit.
8 Prepare the new sending unit for installation by applying a Teflon tape or light coat of sealant to the threads.
9 Unscrew the sending unit from the engine and quickly install the new one to prevent coolant loss.
10 Tighten the sending unit securely and connect the wire.
11 Refill the cooling system and run the engine. Check for leaks and proper gauge operation.

10 Blower unit - removal and installation

Warning: *Some models covered by this manual are equipped with an airbag. Always disable the airbag system before working in the vicinity of the steering column, impact sensors or dashboard to avoid the accidental deployment of the airbag, which could cause personal injury. See Chapter 12 for the airbag disarming procedure.*

Sprint models

Refer to illustrations 10.2 and 10.5
1 Disconnect the negative cable from the battery.
2 The blower unit is located above and to the right of the accelerator pedal. Disconnect the wiring from the blower and remove the rubber air duct running between the motor and the blower housing **(see illustration)**.

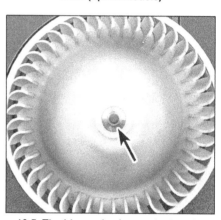

10.5 The blower fan is attached to the motor shaft with a nut (arrow) (Sprint models)

3 Remove the screws holding the blower unit to the vehicle and lower the unit from the dash.
4 Remove the three mounting screws and lift the motor out of the housing.
5 If you're replacing the motor, detach the fan **(see illustration)** and transfer it to the new motor.
6 Installation is the reverse of removal. Run the blower and check for proper operation.

Metro models

1994 and earlier models

Refer to illustrations 10.10, 10.12 and 10.13
7 Disconnect the negative cable from the battery.
8 The blower unit is located on the passenger's side of the dash, behind the glove box. Remove the two screws from the glove box striker and detach the striker.
9 Detach the rear upper glove box panel.
10 Disconnect the wire from the blower and remove the rubber air hose running between the motor and the blower housing **(see illustration)**.
11 Disconnect the fresh/recirculated control cable from the blower case assembly.

10.10 Disconnect the electrical connector to the blower motor (arrow) (1994 and earlier Metro models)

10.12 Use a screwdriver to punch out the plastic cover (arrow) to gain access to the third mounting screw (1994 and earlier Metro models)

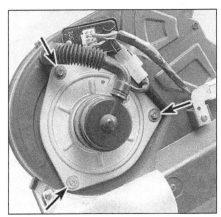

10.13 Blower motor mounting screw locations (arrows) (1994 and earlier Metro models)

10.19 To replace the blower motor resistor, unplug the electrical connector (left arrow) and then remove the resistor mounting screw (right arrow) (1995 and later Metro models)

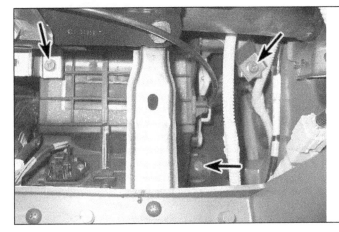

10.21 Remove the three blower case mounting screws (arrows) and then remove the blower case from below (1995 and later Metro models)

12   Remove the three blower case mounting screws **(see illustration)** and lower the case from the vehicle.

13   Remove the three mounting screws **(see illustration)** and lift the motor out of the housing.

14   If you're replacing the motor, detach the fan **(see illustration 10.5)** and transfer it to the new motor.

15   Installation is the reverse of removal. Run the blower and check for proper operation.

1995 and later models

16   Disconnect the negative cable from the battery.

17   Remove the glove box (see Chapter 11).

18   Remove the Powertrain Control Module (PCM) (see Chapter 6).

Blower motor resistor

Refer to illustration 10.19

19   Unplug the electrical connector from the blower motor resistor **(see illustration)**, remove the resistor retaining screw and then remove the resistor. Installation is the reverse of removal.

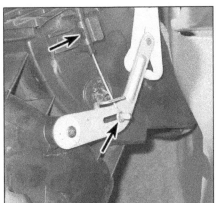

10.22 Disengage the fresh/recirculated air control cable housing from the cable retaining clip (upper arrow) and then disconnect the cable wire loop (lower arrow) from the lever pin (1995 and later Metro models)

Blower unit

Refer to illustrations 10.21, 10.22, 10.23, 10.24 and 10.25

20   Disconnect the electrical connector to the blower resistor **(see illustration 10.19)**.

21   Remove the three blower case mounting screws **(see illustration)** and then reach up

10.23 Remove the air hose (arrow) between the blower motor and the blower case (1995 and later Metro models)

and pull down the blower case from below.

22   Disengage the fresh/recirculated air control cable housing **(see illustration)** from the clip on the side of the blower case and then disconnect the cable loop that's pushed onto the lever pin. To disengage the cable loop from the lever pin, pry it off with a small screwdriver.

23   Remove the air hose between the blower motor and the blower case **(see illustration)**.

10.24 Remove the three mounting screws (arrows) and then remove the motor from the case (1995 and later Metro models)

10.25 To remove the fan from the fan motor shaft, remove the circlip that secures the fan to the shaft (1995 and later Metro models)

11.3 Disconnect the heater hoses at the firewall

24   Remove the three mounting screws **(see illustration)** and then remove the motor from the case.
25   To remove the fan from the fan motor shaft, remove the circlip that secures the fan to the shaft **(see illustration)**.
26   Installation is the reverse of removal.

---

11   Heater core - removal and installation

Refer to illustration 11.3
Note: *The photos accompanying the following procedure depict a typical heater core removal procedure. Heater core removal on all models covered by this manual is similar, although dashboards and the various trim pieces vary somewhat from model to model.*
1   Disconnect the negative cable from the battery.
2   Drain the cooling system (see Chapter 1).
3   Working in the engine compartment, disconnect the heater hoses where they enter the firewall **(see illustration)**.
4   Remove the instrument cluster (see Chapter 12).
5   Remove the air conditioner and heater control assembly (see Section 12). Be sure to mark the locations of the cable clamps on the

cables to ensure correct adjustment upon reinstallation.
6   Remove the glove box and the instrument panel assembly (see Chapter 11).

1994 and earlier models
Refer to illustrations 11.7, 11.8, 11.9a, 11.9b and 11.10
Warning: *Later models covered by this manual are equipped with a driver's side airbag; some late-model vehicles have a passenger airbag as well. Always disable the airbag sys-*

tem before working in the vicinity of the steering column, impact sensors or dashboard to avoid the accidental deployment of the airbag(s), which could cause personal injury. See Chapter 12 for the airbag disarming procedure.
7   Label and detach the air ducts, wiring and controls still attached to the heater housing **(see illustration)**.
8   Unbolt the heater assembly from the firewall **(see illustration)**.
9   Remove the screws and clips and separate the two halves of the heater assembly **(see illustrations)**.

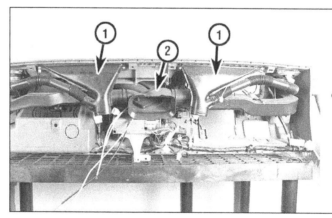

11.7 Heater system components - exploded view (Sprint)

1   *Side vent duct*
2   *Center vent louver*

11.8 Remove the heater assembly mounting nuts (arrow shows one lower nut)

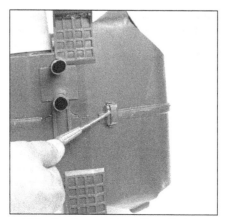

11.9a Pry the retaining clips off of the case with a screwdriver

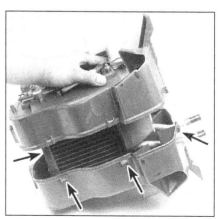

11.9b Separate the two case halves (arrows show clip locations)

11.10 Once the cases are separated, the heater core can be slipped out

1 Heater case
2 Heater core
3 Blower motor

10 Take out the old heater core (see illustration) and install the new one.
11 Reassemble the heater unit and check the operation of the control flaps. If any parts bind, correct the problem before installation.
12 Reinstall the remaining parts in the

reverse order of removal.
13 Refill the cooling system, reconnect the battery and run the engine. Check for leaks and proper system operation.

1995 and later models

14 Disconnect the air duct from the heater case.
15 Remove the two heater case mounting bolts and the two heater mounting nuts and then remove the heater case.
16 Remove the heater case retaining clips.
17 Remove the two screws from the heater case.
18 Separate the heater case halves.
19 Remove the heater core from the heater case.
20 Installation is the reverse of removal.
21 Refill the cooling system, reconnect the battery and run the engine. Check for leaks and proper system operation.

12 Air conditioner and heater control assembly - removal, installation and cable adjustment

1 Disconnect the negative cable from the battery.

1994 and earlier models

Refer to illustrations 12.3a, 12.3b, 12.6, 12.7 and 12.8

Warning: *Some models covered by this manual are equipped with an airbag. Always disable the airbag system before working in the vicinity of the steering column, impact sensors or dashboard to avoid the accidental deployment of the airbag, which could cause personal injury. See Chapter 12 for the airbag disarming procedure.*

2 Remove the ashtray assembly (if necessary) by removing the two upper mounting screws.
3 Pull off the control knobs and detach the trim panels which surround the radio and heater control assembly, as necessary (see illustrations).
4 Remove the radio (see Chapter 12).
5 Remove the glove compartment.
6 Remove the control mounting screws (see illustration).
7 Carefully pull and tilt the unit out of the dash as far as the cables allow (see illustration).
8 Check the cable housings for indentations where the clamps grip them. Mark the cable housing with paint if no indentation is visible. Remove the clamps and detach the

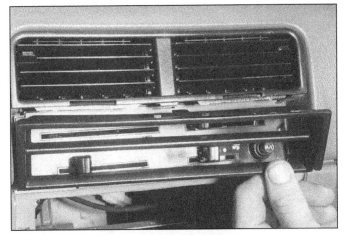

12.3a Remove the trim panels . . .

12.3b . . . and pull off the knobs (1994 and earlier models)

12.6 Remove the control panel mounting screws (arrows) (1994 and earlier models)

12.7 Carefully pull the control unit out for access to the cables (1994 and earlier models)

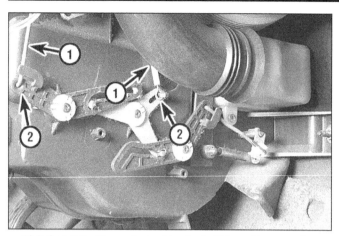

12.8 Separate the control cables/rods (1) from the operating levers (2) (1994 and earlier models)

12.13 Grasp each heater control lever knob firmly and then pull it off (1995 and later models)

control cables. If necessary, disconnect the control cables at the ends opposite from the control by detaching the cable clamps and separating the cables from the operating levers **(see illustration)**.

9   Unplug the wiring connectors and lift the control from the vehicle.

10   Installation is the reverse of removal.

11   Normally, if a control cable is reattached where the clamp mark is, the lever will be adjusted properly. If the lever requires adjustment, remove the clamp and move the cable until the control lever moves freely through its normal range of motion.

1995 and later models

Refer to illustrations 12.13, 12.15, 12.16, 12.17, 12.18, 12.19a, 12.19b, 12.20 and 12.22

12   Remove the center console trim bezel (see Chapter 11).

13   Grasp the knobs of the heater control lever firmly and then pull off the knobs **(see illustration)**.

12.15 Carefully pry off and then remove the heater face plate (1995 and later models)

14   Remove the right instrument panel cluster lower cover (see "Instrument panel - removal and installation in Chapter 11).

15   Carefully pry off and then remove the

12.16 Remove the face plate illumination light socket (1995 and later models)

heater face plate **(see illustration)**.

16   Remove the face plate illumination lamp socket **(see illustration)**.

17   Working under the left side of the dash, and using a flashlight, disconnect the tem-

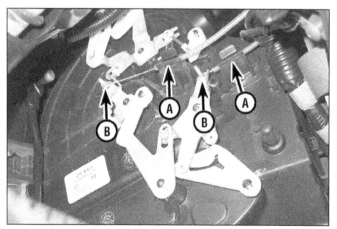

12.17 To disconnect the mode control cable (upper cable) and the temperature control cable (lower cable) from the heater case, pull the cable housings out of their respective clamps (A) and then pry the cable loops off the lever pins (B) (1995 and later models)

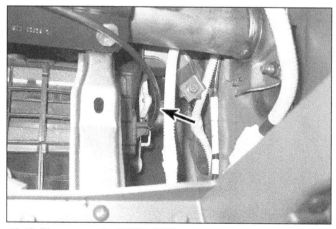

12.18 Disconnect the FRESH/RECIRC control cable (arrow) from the right side of the heater blower unit case the same way you disconnected the two cables in the previous illustration: pull the cable housing out of its clamp and then pry the cable loop off the lever pin (see illustration 10.22 for a closer view of the clamp and cable end) (1995 and later models)

12.19a To remove the heater and A/C control assembly, remove these two mounting screws (arrows) . . .

12.19b . . . and then pull out the assembly just far enough to unplug the blower switch electrical connector (1995 and later models)

perature control cable and the mode control cable from the left side of the heater case **(see illustration)**.

18   Remove the glove box (see Chapter 11), the Powertrain Control Module (PCM) (see Chapter 6) and then, using a flashlight and a small screwdriver, disconnect the FRESH/RECIRC control cable **(see illustration)** from the right side of the heater blower unit case. This cable is installed just like the other two cables shown in the previous step. The cable housing is secured by a clamp and the cable end is looped over a pin on the control lever. For a better look at the FRESH/RECIRC cable housing clamp and the cable wire end, refer to illustration 10.22. But note that even though the blower unit case has been lowered from the dash in that photo, the FRESH/RECIRC cable *can* be disconnected without removing the blower unit case - it's just a little trickier.

19   Remove the heater and A/C control assembly mounting screws and then pull out the assembly **(see illustrations)**.

20   Unplug the blower switch electrical connector **(see illustration)**.

21   Remove the heater and air conditioning control assembly.

22   If you're replacing the blower switch, pry open the locking tangs on each side **(see illustration)** and pull the switch from the heater and air conditioning control assembly.

23   Installation is the reverse of removal. If each cable is reattached where the clamp mark is, the lever should be adjusted correctly. If a lever requires adjustment, remove the cable from the clamp, move the cable until the control lever moves freely through its normal range of motion and then re-clamp the cable.

---

13   Air conditioning system - check and maintenance

Refer to illustrations 13.7 and 13.8

Warning: *The air conditioning system is under high pressure. Do not loosen any hose fittings or remove any components until after the system has been discharged. Air conditioning refrigerant must be correctly dis-*

charged into an EPA-approved recovery/recycling unit at a dealer service department or an automotive air conditioning repair facility. Always wear eye protection when disconnecting air conditioning system fittings.

Caution: *Two different types of air conditioning refrigerant are used on the models covered by this manual. 1993 and earlier models use R-12 refrigerant, while 1994 and later models use the non-ozone-depleting R-134a refrigerant. The R-134a refrigerant and its lubricating oil are not compatible with the R-12 system and under no circumstances should the two different types of refrigerant or lubricating oil be intermixed. The system charging fittings are different to prevent accidental connection of the dissimilar system charging hoses.*

Note: *Because of Federal regulations proposed by the Environmental Protection Agency, R-12 refrigerant is no longer available to the home mechanic. Models with R-12 systems should be serviced at a dealership or at an automotive air conditioning shop.*

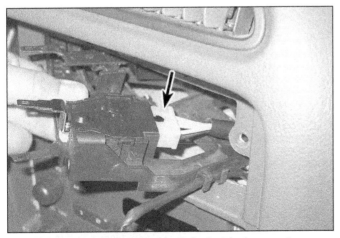

12.20 Unplug the blower switch electrical connector (1995 and later models)

12.22 If you're replacing the blower switch, pry open the locking tangs on each side and then pull the switch out of the heater and air conditioning control assembly (1995 and later models)

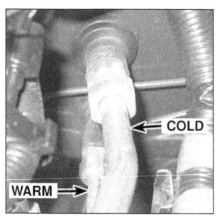

13.7 With the air conditioning system operating, feel the two pipes connected to the evaporator at the firewall: The lower pipe with the thinner diameter, which connects the condenser outlet to the evaporator, should be warm, and the thicker diameter pipe, which connects the evaporator to the compressor, should be cold

13.8 Check the sight glass (arrow) in the top of the receiver/drier

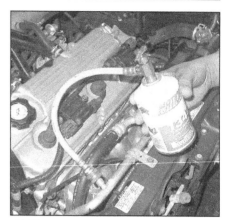

13.12 Hook up an R-134a refrigerant can to the system in accordance with the manufacturer's instructions

1   The following maintenance checks should be performed on a regular basis to ensure that the air conditioner continues to operate at peak efficiency.

a) *Check the compressor drivebelt. If it's worn or deteriorated, replace it (see Chapter 1).*

b) *Check the drivebelt tension and, if necessary, adjust it (see Chapter 1).*

c) *Check the system hoses. Look for cracks, bubbles, hard spots and deterioration. Inspect the hoses and all fittings for oil bubbles and seepage. If there's any evidence of wear, damage or leaks, replace the hose(s).*

d) *Inspect the condenser fins for leaves, bugs and other debris. Use a -fin comb" or compressed air to clean the condenser.*

e) *Make sure the system has the correct refrigerant charge.*

2   It's a good idea to operate the system for about 10 minutes at least once a month, particularly during the winter. Long term non-use can cause hardening, and subsequent failure, of the seals.

3   Because of the complexity of the air conditioning system and the special equipment necessary to service it, in-depth troubleshooting and repairs are not included in this manual. However, simple checks and component replacement procedures are provided in this Chapter.

4   The most common cause of poor cooling is simply a low system refrigerant charge. If a noticeable drop in cool air output occurs, one of the following quick checks will help you determine whether the refrigerant level is low.

5   Warm up the engine to its normal operating temperature.

6   Place the air conditioning temperature selector at the coldest setting and put the

blower at the highest setting. Open the doors (to make sure the air conditioning system doesn't cycle off as soon as it cools the passenger compartment).

7   After the system reaches operating temperature, feel the two pipes connected to the evaporator at the firewall (see illustration). The pipe with the thinner diameter, which connects the condenser outlet to the evaporator, should be warm, and the thicker diameter pipe, which connects the evaporator to the compressor, should be cold. If the two pipes are the same temperature (or close to the same temperature), the system charge is low.

8   If the system is equipped with a sight glass, on top of the receiver/drier (see illustration), check for the presence of air bubbles in the refrigerant. If the refrigerant passing through the sight glass looks foamy, the system charge is low.

9   Further inspection or testing of the system is beyond the scope of the home mechanic and should be left to a professional.

Adding refrigerant

Refer to illustration 13.12

Caution: *Make sure any refrigerant, refrigerant oil or replacement component your purchase is designated as compatible with environmentally-friendly R-134a systems.*

10   1993 and earlier models use R-12 refrigerant. Because of federal restrictions on the sale of R-12 refrigerant, it isn't practical for refrigerant to be added by the home mechanic. When the system needs recharging, take the vehicle to a dealer service department or professional air conditioning shop for evacuation, leak testing and recharging. On 1994 and later models using R-134a refrigerant, make sure any refrigerant, oil or replacement component is designated for R-134a systems.

11   Buy an R-134a automotive charging kit at an auto parts store. A charging kit includes a 12-ounce can of refrigerant, a tap valve and a short section of hose that can be attached between the tap valve and the system low

side service valve. One can of refrigerant might be insufficient to bring the system charge up to the proper level, so it's a good idea to buy an additional can. **Warning:** *Never add more than two cans of refrigerant to the system.*

12   Hook up the charging kit (see illustration) in accordance with the manufacturer's instructions. **Warning:** *DO NOT hook the charging kit hose to the system high side!* The fittings on the charging kit are designed to fit only on the low side of the system.

13   Back off the valve handle on the charging kit and screw the kit onto the refrigerant can, making sure first that the O-ring or rubber seal inside the threaded portion of the kit is in place. **Warning:** *Wear protective eyewear when dealing with pressurized refrigerant cans.*

14   Remove the dust cap from the low-side charging valve and then attach the quick-connect fitting on the kit hose.

15   Warm up the engine and turn on the air conditioning. Keep the charging kit hose away from the fan and other moving parts. **Note:** *The charging process requires the compressor to be running. If the clutch cycles off, you can put the air conditioning switch on High and leave the car doors open to keep the clutch on and compressor working.*

16   Turn the valve handle on the kit until the stem pierces the can, then back the handle out to release the refrigerant. You should be able to hear the rush of gas. Add refrigerant to the low side of the system, keeping the can upright at all times, but shaking it occasionally. Allow stabilization time between each addition. **Note:** *The charging process will go faster if you wrap the can with a hot-water-soaked shop rag to keep the can from freezing up.*

17   If you have an accurate thermometer, you can place it in the center air conditioning duct inside the vehicle and keep track of the output air temperature. A charged system that is working correctly should cool down to approximately 40-degrees F. If the ambient (outside) air temperature is very high, say 110 degrees F, the duct air temperature may be

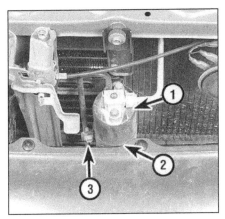

5   Unbolt the receiver/drier **(see illustrations)** and lift it out of the vehicle.
6   Install new O-rings in the refrigerant line fittings and lubricate them with clean refrigerant oil.
7   Installation is the reverse of removal. **Note:** *Do not remove the sealing caps until you're ready to reconnect the lines. Do not mistake the inlet (marked IN) and the outlet (marked OUT) connections.*
8   If a new receiver/drier is installed, add 0.4 US fluid ounces (10 cc) of refrigerant oil to the system.
9   Have the system evacuated, charged and leak tested by the shop that discharged it.

14.3 Disconnect the refrigerant lines (arrows) (Sprint shown, Metro models similar)

14.5a Typical front-mounted receiver/drier details (1994 and earlier Metro models)

1   Receiver/drier   3   Bolts
2   Bracket

as high as 60 degrees F, but generally the air conditioning is 30-40 degrees F cooler than the ambient air.
18   When the can is empty, turn the valve handle to the closed position and release the connection from the low-side port. Replace the dust cap.
19   Remove the charging kit from the can and store the kit for future use with the piercing valve in the UP position, to prevent inadvertently piercing the can on the next use.

14   Air conditioning receiver/drier - removal and installation

Refer to illustrations 14.3, 14.5a, 14.5b and 14.5c
Warning 1: *The air conditioning system is under high pressure. Do not loosen any fittings or remove any components until after the system has been discharged by a dealer service department or service station. Always*

wear eye protection when disconnecting air conditioning system fittings.
Warning 2: *Some models covered by this manual are equipped with an airbag. Always disable the airbag system before working in the vicinity of the steering column, impact sensors or dashboard to avoid the accidental deployment of the airbag, which could cause personal injury. See Chapter 12 for the airbag disarming procedure.*
1   The receiver/drier, which acts as a reservoir and filter for the refrigerant, is located either between the grille and radiator on the driver's side or adjacent to the timing belt end of the engine.
2   Remove the grille for access, if necessary (see Chapter 11).
3   Detach the two refrigerant lines from the receiver/drier **(see illustration)**. Remove and discard the old O-rings from the fittings.
4   Immediately cap the open fittings to prevent the entry of dirt and moisture.

15   Air conditioning compressor - removal and installation

Refer to illustrations 15.6a, 15.6b and 15.7
Warning: *The air conditioning system is under high pressure. Do not loosen any fittings or remove any components until after the system has been discharged by a dealer service department or service station. Always wear eye protection when disconnecting air conditioning system fittings.*
1   Disconnect the negative cable from the battery.
2   Set the parking brake and block the rear tires.
3   Raise the front of the vehicle and support it securely on jackstands.
4   Unbolt the lower splash shield and remove the compressor drivebelt (see Chapter 1).
5   Disconnect the wiring from the compressor.
6   Detach the refrigerant lines from the compressor **(see illustrations)** and immediately cap the open fittings to prevent the entry of dirt and moisture.

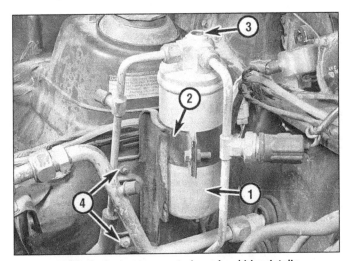

14.5b Typical rear-mounted receiver/drier details (1994 and earlier Metro models)

1   Receiver/drier   3   Sight glass
2   Bracket   4   Mounting bracket nuts

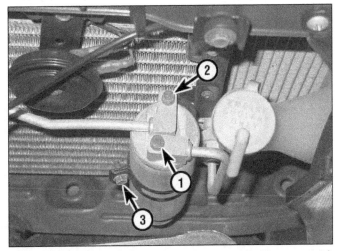

14.5c Typical receiver/drier details (1995 and later Metro models)

1   Condenser-to-receiver/drier line fitting
2   Receiver/drier-to-evaporator line fitting
3   Bracket bolt

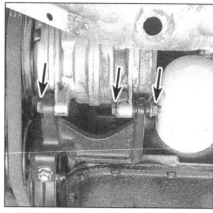

15.6a Unbolt the refrigerant lines (arrows) from the compressor (early Metro shown, Sprint similar)

15.6b Typical refrigerant line fittings (arrows) on late-model Metro

15.7 Locations of the air conditioning compressor lower mounting fasteners

16.2 Detach the vertical brace

16.3 Auxiliary electric fan mounting details - the fan can be removed along with the condenser or by itself

7   Remove the mounting bolts **(see illustration)** and lower the compressor from the engine compartment. **Note:** *Keep the compressor level during handling and storage. If the compressor seized or you find metal particles in the refrigerant lines, the system must be flushed out by an air conditioning technician and the receiver/drier must be replaced.*

8   Prior to installation, turn the center of the clutch six times to disperse any oil that has collected in the head.

9   Install the compressor in the reverse order of removal. Use Loctite (or equivalent) on the bolt threads.

10   If you're installing a new compressor, refer to the manufacturer's instructions for adding refrigerant oil to the system.

11   Have the system evacuated, charged and leak tested by the shop that discharged it.

16 Air conditioning condenser - removal and installation

Warning 1: *The air conditioning system is under high pressure. Do not loosen any fittings or remove any components until after the system has been discharged by a dealer*

service department or service station. *Always wear eye protection when disconnecting air conditioning system fittings.*

Warning 2: *Some models covered by this manual are equipped with an airbag. Always disable the airbag system before working in the vicinity of the steering column, impact sensors or dashboard to avoid the accidental deployment of the airbag, which could cause personal injury. See Chapter 12 for the airbag disarming procedure.*

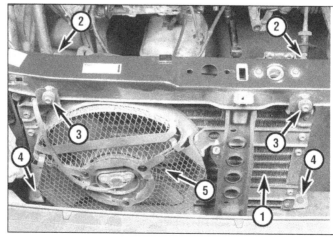

Condenser

1994 and earlier models

Refer to illustrations 16.2, 16.3 and 16.6

1   Remove the grille (see Chapter 11).

2   Remove the vertical brace **(see illustration)**.

3   Remove the auxiliary electric fan **(see illustration)**.

4   Disconnect the refrigerant lines from the condenser assembly. Be sure to use a back-

16.6 Typical condenser mounting details

1 *Condenser*
2 *Nuts*
3 *Bolts*
4 *Studs*
5 *Cooling fan*

16.11a Before the condenser fan or the condenser can be removed on 1995 and later models, these components must be removed first:

A   *Hood latch bolts*
B   *Forward discriminating sensor (airbag sensor) bolts*
C   *Hood latch bracket bolts*
D   *Condenser fan left mounting bolt (other left mounting bolt not visible in this photo, but you'll see it after you remove the previous three components)*

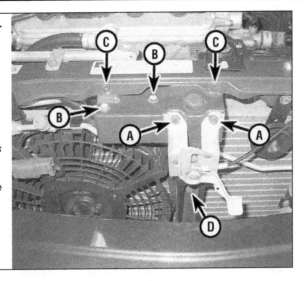

16.11b To detach the vertical brace from the lower crossmember, remove this bolt (arrow) and a screw (not visible in this photo) from underneath the crossmember

16.12 Unplug the condenser fan electrical connector

16.13 To detach the right side of the condenser fan assembly, remove these two screws (arrows)

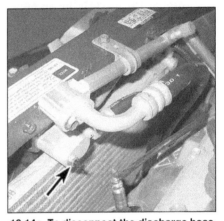

16.14a To disconnect the discharge hose from the condenser, remove this nut (arrow)

up wrench to avoid twisting the lines.
5   Immediately cap the open fittings to prevent the entry of dirt and moisture.
6   Unbolt the condenser **(see illustration)** and lift it out of the vehicle. Store it upright to prevent oil loss.
7   Installation is the reverse of removal.
8   If a new condenser was installed, add 0.7 to 1.0 ounces (20 to 30 cc) of refrigerant oil to the system.
9   Have the system evacuated, charged and leak tested by the shop that discharged it.

1995 and later models

Refer to illustrations 16.11a, 16.11b, 16.12, 16.13, 16.14a and 16.14b
10   Disable the airbag system.
11   Unbolt the forward discriminating sensor (a component of the airbag system) and the hood latch assembly, and then remove the vertical brace **(see illustrations)**.
12   Unplug the condenser fan electrical connector **(see illustration)**.
13   Remove the two condenser fan mounting screws on the right side of the fan assembly **(see illustration)** and remove the remaining bolt on the left side of the fan assembly

(you removed the other left side bolt, from the vertical brace, in Step 11). Then remove the condenser fan assembly by lifting it straight up, between the receiver/drier-to-evaporator line and the radiator (carefully pull the refrigerant line forward to clear the condenser fan assembly). If you're replacing the fan blades or the motor, refer to Steps 20 through 23 below.

14   Disconnect the refrigerant line fittings from the condenser **(see illustration)**.
15   Remove the horn (see Chapter 12).
16   Remove the two condenser mounting bolts **(see illustration 16.14b)** and then remove the condenser.
17   Installation is the reverse of removal.
18   If a new condenser was installed, add 0.7 to 1.0 ounces (20 to 30 cc) of refrig-

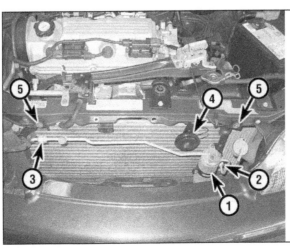

16.14b To remove the condenser, remove the following components

1   *Receiver/drier (see Section 14)*
2   *Condenser-to-receiver/drier line*
3   *Receiver/drier-to-evaporator line (not the entire line, just the section between this fitting and the receiver/drier)*
4   *Horn (see Chapter 12)*
5   *Condenser mounting bolts*

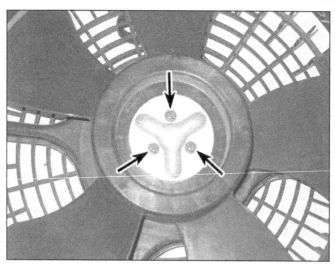

16.21 To remove the fan blades, remove these three screws (arrows) (late-model Metro shown, earlier models similar)

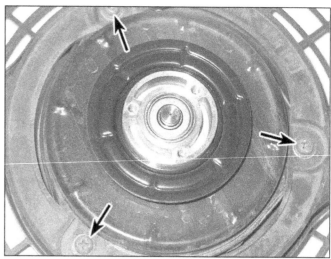

16.22 To remove the fan motor, remove these three screws (arrows) (late-model Metro shown, earlier models similar)

erant oil to the system.

19   Have the system evacuated, charged and leak tested by the shop that discharged it.

Condenser fan motor

Refer to illustrations 16.21 and 16.22

20   Remove the condenser fan (see Steps 1 through 3, or Steps 10 through 13).

21   Remove the fan mounting screws **(see illustration)**.

22   Remove the fan motor mounting screws **(see illustration)**.

23   Installation is the reverse of removal.

Chapter 4
Fuel and exhaust systems

Contents

Specifications

Accelerator cable freeplay

  Sprint

    When carburetor and engine coolant are cold .... 0.40 to 0.59 inch (10 to 15 mm)

    When carburetor and engine coolant are warm .... 0.12 to 0.19 inch (3 to 5 mm)

  Metro

    1989 through 1992 .... 0.0 inch (0.0 mm)

    1993 and 1994 .... 0.4 to 0.6 inch (10 to 15 mm)

    1995 .... 0.12 to 0.20 inch (3 to 5 mm)

    1996 and 1997 .... 0.4 to 0.6 inch (10 to 15 mm)

    1998 on

      Three-cylinder engine .... 0.4 to 0.6 inch (10 to 15 mm)

      Four-cylinder engine .... 0.08 to 0.27 inch (2 to 7 mm)

Choke valve-to-carburetor bore clearance (Sprint)

  At 77-degrees F (25-degrees C) .... 0.004 to 0.019 in (0.1 to 0.5 mm)

  At 95-degrees F (35-degrees C) .... 0.030 to 0.060 in (0.7 to 1.7 mm)

Fuel injector resistance (Metro)

  TBI systems .... 0.5 to 1.5 ohms at 68 degrees F (20 degrees C)

  SFI systems (1998 and later four-cylinder models) .... 12 to 13 ohms at 68 degrees F (20 degrees C)

Fuel pressure (Metro)

  TBI systems

    Engine off and ignition on .... 23 to 31 psi

    Engine at idle .... 13 to 20 psi

  SFI systems (1998 and later four-cylinder models)

    Engine off and ignition on .... 38.5 to 44 psi

    Engine at idle .... 28.5 to 35 psi

Torque specifications   Ft-lbs

Carburetor mounting bolts	18
Throttle Body Injection (TBI) system	
TBI unit mounting bolts	17
Multiport Sequential Fuel Injection (SFI) system	
Throttle body mounting bolts/nuts..................................	14 to 17
Fuel pressure regulator mounting bolts	71 to 106 in-lbs
Fuel rail retaining bolts	13 to 20

1   General information

The fuel system consists of a rear mounted tank, combination metal and rubber fuel hoses, an engine mounted mechanical fuel pump or an electric fuel pump, and either a two-stage, twin-venturi carburetor (Sprint models) or an electronic fuel injection system (Metro models). All Metro three-cylinder models and 1995 through 1997 Metro four-cylinder models are equipped with a Throttle Body Injection (TBI) system (see Section 14). All 1998 and later Metro four-cylinder models are equipped with a multiport Sequential Fuel Injection (SFI) system (see Section 16).

The exhaust system is composed of an exhaust manifold, the catalytic converter and a combination muffler and tailpipe assembly.

The emission control systems modify the functions of both the exhaust and fuel systems. There may be some cross-references throughout this Chapter to sections in Chapter 6 because the emissions control systems are integral with the induction and exhaust systems.

Extreme caution should be exercised when dealing with either the fuel or exhaust system. Fuel is a primary element for combustion. Be very careful! The exhaust system is also an area for exercising caution as it operates at very high temperatures. Serious burns can result from even momentary contact with any part of the exhaust system and the fire potential is ever present.

2   Fuel pressure relief procedure

Warning: *Gasoline is extremely flammable, so take extra precautions when you work on any part of the fuel system. Don't smoke or allow open flames or bare light bulbs near the work area, and don't work in a garage where a natural gas-type appliance (such as a water heater or clothes dryer) with a pilot light is present. If you spill any fuel on your skin, rinse it off immediately with soap and water. When you perform any kind of work on the fuel system, wear safety glasses and have a Class B type fire extinguisher on hand.*

All models

1   Always relieve the fuel pressure before disconnecting any fuel system component to minimize the risk of fire and personal injury.
2   Unscrew the fuel filler cap to release the pressure caused by fuel vapor.

Carbureted models

Refer to illustration 2.4
3   Place rags under the connector or clamp that unites the fuel filter and the fuel line.
4   Loosen the clamp and remove the fuel line. Gently twist the fuel line and allow the fuel to drip into the rags **(see illustration)**. Cover the lines completely with the rags while simultaneously twisting the fuel line.

Fuel-injected models

Refer to illustrations 2.6a and 2.6b
5   Position the shifter in Park (automatic transaxle) or Neutral (manual transaxle) and block the rear wheels.
6   Remove the fuel pump relay **(see illustrations)** from the main fuse panel in the engine compartment.
7   Start the engine, allowing it to run until it stalls. Crank it over five or six times. Reinstall the relay, but don't turn the ignition key on until after all work to the fuel system has been performed, otherwise the fuel system will pressure-up again.
8   Even though fuel pressure should now be safely relieved, it's always a good idea to place a rag over any fuel fitting before loosening it.

3   Fuel lines and fittings - inspection and replacement

Warning: *Gasoline is extremely flammable, so take extra precautions when you work on any part of the fuel system. Don't smoke or allow open flames or bare light bulbs near the work area, and don't work in a garage where a natural gas-type appliance (such as a water heater or clothes dryer) with a pilot light is present. If you spill any fuel on your skin, rinse it off immediately with soap and water. When you perform any kind of work on the fuel system, wear safety glasses and have a Class B type fire extinguisher on hand.*

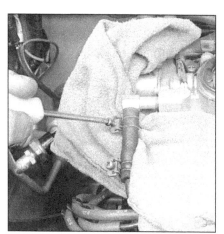

2.4 Place a rag directly under and surrounding the fuel line connection hat will be removed

2.6a Fuel pump relay location (1994 and earlier models)

2.6b Fuel pump relay location (1995 and later models)

Inspection

1 Once in a while, you will have to raise the vehicle to service or replace some component. Whenever you work under the vehicle, always inspect fuel lines and all fittings and connections for damage or deterioration.
2 Check all hoses and pipes for cracks, kinks, deformation or obstructions.
3 Make sure all hoses and pipe clips attach their associated hoses or pipes securely to the underside of the vehicle.
4 Verify all hose clamps attaching rubber hoses to metal fuel lines or pipes are snug enough to assure a tight fit between the hoses and pipes.

Replacement

Refer to illustrations 3.6, 3.8a and 3.8b
5 If you must replace any damaged sections, use original equipment replacement hoses or pipes constructed from exactly the same material as the section you are replacing. Do not install substitutes constructed from inferior or inappropriate material or you could cause a fuel leak or a fire.
6 Always, before detaching or disassembling any part of the fuel line system, note the routing of all hoses and pipes and the orientation of all clamps and clips to assure that replacement sections are installed in exactly the same manner **(see illustration)**.
7 Before detaching any part of the fuel system, be sure to relieve the fuel line and tank pressure (see Section 2).
8 To disconnect a quick-connect fitting (1998 and later models), simply squeeze the two locking tabs on the sides of the fitting and pull the two halves of the fitting apart **(see illustrations)**. Inspect the O-ring seal inside the female half of the fitting. If it's cracked or deteriorated, replace it. To reconnect the fitting, push the two halves together until it "clicks."
9 While you're under the vehicle, check

the condition of the fuel filter; make sure that it's not clogged or damaged (see Chapter 1).

---

4 Fuel pump/fuel system pressure - check

Warning: *Gasoline is extremely flammable, so take extra precautions when you work on any part of the fuel system. Don't smoke or allow open flames or bare light bulbs near the work area, and don't work in a garage where a natural gas-type appliance (such as a water heater or clothes dryer) with a pilot light is present. If you spill any fuel on your skin, rinse it off immediately with soap and water. When you perform any kind of work on the fuel system, wear safety glasses and have a Class B type fire extinguisher on hand.*
Note: *The following checks assume the fuel filter is in good condition. If you doubt it's condition, install a new one (see Chapter 1).*
1 Check that there is adequate fuel in the fuel tank. If you doubt the reading on the gauge, insert a long wooden dowel at the filler opening; it will serve as a dipstick.

Carbureted models
Preliminary check
2 If you suspect insufficient fuel delivery, first inspect all fuel lines to ensure that the problem is not simply a leak in a line.
3 If there are no leaks evident in the fuel lines, inspect the fuel pump itself. The following checks will tell you if the fuel pump is leaking and whether it is pumping fuel.
4 Remove the air cleaner housing.

Fuel pump output check
5 Hook up a remote starter switch in accordance with the manufacturer's instructions. If you don't have a remote starter switch, you will need an assistant to help you with this and the following procedure.

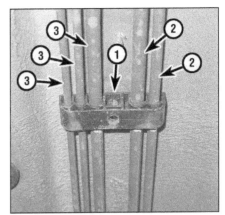

3.6 Various types of clamps are used to secure fuel, vapor and brake lines underneath the vehicle; this is a typical late-model fuel and brake line clamp

1 *Pry here, between the body and the clamp, with a screwdriver to pry the clamp loose from the vehicle*
2 *Brake lines*
3 *Fuel and vapor lines*

6 Trace the fuel outlet hose from the pump to the carburetor and detach it at the carburetor.
7 Attach the cable to the negative terminal of the battery.
8 Detach the wires from the primary terminals of the ignition coil (see Chapter 5).
9 Place a metal container under the open end of the fuel pump outlet hose.
10 Direct the fuel pump outlet hose into the container while cranking the engine for a few seconds with the remote starter (or while an assistant cranks the engine with the ignition key).
11 If fuel is emitted in well defined spurts, the pump is operating satisfactorily. If fuel dribbles or trickles out the hose, the pump is defective. Replace it (see Section 6).

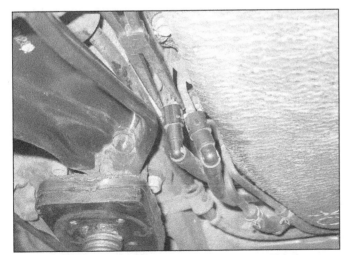

3.8a Quick-connect fuel line fittings are used on 1998 and later models; to disconnect one of these fittings, squeeze the two locking tabs together with a pair of needle-nose pliers . . .

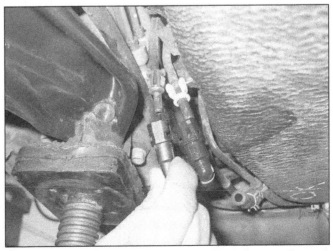

3.8b . . . and then pull the two halves of the fitting apart. Be sure to inspect the O-ring seal inside the female half of the fitting and replace it if it's damaged. To reconnect the fitting, simply push the two halves together until it clicks

4.15 Connect a fuel pressure gauge onto the inlet side
of the TBI fuel line and the TBI unit

4.16a On 1998 and later four-cylinder Metro models, disconnect
the fuel supply hose (arrow) from the fuel rail . . .

Fuel-injected models

Fuel pump operational check

Note: *On 1989 and later models, the fuel pump is located inside the fuel tank.*

12   Set the parking brake and have an assistant turn the ignition switch to the On position while you listen at the fuel pump. You should hear a whirring sound, lasting for a couple of seconds. Start the engine. The whirring sound should now be continuous (although harder to hear with the engine running). If there is no whirring sound, either the fuel pump or the fuel main relay circuit is defective.

Pressure check

Refer to illustrations 4.15, 4.16a, 4.16b and 4.16c

13   Relieve the fuel pressure (see Section 2).
14   Remove the air cleaner assembly.
15   On Metro three-cylinder models and on 1995 through 1997 Metro four-cylinder models, remove the fuel feed line from the throttle body and then attach a fuel pressure gauge between the fuel feed line and the throttle body **(see illustration)**. Clamp the lines securely to make sure there will not be any

leaks.
16   On 1998 and later four-cylinder Metro models, disconnect the fuel feed hose from the fuel rail **(see illustration)** and then, using a suitable T-fitting and hose **(see illustration)**, connect a fuel pressure gauge between the fuel feed hose and the fuel rail **(see illustration)**.
17   Start the engine. With the engine idling, measure the fuel pressure and then compare your measurement to the fuel pressure listed in this Chapter's Specifications. Shut off the engine, turn the ignition key to the ON position, measure the fuel pressure and then compare your measurement to the fuel pressure listed in this Chapter's Specifications. If the fuel pressure is not within specification, check for a pinched or clogged fuel return hose or pipe.
18   If the pressure is lower than specified, inspect the fuel filter - make sure it's not clogged. Look for a pinched or clogged fuel hose between the fuel tank and the fuel pump.
19   If there are no problems with any of the above-listed components, check the fuel pump (see below).

Fuel pump check

Refer to illustration 4.23

20   If you suspect a problem with the fuel pump, verify that the pump actually runs. Have an assistant turn the ignition switch to ON - you should hear a brief whirring noise as the pump comes on and pressurizes the system. Have the assistant start the engine. This time you should hear a constant whirring sound from the pump (but it's more difficult to hear with the engine running).
21   If the pump does not come on (makes no sound), proceed to the next step.
22   Remove the rear seat (see Chapter 11). Remove the fuel pump/fuel sending unit access cover.
23   Disconnect the fuel pump electrical connector **(see illustration)**. Make sure the ignition switch is turned Off before disconnecting the wires.
24   Touch the positive probe of a voltmeter to the red or pink wire and the negative probe to the black wire, then turn on the ignition switch and verify there is voltage available.
25   If voltage is available, replace the fuel pump (see Section 6).
26   If no voltage is available, check the main

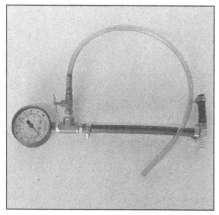

4.16b . . . and then, using a suitable
T-fitting and hose . . .

4.16c . . . connect a fuel pressure gauge
between the fuel feed hose
and the fuel rail

4.23 Do not mistake the fuel sender
connector (yellow/red and black
wires) with the fuel pump connector
(red and black wires)

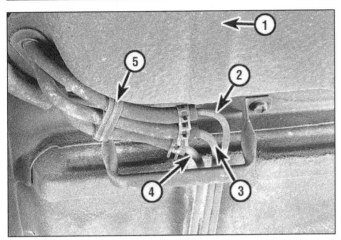

5.6 On Sprint models, remove the three fuel hoses from the lines

1	Fuel tank	3	Vapor vent hose (to
2	Fuel return hose (to		canister)
	carburetor)	4	Fuel hose (to fuel filter)
		5	Clamp

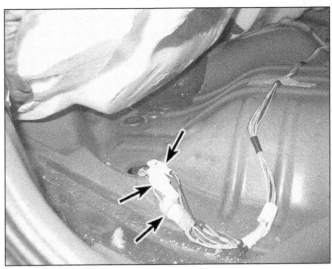

5.7 On Metro models, remove the rear seat cushion and then unplug the electrical connectors (arrows) for the fuel pump, fuel gauge sending unit, etc.

relay (see Chapter 12). If the main relay is okay, check the wiring between the relay and the fuel pump (see Wiring Diagrams at the end of Chapter 12).

5   Fuel tank - removal and installation

Removal

Refer to illustrations 5.6, 5.7, 5.8a, 5.8b, 5.8c, 5.8d and 5.10

Warning: *Gasoline is extremely flammable, so take extra precautions when you work on any part of the fuel system. Don't smoke or allow open flames or bare light bulbs near the work area, and don't work in a garage where a natural gas-type appliance (such as a water heater or clothes dryer) with a pilot light is present. If you spill any fuel on your skin, rinse it off immediately with soap and water. When you perform any kind of work on the fuel system, wear safety glasses and have a Class B type fire extinguisher on hand.*

Note: *The following procedure is much easier to perform if the fuel tank is empty. Some tanks have a drain plug for this purpose. If the tank does not have a drain plug, drive the vehicle until the tank is nearly empty (if possible) or siphon the fuel from the tank using a siphoning kit (available at most auto parts stores).*

1   Remove the fuel filler cap to relieve fuel tank pressure.
2   Relieve the fuel pressure (see Section 2).
3   Detach the cable from the negative terminal of the battery.
4   If the tank has a drain plug, remove it and allow the fuel to drain into an approved gasoline container.
5   Raise the vehicle and place it securely on jackstands.
6   On Sprint models, remove the three fuel hoses from the pipes **(see illustration)**.

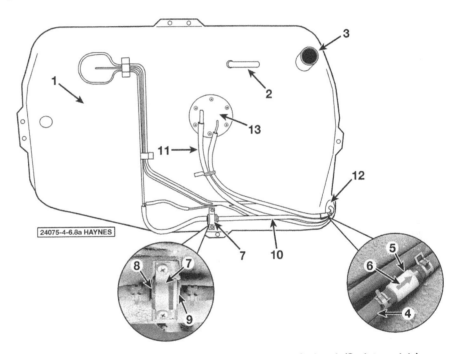

5.8a Typical routing of fuel hoses and lines above fuel tank (Sprint models)

1	Fuel tank	8	Black nozzle
2	Breather inlet	9	Orange nozzle
3	Fuel filler inlet	10	Vapor vent hose (to canister)
4	Fuel return hose	11	Fuel hose (to fuel filter)
5	Fuel return check valve	12	Fuel hose grommet
6	Arrow mark	13	Fuel level gauge and intake filter
7	Two way check valve		

7   On Metro models, remove the rear seat (see Chapter 11). Disconnect the fuel pump electrical connectors **(see illustration)** and then pull the harness through the floor pan so that it's out of the way.
8   Disconnect the fuel lines, the vapor return lines and the fuel filler neck **(see illustrations)**. **Note:** *The fuel lines are usually dif-* ferent diameters, so reattachment is simplified. If you have any doubts, however, clearly label the lines and the fittings. Be sure to plug the hoses to prevent leakage and contamination of the fuel system. On 1998 and later models, remove the fuel tank check valve from the fuel filler neck pipe **(see illustration)**.

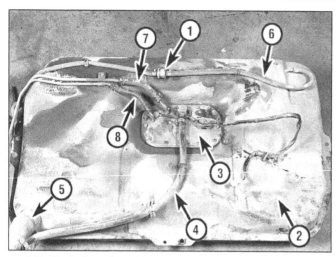

5.8b Typical routing of fuel hoses and lines above fuel tank
(1994 and earlier Metro models)

1	2-way check valve	5	Fuel filler hose
2	Fuel tank	6	Fuel vapor line
3	Fuel pump and sending unit	7	Fuel feed line
4	Breather hose	8	Fuel return line

5.8c Typical routing of fuel hoses and lines above fuel tank
(1995 and later Metro models)

1	Fuel breather hose	4	Fuel return line
2	Fuel filler hose	5	Fuel feed line
3	Air inlet hose		

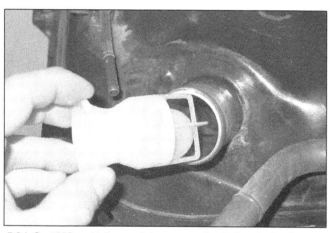

5.8d On 1998 and later models, remove the fuel tank check valve
from the fuel filler neck pipe (and don't forget to put it back in the
filler neck pipe when reinstalling the tank)

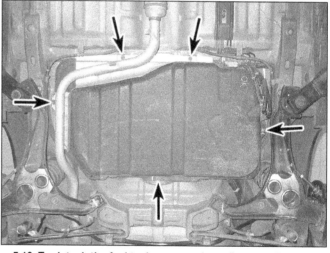

5.10 To detach the fuel tank, remove these five bolts (arrows)

9   Support the fuel tank with a floor jack. Position a piece of wood between the jack head and the fuel tank to protect the tank.
10   Remove the fuel tank retaining bolts **(see illustration)**.
11   Lower the tank enough to verify that all electrical wiring and fuel and vapor hoses and lines have been disconnected.
12   Remove the tank from the vehicle.

Fuel tank cleaning and repair

13   Any repairs to the fuel tank or filler neck should be carried out by a professional who has experience in this critical and potentially dangerous work. Even after cleaning and flushing of the fuel system, explosive fumes can remain and ignite during repair of the tank.
14   If the fuel tank is removed from the vehicle, it should not be placed in an area where sparks or open flames could ignite the fumes

coming out of the tank. Be especially careful inside garages where a natural gas-type appliance is located, because the pilot light could cause an explosion.

Installation

15   Installation is the reverse of removal. On 1998 and later models, don't forget to install the fuel tank check valve in the fuel filler neck pipe before reconnecting the fuel filler hose.

6   Fuel pump/fuel gauge sending unit - removal and installation

Warning: *Gasoline is extremely flammable, so take extra precautions when you work on any part of the fuel system. Don't smoke or allow open flames or bare light bulbs near the*

work area, and don't work in a garage where a natural gas-type appliance (such as a water heater or clothes dryer) with a pilot light is present. If you spill any fuel on your skin, rinse it off immediately with soap and water. When you perform any kind of work on the fuel system, wear safety glasses and have a Class B type fire extinguisher on hand.
1   Disconnect the cable from the negative terminal of the battery.
2   Relieve the fuel system pressure (see Section 2).
3   Relieve the fuel tank pressure by removing the fuel filler cap.

Carbureted models

Mechanical pump
Refer to illustration 6.6
4   Locate the fuel pump mounted on the left rear corner of the cylinder head. Place

6.6 Remove the fuel pump nuts (arrows) from the base of the pump (the other nut is not visible)

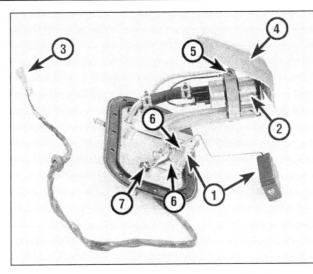

6.14 Typical fuel pump and sending unit assembly (1994 and earlier Metro models)

1   Sending unit
2   Fuel pump
3   Electrical connector
4   Fuel filter
5   Fuel pump clamp screw
6   Sending unit retaining screws
7   Sending unit electrical lead terminal nut

rags underneath the pump to catch any spilled fuel.
5   Loosen hose clamps and slide them down the hoses, past the fittings. Disconnect the hoses from the pump, using a twisting motion as you pull them from the fittings. Immediately plug the hoses to prevent leakage of fuel and the entry of dirt.
6   Unscrew the fasteners that retain the pump to the cylinder head **(see illustration)**, then detach the pump from the head. Inspect the fuel pump arm for wear. Coat it with clean engine oil before installing it.
7   Using a gasket scraper or putty knife, remove all traces of old gasket material from the mating surfaces on the cylinder head (and the fuel pump, if the same one will be reinstalled). While scraping, be careful not to gouge the soft aluminum surfaces.
8   Installation is the reverse of the removal procedure. Be sure to use a new gasket and tighten the mounting fasteners securely.

Fuel gauge sending unit

9   Remove the fuel tank (see Section 5).
10   Remove the fuel gauge sending unit flange retaining screws **(see illustration 5.8a)** and then pull out the fuel gauge sending unit.
11   Installation is the reverse of removal.

Fuel-injected models

12   Remove the fuel tank (see Section 5).

1994 and earlier models

Electric pump
Refer to illustration 6.14
13   Remove the fuel feed lines and the clamps and hoses from the fuel return lines.
14   Remove the retaining screws, then lift the assembly out of the fuel tank **(see illustration)**.
15   Remove the clamp screw that secures the fuel pump motor assembly **(see illustration 6.14)**.
16   Remove the two fuel pump electrical connectors and remove the fuel pump.
17   Installation is the reverse of the removal procedure. Be sure to replace the sending unit gasket if it shows any signs of deterioration.

Fuel gauge sending unit

18   Remove the electric pump from the fuel tank (see Steps 13 through 17).
19   Remove the nut that secures the sending unit electrical lead to the stud on the flange, remove the sending unit retaining screws **(see illustration 6.14)** and then remove the sending unit from its mounting bracket.
20   Installation is the reverse of removal.

1995 and later models
Refer to illustrations 6.21, 6.22, 6.23, 6.24, 6.25, 6.26a, 6.26b, 6.28, 6.29, 6.30 and 6.31
Note: *This subsection includes instructions for replacing the fuel pump and the fuel gauge sending unit.*
21   Unplug the fuel pump/fuel gauge sending unit electrical connector and disconnect the fuel vapor hose, fuel feed line and fuel return line **(see illustration)** from the fuel pump/fuel gauge sending unit.
22   Remove the retaining screws **(see illustration)** and then remove the fuel pump from the fuel tank.
23   To remove the fuel pump/fuel gauge sending unit from the fuel tank, lift it straight up **(see illustration)**.
24   Unplug the fuel gauge sending unit elec-

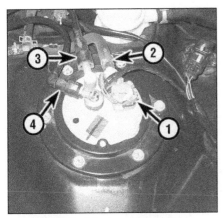

6.21 Typical fuel pump/fuel gauge sending unit assembly (1995 and later models)

1   Fuel pump/fuel gauge sending unit electrical connector
2   Fuel vapor hose
3   Fuel feed line
4   Fuel return line

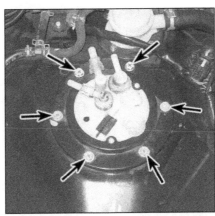

6.22 To detach the fuel pump/fuel gauge sending unit assembly from the fuel tank, remove these retaining screws (arrows)

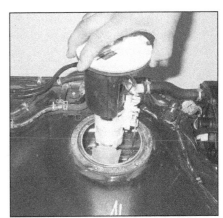

6.23 To remove the fuel pump/fuel gauge sending unit from the fuel tank, lift it straight up

6.24 Unplug the fuel gauge sending unit electrical connector

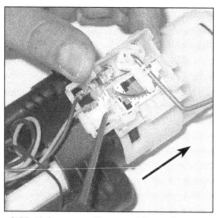

6.25 Using a small screwdriver, depress the fuel gauge sending unit retaining tab and then remove the sending unit by sliding it in the direction indicated by the arrow

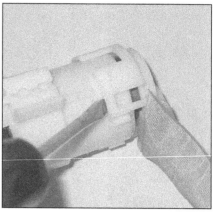

6.26a Using a small screwdriver, depress the lower end cap retaining tab . . .

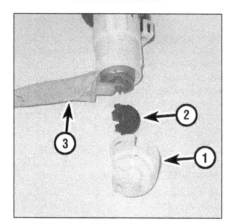

6.26b . . . and then remove the end cap (1) and the cushion (2); if the strainer (3) is damaged, remove it from the pump and replace it

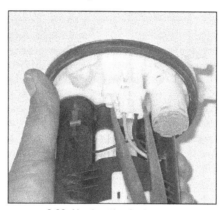

6.28 Unplug the fuel pump electrical connector

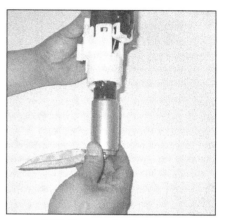

6.29 Remove the fuel pump from the housing

trical connector **(see illustration)**.

25   Using a small screwdriver, depress the fuel gauge sending unit retaining tab **(see illustration)** and then remove the sending unit.

26   Using a small screwdriver, depress the lower end cap retaining tab and then remove the end cap and the cushion **(see illustrations)**.

27   If the fuel sock (strainer) is damaged, or if it needs to be cleaned, or if you're swapping pumps, remove the fuel sock **(see illustration 6.26b)**.

28   Unplug the fuel pump electrical connector **(see illustration)**.

29   Remove the fuel pump from the housing **(see illustration)**.

30   Using a small screwdriver, depress the retaining tab **(see illustration)** and then remove the housing.

31   Remove the fuel tube **(see illustration)**. Inspect the fuel tube grommets at both ends of the fuel tube. If they're cracked or deteriorated, replace them.

32   Reassembly is the reverse of disassembly.

33   Installation is the reverse of removal.

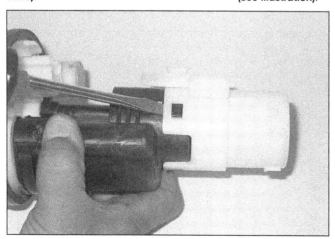

6.30 Using a small screwdriver, depress the retaining tab and then remove the housing.

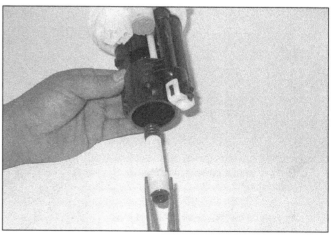

6.31 Remove the fuel tube and then inspect the fuel tube grommets on the upper and lower ends of the fuel tube; if they're cracked or deteriorated, replace them

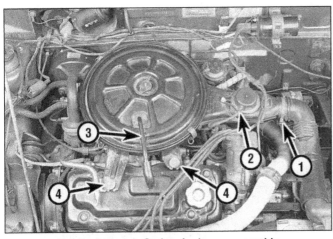

**7.1 Typical early Sprint air cleaner assembly
(Metro three-cylinder air cleaners similar)**

1   Air intake duct hose clamp
2   Thermostatically Controlled Air Cleaner (TAC) hose
3   PCV hose
4   Air cleaner assembly bracket bolts (Later Sprint models and all
    Metro models have one bolt)

**7.5 To remove the air cleaner assembly from the carburetor or
throttle body on later Sprint and all three-cylinder Metro models,
remove this bolt (arrow) that attaches the air cleaner to the
cylinder head**

7   Air cleaner assembly - removal and installation

Sprint models and Metro three-cylinder models

Refer to illustrations 7.1 and 7.5

1   Loosen the clamp that secures the air intake duct to the air cleaner snorkel **(see illustration)**.
2   Remove the wing nut that secures the air filter assembly to the carburetor or throttle body. **Note:** *It's not necessary to actually remove the upper part of the air filter housing and the air filter element in order to remove the air cleaner assembly. The only thing you have to remove is the wing nut on top that secures the assembly to the carburetor or throttle body. However, as long as you're removing the air cleaner assembly, this is a good time to inspect the condition of the element and the inside of the air cleaner assem-*

bly *(see "Air filter replacement" in Chapter 1)*.
3   Unplug the electrical connector from the Intake Air Temperature (IAT) sensor (see Chapter 6).
4   Disconnect the Positive Crankcase Ventilation (PCV) hose from the air cleaner assembly (for information on how to check the PCV valve, see Chapter 6).
5   To detach the lower part of the air filter assembly from the cylinder head, remove the bolt that attaches the air cleaner bracket to the cylinder head **(see illustration)**. **Note:** *Some early Sprint models have two bracket bolts* **(see illustration 7.1)**.
6   Installation is the reverse of removal.

Metro four-cylinder models

Refer to illustration 7.7

7   Loosen the clamp that secures the air intake duct to the air cleaner assembly, remove the duct retaining bolt **(see illustration)** and then detach the air intake duct from the air cleaner assembly. It's not necessary to

disconnect the air intake duct from the throttle body; just pull it off the air cleaner and push it aside.
8   Remove the air cleaner assembly mounting bolt.
9   Remove the air cleaner assembly.
10   Installation is the reverse of removal.

8   Accelerator cable - removal, installation and adjustment

Sprint models
Removal

Refer to illustrations 8.2, 8.3 and 8.5

1   Detach the cable from the negative terminal of the battery.
2   Unscrew the locknut on the threaded portion of the accelerator cable at the carburetor **(see illustration)**.

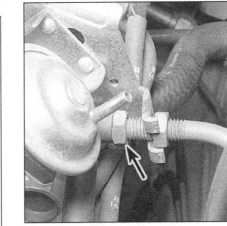

**7.7 To disconnect
the air intake duct
from the air
cleaner assembly
on four-cylinder
Metro models,
remove this bolt
and loosen the
hose clamp (upper
arrows); to detach
the air cleaner
assembly from the
vehicle body,
remove this bolt
(lower arrow)**

**8.2 Unscrew the locknut (arrow)
on the threaded portion of the
cable (Sprint models)**

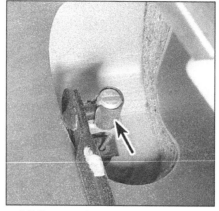

8.3 Grasp the throttle lever arm with a pair of needle nose pliers and rotate it to put some slack in the cable, then slip the cable end (arrow) out of the slot in the arm (Sprint models)

8.5 Remove the retaining clip (arrow) from the end of the accelerator cable by sliding it out (Sprint models)

8.13 Rotate the throttle lever arm to put some slack in the accelerator cable and then slip the cable end out of its slot in the arm (Metro models)

3   Grasp the throttle lever arm and rotate it to put some slack in the accelerator cable, then slip the cable end out of its slot in the arm **(see illustration)**.
4   Trace the accelerator cable to the firewall, detaching it from all brackets.
5   The cable is secured to the firewall with a retaining clip **(see illustration)** that must be removed from inside the vehicle, directly above the throttle pedal.
6   Detach the accelerator cable from the accelerator pedal.
7   From inside the vehicle, pull the cable through the firewall.

Installation and adjustment

8   Installation is the reverse of removal.
9   To adjust the cable, fully depress the accelerator pedal and check that the throttle is fully opened.
10   If not fully opened, loosen the locknuts, depress the accelerator pedal and adjust the cable so full throttle can be attained.

11   Tighten the locknuts and recheck the adjustment. Make sure the throttle closes fully when the pedal isn't depressed.

Metro models

Removal

Refer to illustrations 8.13, 8.14 and 8.16

12   Detach the cable from the negative terminal of the battery.
13   Grasp the throttle lever arm and rotate it to put some slack in the accelerator cable **(see illustration)**, then slip the cable end out of its slot in the arm.
14   Using a back-up wrench to hold the adjustment nut, unscrew the locknut at the accelerator cable bracket **(see illustration)**.
15   Trace the accelerator cable to the firewall and detach it from all brackets, clips, clamps and cable guides.
16   Detach the accelerator cable from the accelerator pedal **(see illustration)**.
17   From inside the vehicle, pull the cable through the firewall.

8.14 Using a back-up wrench to hold the adjustment nut, unscrew the locknut at the accelerator cable bracket (Metro models)

8.16 To disengage the accelerator cable from the accelerator pedal, push the upper end of the pedal forward to remove tension from the cable and then slide the cable end (arrow) out of the slot in the left side of the pedal (Metro models)

8.19 To measure the accelerator pedal freeplay, place a ruler between the lower end of the pedal and the carpet on the lower firewall as shown, and then push down on the pedal and measure the distance the pedal travels before you feel any resistance (this is the point at which the cable begins to open the spring-loaded throttle lever arm) (Metro models)

8.22 To measure the clearance between the throttle lever and the throttle stop, have an assistant depress the accelerator pedal to the floor (or push it down yourself and hold it down with a weight), and then measure the throttle lever-to-throttle stop clearance with a feeler gauge (Metro models)

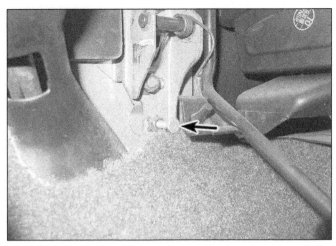

8.23 To bring the throttle lever-to-throttle stop clearance within the specified range, screw the pedal stopper bolt (arrow) in or out until the throttle lever-to-throttle stop clearance is within the specified range (Metro models)

Installation and adjustment

Refer to illustrations 8.19, 8.22 and 8.23

18   Installation is the reverse of removal.

19   Inside the vehicle, measure the accelerator pedal freeplay **(see illustration)** and then compare your measurement with the freeplay listed in this Chapter's Specifications.

20   If the freeplay is within the specified range, skip the next Step and proceed to Step 22. If it isn't, go to the next Step.

21   Loosen the accelerator cable locknut **(see illustration 8.14)** and then turn the accelerator cable adjustment nut until the accelerator pedal freeplay is within the specified range. Once the freeplay is correct, tighten the accelerator cable locknut and then recheck the accelerator pedal freeplay. If the freeplay is still incorrect, repeat this procedure. Continue to do so until the freeplay is correct.

22   Have an assistant depress the accelerator pedal to the floor (or push it down yourself and hold it down with a weight), and then measure the clearance between the throttle lever and the throttle stop **(see illustration)**. Compare your measurement to the throttle lever-to-throttle stop clearance listed in this Chapter's Specifications. If the clearance is within the specified range, proceed to Step 24. If it isn't, go to the next Step.

23   Adjust the height of the pedal stopper bolt **(see illustration)** until the throttle lever-to-throttle stop clearance is within the specified range.

24   Depress the accelerator pedal and verify that the accelerator cable operates smoothly, with no binding.

---

9   Carburetor (Sprint models) - diagnosis and overhaul

Warning: *Gasoline is extremely flammable, so take extra precautions when you work on any part of the fuel system. Don't smoke or allow open flames or bare light bulbs near the* work area, and don't work in a garage where a natural gas-type appliance (such as a water heater or clothes dryer) with a pilot light is present. If you spill any fuel on your skin, rinse it off immediately with soap and water. When you perform any kind of work on the fuel system, wear safety glasses and have a Class B type fire extinguisher on hand.

General diagnosis

1   A thorough road test and check of carburetor adjustments should be done before any major carburetor service work. Specifications for some adjustments are listed on the Vehicle Emissions Control Information (VECI) label found in the engine compartment.

2   Carburetor problems usually show up as flooding, hard starting, stalling, severe backfiring and poor acceleration. A carburetor that's leaking fuel and/or covered with wet looking deposits definitely needs attention.

3   Some performance complaints directed at the carburetor are actually a result of loose, out-of-adjustment or malfunctioning engine or electrical components. Others develop when vacuum hoses leak, are disconnected or are incorrectly routed. The proper approach to analyzing carburetor problems should include the following items:

a) Inspect all vacuum hoses and actuators for leaks and correct installation (see Chapters 1 and 6).
b) Tighten the intake manifold and carburetor mounting nuts/bolts evenly and securely.
c) Perform a cylinder compression test (see Chapter 2).
d) Clean or replace the spark plugs as necessary (see Chapter 1).
e) Check the spark plug wires (see Chapter 1).
f) Inspect the ignition primary wires.
g) Check the ignition timing (follow the instructions printed on the Emissions Control Information label).

h) Check the fuel pump (see Section 4).
i) Check the heat control valve in the air cleaner for proper operation (see Chapter 1).
j) Check/replace the air filter element (see Chapter 1).
k) Check the PCV system (see Chapters 1 and 6).
l) Check/replace the fuel filter (see Chapter 1). Also, the strainer in the tank could be restricted.
m) Check for a plugged exhaust system.
n) Check EGR valve operation (see Chapter 6).
o) Check the choke - it should be completely open at normal engine operating temperature (see Chapter 1).
p) Check for fuel leaks and kinked or dented fuel lines (see Chapters 1 and 4)
q) Check accelerator pump operation with the engine off (remove the air cleaner cover and operate the throttle as you look into the carburetor throat - you should see a stream of gasoline enter the carburetor).
r) Check for incorrect fuel or bad gasoline.
s) Check the valve clearances (if applicable) and camshaft lobe height (see Chapters 1 and 2)
t) Have a dealer service department or repair shop check the electronic engine and carburetor controls.

4   Diagnosing carburetor problems may require that the engine be started and run with the air cleaner off. While running the engine without the air cleaner, backfires are possible. This situation is likely to occur if the carburetor is malfunctioning, but just the removal of the air cleaner can lean the fuel/air mixture enough to produce an engine backfire. **Warning:** *Don't position any part of your body, especially your face, directly over the carburetor during inspection and servicing procedures. Wear eye protection!*

9.5 Exploded view of the MR08 carburetor

1   Screw & lockwasher (2) - actuator
2   Actuator - idle up
3   Spring - aux. throttle return
4   E-clip - secondary diaphragm rod
5   Screw & lockwasher (2) - secondary diaphragm
      assembly
6   Secondary diaphragm assembly
7   Screw & lockwasher - bracket
8   Screw & lockwasher - bracket
9   Bracket - cable
10   Screw - pump lever
11   Lever - pump
12   Washer - pump lever screw
13   Lockwasher - pump lever screw
14   Screw & lockwasher - bracket
15   Screw & lockwasher - bracket
16   Bracket
17   Spring - switch return (long)
18   Spring - switch return (short)
19   Screw & lockwasher (2) - bracket micro switch
20   Bracket & micro switch assembly
21   Gasket - window level gauge
22   Window - level gauge
23   O-ring - window level gauge
24   E-clip - choke pull-off link
25   Screw & lockwasher - choke pull-off
26   Screw & lockwasher - choke pull-off
27   Choke pull-off assembly
28   Screw & lockwasher (2) - thermo element
29   Thermo element assembly
30   Screw & lockwasher (2) - bowl vent solenoid
31   Bowl vent solenoid assembly
32   O-ring - bowl vent solenoid (inner)
33   O-ring - bowl vent solenoid (outer)
34   Screw & lockwasher - fuel fitting bracket
35   Bracket - fuel fitting
36   Screw & lockwasher (2) - bowl cover
37   Bowl cover assembly
38   Pin - float
39   Float assembly
40   Needle & seat assembly
41   Filter screen - needle seat
42   Pump assembly
43   Boot - pump
44   Screw & lockwasher (3) - mixture solenoid
45   Solenoid assembly
46   O-ring - solenoid (upper)
47   O-ring - solenoid (lower)
48   Gasket - bowl cover
49   Spring - pump return
50   Ball - pump intake check
51   Weight (2) - pump disc ball
52   Spring - pump disc ball weight
53   Spring - pump discharge check
54   Jet - slow air bleed
55   Jet - slow
56   Jet - primary main air bleed
57   Jet - secondary main air bleed
58   Plug - secondary slow
59   Jet - secondary slow
60   Solenoid assembly - idle shut-off
61   Gasket - solenoid
62   Screw & lockwasher - throttle body
63   Screw & lockwasher (3) - throttle body
64   Bowl assembly
65   Retainer - plug

66   Plug - primary main jet
67   Gasket - plug
68   Jet - primary main
69   Plug - secondary main jet
70   Gasket - plug
71   Jet - secondary main
72   Venturi assembly - primary
73   Locking pin - primary venturi
74   O-ring - primary venturi seal

75   Venturi assembly - secondary
76   Locking pin - secondary venturi
77   O-ring - secondary venturi seal
78   Gasket - throttle body
79   Needle - idle adjusting
80   Spring - idle needle
81   Washer - idle needle
82   Seal - idle needle
83   Throttle body assembly

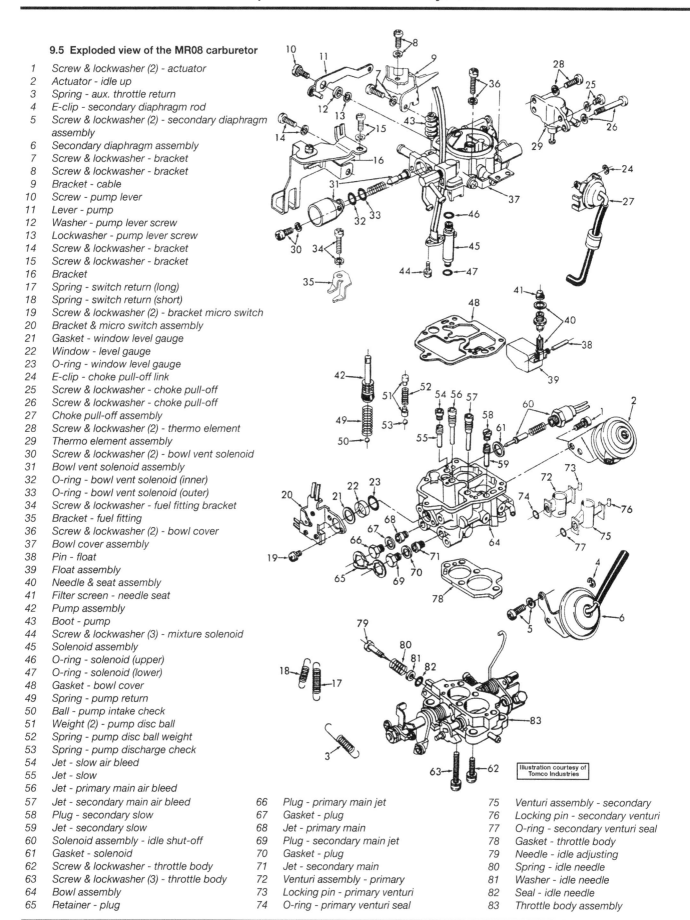

Illustration courtesy of
Tomco Industries

10.5 Clearly label all vacuum hoses before
disconnecting them

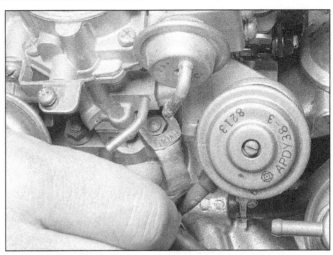

10.8 Because of the difficult angles and the lack of room at the
base of the carburetor, a curved, open end wrench is
the best tool for reaching the carburetor nuts.

Overhaul

Refer to illustration 9.5

5   Once it's determined that the carburetor needs an overhaul **(see illustration)**, several options are available. If you're going to attempt to overhaul the carburetor yourself, first obtain a good quality carburetor rebuild kit (which will include all necessary gaskets, internal parts, instructions and a parts list). You'll also need some special solvent and a means of blowing out the internal passages of the carburetor with air.

6   An alternative is to obtain a new or rebuilt carburetor. They're readily available from dealers and auto parts stores. Make absolutely sure the exchange carburetor is identical to the original. A tag is usually attached to the top of the carburetor or a number is stamped on the float bowl. It will help determine the exact type of carburetor you have. When obtaining a rebuilt carburetor or a rebuild kit, make sure the kit or carburetor matches your application exactly. Seemingly insignificant differences can make a large difference in engine performance.

7   If you choose to overhaul your own carburetor, allow enough time to disassemble it carefully, soak the necessary parts in the cleaning solvent (usually for at least one-half day or according to the instructions listed on the carburetor cleaner) and reassemble it, which will usually take much longer than disassembly. When disassembling the carburetor, match each part with the illustration in the carburetor kit and lay the parts out in order on a clean work surface. Overhauls by inexperienced mechanics can result in an engine which runs poorly or not at all. To avoid this, use care and patience when disassembling the carburetor so you can reassemble it correctly.

8   Because carburetor designs are constantly modified by the manufacturer in order to meet increasingly more stringent emissions regulations, the overhaul procedures in this Chapter may not apply exactly to your

vehicle. You'll receive a detailed, well illustrated set of instructions with any carburetor overhaul kit; they'll apply in a more specific manner to the carburetor on your vehicle.

---

10 Carburetor (Sprint models) - removal and installation

Warning: *Gasoline is extremely flammable, so take extra precautions when you work on any part of the fuel system. Don't smoke or allow open flames or bare light bulbs near the work area, and don't work in a garage where a natural gas-type appliance (such as a water heater or clothes dryer) with a pilot light is present. If you spill any fuel on your skin, rinse it off immediately with soap and water. When you perform any kind of work on the fuel system, wear safety glasses and have a Class B type fire extinguisher on hand.*

Removal

Refer to illustrations 10.5 and 10.8

1   Remove the fuel tank cap to relieve the tank pressure.

2   Remove the air cleaner from the carburetor. Be sure to label all vacuum hoses attached to the air cleaner housing.

3   Disconnect the accelerator cable from the throttle lever (see Section 8).

4   If the vehicle is equipped with an automatic transaxle, disconnect the TV cable from the throttle lever.

5   Clearly label all vacuum hoses and fittings, then disconnect the hoses **(see illustration)**.

6   Disconnect the fuel line from the carburetor.

7   Label the wires and terminals, then unplug all wire harness connectors.

8   Remove the mounting fasteners **(see illustration)**, remove the EGR modulator bracket (see Chapter 6) and lift the carburetor from the intake manifold. Remove the carbu-

retor mounting gasket. Stuff a shop rag into the intake manifold openings.

Installation

9   Use a gasket scraper to remove all traces of gasket material and sealant from the intake manifold, being careful not to damage the surface (and the carburetor, if it's being reinstalled). Clean the mating surfaces with lacquer thinner or acetone.

10   Place a new gasket on the intake manifold and the carburetor baseplate.

11   Position the carburetor on the gasket and install the mounting fasteners.

12   To prevent carburetor distortion or damage, tighten the fasteners, in a criss-cross pattern, 1/4 turn at a time, until the torque listed in this Chapter's Specifications is reached.

13   The remaining installation steps are the reverse of removal.

14   Check and, if necessary, adjust the idle speed (see Chapter 1).

15   If the vehicle is equipped with an automatic transaxle, refer to Chapter 7, Part B for the TV cable adjustment procedure.

16   Start the engine and check carefully for fuel leaks.

---

11 Carburetor (Sprint models) - adjustments

Warning: *Gasoline is extremely flammable, so take extra precautions when you work on any part of the fuel system. Don't smoke or allow open flames or bare light bulbs near the work area, and don't work in a garage where a natural gas-type appliance (such as a water heater or clothes dryer) with a pilot light is present. If you spill any fuel on your skin, rinse it off immediately with soap and water. When you perform any kind of work on the fuel system, wear safety glasses and have a Class B type fire extinguisher on hand.*

11.2 Using a feeler gauge (arrow), check the clearance of the choke plate to the carburetor bore

Note: If your vehicle's engine is hard to start or does not start at all, has an unstable idle or poor driveability and you suspect the carburetor is malfunctioning, it's best to first take the vehicle to a dealer service department that has the equipment necessary to diagnose this highly complicated system. The following procedures are intended to help the home mechanic verify proper operation of components, make minor adjustments, and replace some components. They are not intended as troubleshooting procedures.

Choke check and adjustment

Refer to illustrations 11.2 and 11.5

1   Remove the air cleaner assembly. With the engine stopped and cold, hold the throttle open and depress the choke plate with your finger - it should move smoothly. The choke should be almost fully closed if the ambient air temperature is below 77-degrees F (25-degrees C) and the engine is cold.

2   Check the clearance of the choke plate to the carburetor bore **(see illustration)** and compare your reading with those listed in this Chapter's Specifications section. If the clearances are too small or too large, lubricate the choke linkage and take another measurement. **Note:** *If the air temperature is very warm, use the alternate specification to check the carburetor bore clearance. If the air temperature is at around 85 degrees F, calculate a midpoint value using the two other specifications.*

3   Start the engine and allow it to warm to normal operating temperature. Depress and release the accelerator once. The choke should now be fully open. If it's not, and the plate-to-bore clearance is correct, there's a problem with the choke linkage or operating mechanism.

4   If the plate-to-bore clearance is not as specified, remove the carburetor (see Section 10) and adjust the choke lever as follows.

5   Remove the idle-up actuator from the carburetor, turn the fast idle cam counterclockwise and insert a pin into the cam and bracket to lock them into place **(see illustration)**.

11.5 With the idle-up actuator removed, rotate the fast idle cam to align the hole in the cam with the hole in the bracket - insert a punch through the two holes to hold the cam in this position - the choke lever can now be adjusted with a pair of pliers

6   Using a pair of pliers, bend the choke lever up or down until the choke-to-bore clearance is set to the specified amount. Bending the tab up causes the choke valve to close, while bending it down allows it to open a little more **(see illustration 11.5)**.

7   Reinstall the carburetor and check the choke again.

Accelerator pump

8   Remove the air intake case. With the engine Off, operate the throttle linkage through its full range of travel while looking down the throat of the carburetor (you may have to hold the choke plate open). A healthy stream of fuel should squirt out of the pump discharge nozzle. If fuel just dribbles out or there is no squirt at all, the accelerator pump is defective or the discharge passage is clogged. In either case, an overhaul of the carburetor is required.

Fast idle adjustment

Refer to illustrations 11.10 and 11.11

9   Allow the engine to cool for at least four hours. The ambient temperature should be below 77-degrees F. Drain the coolant (see Chapter 3).

10   Disconnect the hoses from the choke thermo-element holder and plug them **(see illustration)**. Be sure to use clamps to prevent leakage at the seal ends.

11   After the carburetor has cooled, check to see that the mark on the fast idle cam and the center of the cam follower are in alignment **(see illustration)**.

12   Install the radiator drain plug and refill the cooling system (see Chapter 3). Check the hoses for any leaks.

13   With the engine running at idle speed, force the fast idle cam to pivot around until the mark on the cam centers with the cam follower (refer to Step 11) and check the idle speed. It should be 2100 to 2700 rpm. **Note:** *If necessary, pivot the fast idle cam using a*

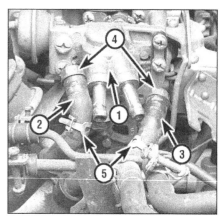

11.10 Plug the hoses on the thermo element holder

1   *Carburetor thermo element holder*
2   *Choke number 1 hose*
3   *Choke number 2 hose*
4   *Plugs*
5   *Clamps*

11.11 With the carburetor cool, make sure the mark on the fast idle cam is aligned with the center of the cam follower

1   *Fast idle cam*
2   *Mark on cam*
3   *Cam follower*
4   *Fast idle adjusting screw*

pair of pliers.

14   Adjust the fast idle by turning the fast idle screw **(see illustration 11.11)**.

15   Allow the engine to cool and install the cooling lines back onto the choke thermo-element holder. Installation is the reverse of removal.

Idle-up adjustment

Refer to illustration 11.17

Note: *The idle-up adjustment must be performed when the cooling fan is not running.*

16   Run the engine until it reaches normal operating temperature. Connect a tachometer and verify that the idle speed is as specified (see Chapter 1).

17   Turn the parking lamps On and check to see that the idle-up actuator rod moves down. Now turn the headlights On and check

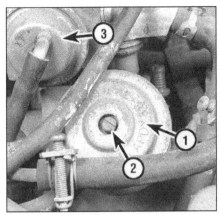

11.17 Idle-up adjusting screw

1 *Idle-up actuator*
2 *Idle-up adjusting screw*
3 *Choke piston*

the engine rpm (the heater fan, rear defogger and air conditioner must be turned Off). It should increase to 750 - 850 rpm. If it doesn't increase, turn the adjusting screw **(see illustration)**. **Note:** *On vehicles equipped with an automatic transaxle, the idle-up actuator also should function with the brake pedal depressed and the shift selector in Drive.*

12   Electronic Fuel Injection (EFI) system (Metro three-cylinder and 1995 through 1997 four-cylinder models) - general information

Refer to illustration 12.2

Electronic fuel injection provides optimum mixture ratios at all stages of combustion and offers immediate throttle response characteristics. It also enables the engine to run at the leanest possible air/fuel mixture ratio, reducing exhaust gas emissions.

A Throttle Body Injection (TBI) unit **(see illustration)** replaces the carburetor. It is controlled by the Electronic Control Module (ECM), which monitors engine performance and adjusts the air/fuel mixture accordingly (see Chapter 6 for a complete description of the fuel control system).

An electric fuel pump - located in the fuel tank with the fuel gauge sending unit - pumps fuel to the fuel injection system through the fuel feed line and an inline fuel filter. A fuel pressure regulator maintains a constant fuel pressure. Excess fuel is returned to the fuel tank through a fuel return line.

The TBI unit consists of the throttle body housing, a fuel injector, the fuel pressure regulator, a throttle opener, a Throttle Position (TP) sensor, an air valve and an idle speed control solenoid valve. The throttle opener controls the throttle valve opening so that it's a little wider when the engine is starting than when it's idling. The air valve allows extra air past the throttle plate during cold engine operation. The idle speed control solenoid valve controls the idle speed according to the

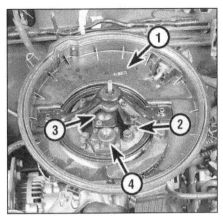

12.2 Throttle Body Injection (TBI) components

1 *Air cleaner*
2 *Throttle body*
3 *Fuel injector*
4 *Fuel pressure regulator*

ECM. For more information on servicing the throttle opener, the air valve and the idle speed control solenoid valve, refer to Chapter 6.

The fuel injector is a solenoid-operated device controlled by the ECM. The ECM turns on the solenoid, which lifts a (normally closed) needle valve off its seat. The fuel, which is under pressure, is injected in a conical spray pattern at the walls of the throttle body bore above the throttle valve. The pressure regulator returns any fuel not used by the injector to the fuel tank.

13   Electronic Fuel Injection (EFI) system (Metro three-cylinder and 1995 through 1997 four-cylinder models) - check

Warning: *Gasoline is extremely flammable, so take extra precautions when you work on any part of the fuel system. Don't smoke or allow open flames or bare light bulbs near the work area, and don't work in a garage where a natural gas-type appliance (such as a water heater or clothes dryer) with a pilot light is present. If you spill any fuel on your skin, rinse it off immediately with soap and water. When you perform any kind of work on the fuel tank, wear safety glasses and have a Class B type fire extinguisher on hand.*

1 Check the ground wire connections on the intake manifold for tightness. Check all electrical connectors that are related to the system. Loose connectors and poor grounds can cause many problems that resemble more serious malfunctions.
2 Check to see that the battery is fully charged, as the control unit and sensors depend on an accurate supply voltage in order to properly meter the fuel.
3 Check the air filter element - a dirty or partially blocked filter will severely impede performance and economy (see Chapter 1).

4 If a blown fuse is found, replace it and see if it blows again. If it does, search for a grounded wire in the harness to the fuel pump.
5 Check the condition of the vacuum hoses attached to the throttle body and intake manifold. Vacuum leaks can result in an excessively lean mixture.
6 Check the fuel system pressure (see Section 4).

TBI systems only

7 Set the parking brake, remove the air cleaner top plate and with the engine idling in Park, observe the operating fuel injector. The spray pattern should be even and conical in shape. The spray should touch the throttle body bore.

   a) *If the spray is weak or uneven, the injector is clogged or faulty. Gasoline additives designed to clean fuel injectors can sometimes clear a clogged injector. If not, a dealer service department has more effective cleaning equipment.*
   b) *If an injector is not operating at all, check its electrical connector. If the connection is good and the injector is receiving voltage, but the injector still doesn't work, the injector is faulty.*

8 The remainder of the system checks should be left to a dealer service department or other qualified repair shop, as there is a chance that the control unit may be damaged if they are not performed properly.

14   Throttle Body Injection (TBI) unit (Metro three-cylinder and 1995 through 1997 four-cylinder models) - removal and installation

Warning: *Gasoline is extremely flammable, so take extra precautions when you work on any part of the fuel system. Don't smoke or allow open flames or bare light bulbs near the work area, and don't work in a garage where a natural gas-type appliance (such as a water heater or clothes dryer) with a pilot light is present. If you spill any fuel on your skin, rinse it off immediately with soap and water. When you perform any kind of work on the fuel tank, wear safety glasses and have a Class B type fire extinguisher on hand.*

Removal

Refer to illustration 14.6a and 14.6b

1 Disconnect the cable from the negative terminal of the battery.
2 Relieve the fuel system pressure (see Section 2).
3 Following the procedure described in Section 8, disconnect the accelerator cable from the throttle lever at the TBI unit.
4 Disconnect the fuel feed and return lines from the TBI unit.
5 Label and disconnect any electrical connectors and vacuum hoses.

14.6a Remove the two long TBI unit attaching bolts (arrows)

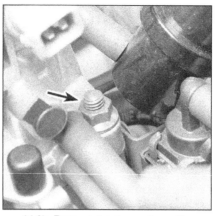

14.6b Remove the two nuts on the backside of the TBI unit (arrow) - the other mounting nut is not visible in this photo

15.2 The fuel injector should emit a strong, conical spray of fuel against the walls of the TBI unit bore - a pulsing timing light can make the spray pattern more visible

6   Remove the attaching bolts and lift the TBI unit from the intake manifold **(see illustrations)**.

Installation

7   Using a gasket scraper or a putty knife, remove all traces of old gasket material and sealant from the intake manifold (and throttle body, if the same one will be installed). While scraping, be careful not to gouge the soft aluminum surfaces.
8   Installation is the reverse of the removal procedure, but be sure to use a new base gasket, and tighten the bolts to the torque listed in this Chapter's Specifications.

15   Throttle Body Injection (TBI) unit (Metro three-cylinder and 1995 through 1997 four-cylinder models) - component check and replacement

Warning: *Gasoline is extremely flammable, so take extra precautions when you work on any part of the fuel system. Don't smoke or*

allow open flames or bare light bulbs near the work area, and don't work in a garage where a natural gas-type appliance (such as a water heater or clothes dryer) with a pilot light is present. If you spill any fuel on your skin, rinse it off immediately with soap and water. When you perform any kind of work on the fuel tank, wear safety glasses and have a Class B type fire extinguisher on hand.
Note: *All TBI components and parts (except IAC valve, pressure regulator diaphragm and any rubber pieces) should be cleaned in a cold immersion cleaner such as Carbon X or equivalent.*

Fuel injector

Refer to illustrations 15.2, 15.4, 15.9a and 15.9b
1   Unbolt the air cleaner assembly from the top of the throttle body and move it aside.
2   Start the engine and carefully peer down into the throttle body (wear safety goggles), checking the injector spray, using a timing light to illuminate the pattern. It should be an even, conical pattern **(see illustration)** - if it isn't, the injector must be replaced with a

new one.
3   Shut off the engine and make sure the injection of fuel stops as well. The injector should not leak more than one drop per minute - if it does, replace it.
4   Disconnect the electrical connector from the injector **(see illustration)**.
5   To replace the fuel injector, begin by relieving the fuel system pressure (see Section 2).
6   Disconnect the cable from the negative terminal of the battery.
7   Disconnect the fuel feed line from the throttle body. Remove the two injector cover screws and lift the cover from the injector.
8   Disconnect the injector electrical connector, release the harness clamp and dislodge the grommet from the throttle body housing.
9   To remove the injector from the throttle body, carefully direct compressed air into the fuel inlet port while pulling up on the injector **(see illustrations)**. **Caution:** *Apply the compressed air gradually, using only enough to ease the injector out of the throttle body. Do not exceed 85 psi, or damage to the injector and other components may occur. Also, once*

15.4 Disconnecting the injector electrical connector

1   Injector
2   Electrical connector

15.9a Apply compressed air to the inlet of the TBI unit while simultaneously holding a rag over the injector to catch any fuel that might blow up past the seals at the moment the injector releases from the bore

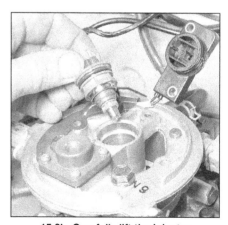

15.9b Carefully lift the injector from the TBI unit

15.17 Use a Torx driver to remove the bolts that secure the fuel pressure regulator to the body of the TBI unit

15.18 Carefully inspect the rubber diaphragm for rips or defects

the injector has been removed, handle it carefully and don't immerse it in solvent to clean it.

10   Check the fuel filters on the injector for dirt particles. If there is any residue, clean the filters and check the fuel tank and lines for contamination.

11   Before installing the injector, lubricate the O-rings with light oil (if you are reinstalling the same injector, use new O-rings). Push the injector firmly into its bore, making sure the wiring harness is pointing toward its slot in the throttle body housing. Push the grommet on the wiring harness into the slot.

12   Install the injector cover and tighten the screws securely.

13   Hook up the cable to the negative battery terminal. Connect the injector electrical connector and pressurize the fuel system by turning the ignition key to the On position. Check the fuel feed line and the injector for leakage.

14   Install the air cleaner assembly.

Fuel pressure regulator

Refer to illustration 15.17 and 15.18

Note: *Checking the fuel pressure requires the use of some special adapters not normally available to the home mechanic. If the necessary tools are not available (see Section 4), the fuel pressure should be checked by a dealer service department or other service facility with the necessary hardware. If the regulator has been determined to be faulty, replace it using the following procedure.*

15   Disconnect the cable from the negative terminal of the battery.

16   Relieve the fuel system pressure (see Section 2).

17   Remove the screws that secure the regulator to the throttle body using a Torx drive (T-20) **(see illustration)**, then pull the regulator straight up.

18   Check the diaphragm **(see illustration)** for cuts or defects and also make sure the spring is intact and not twisted or damaged.

19   Install the new fuel pressure regulator straight into the throttle body. Tighten the screws securely.

20   Install the fuel return line, tightening the hose clamp securely.

21   Connect the cable to the negative battery terminal, pressurize the fuel system by turning the ignition switch to the On position and check around the regulator for fuel leaks.

---

16   Multiport Sequential Fuel Injection (SFI) system (1998 and later Metro four-cylinder models) - general information

---

Refer to illustration 16.1

The multiport Sequential Fuel Injection (SFI) system **(see illustration)** consists of the air intake system, the fuel delivery system, the engine management system and the emission control systems. The engine management system uses a computer, known as the Powertrain Control Module (PCM), and an array of information sensors (coolant temperature sensor, throttle position sensor, mass airflow sensor, oxygen sensor, etc.) to determine the correct air/fuel ratio under all operating conditions. The emission control systems monitor and mitigate emission byprod-

ucts - hydrocarbons, carbon monoxide, oxides of nitrogen, etc. - produced by combustion.

The fuel injection system, the engine control system and the emission control systems are highly integrated For information on the engine management system and the emission control systems, refer to Chapter 6.

Air intake system

The air intake system consists of the air cleaner housing, the air filter element, the air intake duct, the throttle body, the air intake plenum and the intake manifold runners (the plenum and the manifold runners are a one-piece design).

Engine management system and emission control systems

The engine management and emission control systems are described in Chapter 6.

Fuel delivery system

The fuel delivery system consists of the fuel tank, the in-tank fuel pump, the fuel pressure regulator, the fuel rail, the fuel injectors and the hoses and lines connecting all of these components.

The fuel pump is an electric unit located inside the fuel tank. Fuel is drawn through a "sock" (inlet screen) into the pump. From the pump, fuel is pumped under pressure through the fuel filter and then into the fuel rail before being discharged into the intake ports by the fuel injectors. The fuel pressure regulator maintains a constant fuel pressure to the injectors. The pressure regulator is mounted on the fuel rail. Excess fuel is released through the fuel pressure regulator and then through a return hose back to the fuel tank.

The injectors are solenoid-actuated, pintle-type units consisting of a solenoid, plunger, needle valve and housing. When current is applied to the solenoid coil, the needle valve raises and pressurized fuel sprays out the nozzle. The quantity of fuel injected by each injector each time it opens is

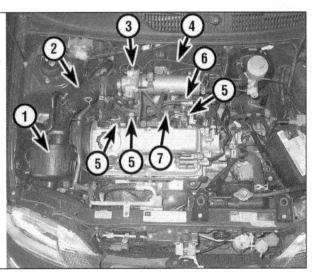

16.1 Typical multiport Sequential Fuel Injection (SFI) system components

1   Air cleaner assembly
2   Air intake duct
3   Throttle body
4   Air intake plenum
5   Fuel injectors (injector for No. 2 cylinder, under air intake plenum support bracket, not visible in this photo)
6   Fuel pressure regulator
7   Fuel rail

determined by the length of time the valve is open (the length of time during which current is supplied to the solenoid coils).

The fuel pump control relay is located on the relay panel in the engine compartment. The fuel pump relay controls the circuit that carries battery voltage to the fuel pump. The PCM controls the fuel pump relay. If the PCM senses that there is no signal from the engine speed sensor, i.e. the engine is not running or cranking, the PCM de-energizes the fuel pump relay.

17   Multiport Sequential Fuel Injection (SFI) system (1998 and later Metro four-cylinder models) - check

Refer to illustrations 17.7 and 17.8
Note: *The following procedure is based on the assumption that the fuel pressure is adequate (see Section 4).*
1   Inspect all electrical connectors related to the SFI system. Check the ground wire connections for tightness (see the wiring diagrams at the end of Chapter 12). Loose connectors or poor grounds can cause problems that resemble more serious malfunctions.
2   Verify that the battery is fully charged, as the control unit and sensors depend on an accurate supply voltage in order to properly meter the fuel.
3   Inspect the air filter element (see Chapter 1). A dirty or partially blocked filter element will severely impede performance and economy.
4   Check all fuses related to the SFI system. If you find a blown fuse, replace it and see if it blows again. If it does, a wire in the harness served by that particular fuse has shorted to ground; refer to the wiring diagrams and then find the short.
5   Inspect the air intake duct to the intake manifold for air leaks, which will result in an excessively lean mixture. Also check the con-

17.7 With the engine running, place an automotive stethoscope against each injector, one at a time, and listen for a clicking sound, indicating operation

dition of all vacuum hoses connected to the intake manifold and/or throttle body.
6   Remove the air intake duct from the throttle body and check for dirt, carbon or other residue build-up. If it's dirty, clean it with carburetor cleaner spray, a toothbrush and a shop towel.
7   With the engine running, place an automotive stethoscope against each injector, one at a time, and listen for a clicking sound, indicating operation **(see illustration)**. If you don't have a stethoscope, place the tip of a screwdriver against the injector and listen through the handle. If you hear the injectors operating, the electrical circuits are functioning, but the injectors may be dirty or fouled from carbon deposits - commercial cleaning products may help or they may require replacement. If one or more injectors are not operating, proceed with the injector check.
8   Turn the ignition key to OFF. Disconnect the injector electrical connectors and measure the resistance of each injector **(see illustration)**. Compare your measurements with the injector resistance listed in this

17.8 Measure the resistance of each injector with an ohmmeter on these terminals (arrows)

Chapter's Specifications. Replace any injector with a resistance value that does not fall within specifications.
9   Testing of the fuel injector *circuits* should be left to a dealer service department.

18   Throttle body (1998 and later Metro four-cylinder models) - removal and installation

Removal

Refer to illustrations 18.2, 18.5, 18.6 and 18.7
1   Drain the engine coolant (see Chapter 1).
2   Remove the breather hose **(see illustration)**.
3   Remove the bolt that secures the air intake duct to the vehicle body **(see illustration 18.2)**, loosen the air intake duct hose clamps and then remove the duct and resonator (it's not necessary to disconnect the resonator from the air intake duct).
4   Disconnect the accelerator cable from the throttle lever arm (see Section 8). On

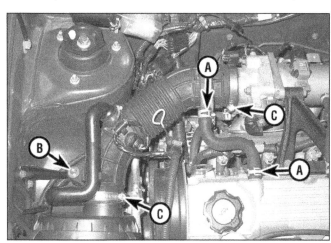

18.2 Before removing the throttle body, loosen the hose clamps (A) and then remove the breather hose; then remove the air intake duct bolt (B), loosen the intake duct hose clamps (C) and then remove the air intake duct

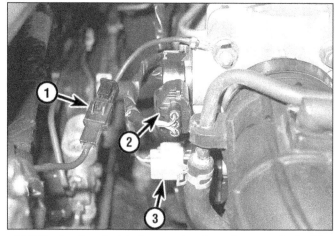

18.5 Unplug the electrical connectors from the ground wire (1), the Throttle Position (TP) sensor (2) and the Idle Air Control (IAC) valve (3)

18.6 Mark the two coolant hoses (arrows) connected to the IAC valve on the underside of the throttle body and then disconnect both hoses from the IAC valve (if the hoses are hard to remove, pull them off after you've removed the throttle body from the air intake plenum)

18.7 To detach the throttle body from the air intake plenum, remove these two nuts (upper arrows) and these two bolts (lower arrows)

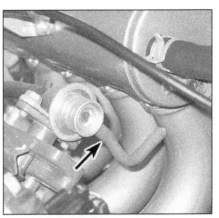

19.2 Disconnect the vacuum hose (arrow) from the port on the fuel pressure regulator

19.3 Loosen the fuel return hose clamp (left arrow) and then disconnect the fuel return hose from the fuel pressure regulator; to detach the fuel pressure regulator from the fuel rail, remove these two bolts (right arrows)

models with cruise control, also disconnect the cruise control cable (the cruise control cable is attached to the throttle linkage and to the cable bracket exactly the same way as the accelerator cable).

5   Disconnect the electrical connectors from the Idle Air Control (IAC) valve and from the Throttle Position (TP) sensor **(see illustration)**.

6   To ensure correct reassembly, clearly mark the two coolant hoses **(see illustration)** connected to the IAC valve on the underside of the throttle body and then loosen the hose clamps and disconnect the hoses from the IAC valve. If the coolant hoses are difficult to remove at this stage, wait until you have removed the throttle body from the air intake plenum to pull them off.

7   Remove the two throttle body retaining nuts and the two retaining bolts **(see illustration)**.

8   Remove the throttle body from the air intake plenum, turn it upside down and then, if you have not already done so, disconnect the coolant hoses from the IAC valve. Remove the throttle body.

9   Remove the throttle body gasket and discard it (you must use a new gasket when reinstalling the throttle body).

10   If the throttle body is going to be washed in solvent, *remove* the IAC valve and the TP sensor (see Chapter 6).

Installation

11   Install the IAC valve and the TP sensor, if you removed them (see Chapter 6).

12   Reconnect the coolant hoses to their respective pipes on the IAC valve. Make sure that the hoses are in good condition and that the hose clamps are tight.

13   Ensure that the gasket mating surfaces of the throttle body and the air intake plenum are clean and dry, place the new gasket in position, install the throttle body and then tighten the retaining nuts and bolts to the

torque listed in this Chapter's Specifications.

14   Reconnect the IAC valve and TP sensor electrical connectors.

15   Reconnect the accelerator cable to the throttle lever arm (see Section 8). On models with cruise control, reconnect the cruise control cable.

16   Reconnect the air intake duct to the air cleaner assembly and to the throttle body and then tighten the hose clamp screws securely.

17   Reattach the air intake duct to the vehicle body with the retaining bolt and tighten the bolt securely.

18   Reconnect the breather hose to the valve cover and to the air intake duct. Make sure that the grommet in the duct is in good condition. If it's cracked or torn or deteriorated, replace it.

19   Check the coolant level and refill as necessary (see Chapter 1).

19   Fuel pressure regulator (1998 and later Metro four-cylinder models) - removal and installation

Refer to illustrations 19.2 and 19.3

Warning: *Gasoline is extremely flammable, so take extra precautions when working on any part of the fuel system. Do not smoke or allow open flames or bare light bulbs in or near the work area. Also, don't work in a garage if a natural gas appliance with a pilot light is present. While performing any work on the fuel tank it is advisable to wear safety glasses and to have a dry chemical (Class B) fire extinguisher on hand. If you spill any fuel on your skin, rinse it off immediately with soap and water.*

1   Relieve the system fuel pressure (see Section 2).

2   Disconnect the vacuum hose **(see illustration)** from the port on the fuel pressure regulator.

3   Disconnect the fuel return hose **(see illustration)** from the fuel pressure regulator.

4   Remove the fuel pressure regulator

retaining bolts **(see illustration 19.3)** and then remove the fuel pressure regulator from the fuel rail. Discard the old O-ring.

5   Installation is the reverse of removal. Be sure to use a new O-ring. Tighten the pressure regulator bolts to the torque listed in this Chapter's Specifications.

6   Start the engine and check for leaks.

20   Fuel rail and injectors (1998 and later Metro four-cylinder models) - removal and installation

Warning: *Gasoline is extremely flammable, so take extra precautions when working on any part of the fuel system. Do not smoke or allow open flames or bare light bulbs in or near the work area. Also, don't work in a garage if a natural gas appliance with a pilot light is present. While performing any work on the fuel tank it is advisable to wear safety glasses and to have a dry chemical (Class B) fire extinguisher on hand. If you spill any fuel on your skin, rinse it off immediately with soap and water.*

Removal

Refer to illustrations 20.3, 20.4, 20.5, 20.7, 20.8 and 20.9

1   Relieve the system fuel pressure (see Section 2).

2   Disconnect the negative battery cable.

3   Remove the three intake manifold bracket bolts **(see illustration)** and then remove the bracket.

4   Disconnect the breather hose and the PCV valve and hose from the air intake plenum and from the valve cover **(see illustration)**.

5   Disconnect the electrical connectors from the four fuel injectors **(see illustration)**.

6   Disconnect the vacuum hose from the fuel pressure regulator **(see illustration 19.2)**.

7   Disconnect the fuel feed hose and the fuel return hose from the fuel rail **(see illustration)**.

8   Clean any debris from around the injec-

tors. Remove the fuel rail mounting bolts **(see illustration)**. Gently rock the fuel rail and injectors to loosen the injectors. Remove the fuel rail and fuel injectors as an assembly. Remove the fuel rail insulators (the black plastic spacers between the fuel rail and the intake manifold). Inspect the insulators for cracks. If the insulators are cracked or otherwise damaged, replace them.

9   Remove the injector(s) from the fuel rail assembly and then remove and discard the O-rings and seals **(see illustration)**. **Note:** *Whether you're replacing an injector or a leaking O-ring, it's a good idea to remove all the injectors from the fuel rail and replace all the O-rings.*

Installation

10   Coat the new O-rings with clean engine oil and install them on the injectors, and then insert each injector into its corresponding bore in the fuel rail.

20.3 To detach the intake manifold bracket, remove these three bolts (arrows)

20.4 Loosen the hose clamps (arrows) and then remove the breather hose (left hose) and the PCV valve and hose (right hose)

20.5 Disconnect the electrical connectors (arrows) from the four fuel injectors

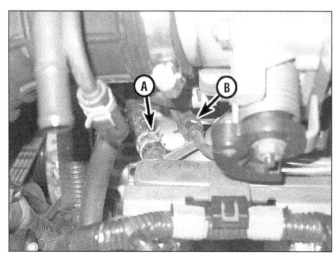

20.7 Disconnect the fuel feed hose (A) and the fuel return hose (B) from the fuel rail

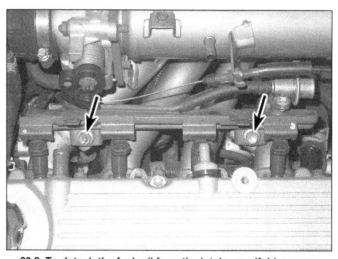

20.8 To detach the fuel rail from the intake manifold, remove these two bolts (arrows). After removing the fuel rail, remove the fuel rail insulators (the black plastic spacers between the fuel rail and the intake manifold) and then inspect them for cracks; if the insulators are damaged, replace them

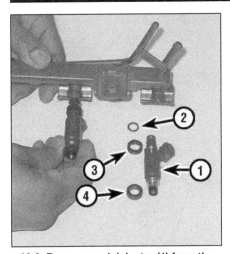

20.9 Remove each injector (1) from the fuel rail assembly and then remove the O-ring (2), the grommet (3) and the cushion (4) from the injector; inspect these parts and make sure that they're neither cracked nor deteriorated

11   Install the fuel rail insulators and then install the injector and fuel rail assembly. Fully seat the injectors, then tighten the fuel rail mounting bolts to the torque listed in this Chapter's Specifications.

12   Connect the fuel feed and return hoses to the fuel rail. Make sure that the hose clamps are very tight. If they're not, replace them.

13   Connect the vacuum line to the fuel pressure regulator.

14   Connect the electrical connectors to each injector, referring to the numbered tags.

15   Install the intake manifold bracket and tighten the three bracket bolts securely.

16   Reconnect the breather hose and the PCV valve and hose between the air intake plenum and the valve cover.

17   Connect the negative battery cable.

18   Turn the ignition switch to ON, but don't start the engine. When the ignition key is turned to ON, it activates the fuel pump for about two seconds, which builds up fuel pressure in the fuel lines and the fuel rail. Repeat this step two or three times and then check the fuel pressure line, fuel return hose,

the fuel rail and the fuel injectors for fuel leakage.

19   Start the engine and then check for leaks.

21 Exhaust system servicing - general information

Refer to illustrations 21.1a, 21.1b, 21.1c, 21.1d, 21.1e and 21.1f

Warning: *Inspection and repair of exhaust system components should be done only after enough time has elapsed after driving the vehicle to allow the system components to cool completely. Also, when working under the vehicle, make sure it is securely supported on jackstands.*

1   The exhaust system consists of the exhaust manifold, catalytic converter, the muffler, the tailpipe and all connecting pipes, brackets, hangers and clamps (**see illustrations**). The exhaust system is attached to the body with mounting brackets and rubber hangers. If any of these parts are damaged or

21.1a The exhaust pipe is connected to the exhaust manifold with two bolts or nuts - there is a gasket under the flange which should be replaced whenever the pipe is unbolted from the manifold

21.1b Typical exhaust manifold and shield

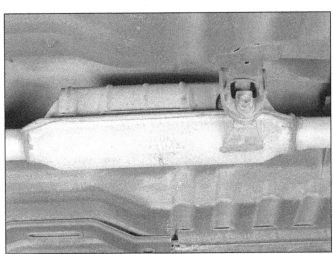

21.1c Typical catalytic converter and center pipe

21.1d Typical muffler

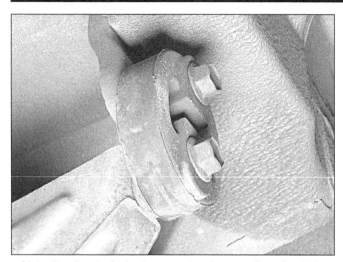

21.1e Here's a typical exhaust system hanger - they should be inspected for cracks and replaced if deteriorated

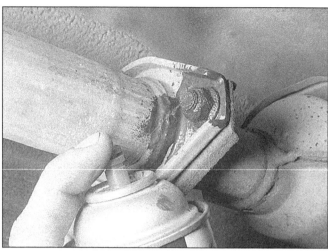

21.1f Exhaust system bolts and nuts, particularly those on the exhaust manifold and catalytic converters, can be very difficult to loosen - spraying them with a penetrant will free up the threads

deteriorated, excessive noise and vibration will be transmitted to the body.

2   Conducting regular inspections of the exhaust system will keep it safe and quiet. Look for any damaged or bent parts, open seams, holes, loose connections, excessive corrosion or other defects which could allow exhaust fumes to enter the vehicle. Deteriorated exhaust system components should not be repaired - they should be replaced with new parts.

3   If the exhaust system components are extremely corroded or rusted together, they will probably have to be cut from the exhaust system. The convenient way to accomplish this is to have a muffler repair shop remove the corroded sections with a cutting torch. If, however, you want to save money by doing it yourself (and you don't have an oxy/acety-lene welding outfit with a cutting torch), simply cut off the old components with a hacksaw. If you have compressed air, special pneumatic cutting chisels can also be used. If you do decide to tackle the job at home, be sure to wear eye protection to protect your eyes from metal chips and work gloves to protect your hands.

4   Here are some simple guidelines to apply when repairing the exhaust system:
  a) *Work from the back to the front when removing exhaust system components.*
  b) *Apply penetrating oil to the exhaust system component fasteners to make them easier to remove.*
  c) *Use new gaskets, hangers and clamps when installing exhaust system components.*
  d) *Apply anti-seize compound to the threads of all exhaust system fasteners during reassembly.*
  e) *Be sure to allow sufficient clearance between newly installed parts and all points on the underbody to avoid overheating the floor pan and possibly damaging the interior carpet and insulation. Pay particularly close attention to the catalytic converter and its heat shield.* **Warning:** *The catalytic converter operates at very high temperatures! Wait until it cools before attempting to remove the converter. Failure to do so could result in serious burns.*

Chapter 5
Engine electrical systems

Contents

Specifications

Ignition coil

Chevrolet Sprint
    Primary resistance .... 1.06 to 1.43 ohms
    Secondary resistance .... 10.8 to 16.2 k-ohms
Geo Metro
  1992 and earlier
    Primary resistance .... 1.33 to 1.55 ohms
    Secondary resistance .... 10.7 to 14.5 K-ohms
  1993
    Hardtop models
      Primary resistance .... 1.35 to 1.65 ohms
      Secondary resistance .... 22.1 to 29.9 K-ohms
    Convertible models
      Primary resistance .... 1.33 to 1.55 ohms
      Secondary resistance .... 10.7 to 14.5 K-ohms
  1994
    Standard emission
      Primary resistance .... 1.35 to 1.65 ohms
      Secondary resistance .... 22.1 to 29.9 K-ohms
    Upgraded emissions
      Primary resistance .... 1.08 to 1.32 ohms
      Secondary resistance .... 22.1 to 29.9 K-ohms
  1995 through 1998
    Primary resistance .... 1.35 to 1.65 ohms
    Secondary resistance .... 22.1 to 29.9 K-ohms
  1999 on
    Three-cylinder
      Primary resistance .... 1.35 to 1.65 ohms
      Secondary resistance .... 22.1 to 29.9 K-ohms
    Four-cylinder
      Primary resistance .... N/A
      Secondary resistance .... 8.5 to 11.5 K-ohms

Distributor

Pick-up coil resistance (vacuum/centrifugal advance distributor)...........   130 to 190 ohms
Pick-up coil/camshaft Position (CMP) sensor resistance
    (electronic spark control distributor) ....   140 to 180 ohms
Signal rotor air gap
    1996 and earlier...   0.008 to 0.016 in (0.2 to 0.4 mm)
    1997 on ....   0.008 in (0.2 mm)

Ignition timing

Chevrolet Sprint
    Manual transaxle ....   10-degrees BTDC at 700 to 800 rpm
    Automatic transaxle ....   6-degrees BTDC at 800 to 900 rpm
Geo Metro...   Refer to VECI label under hood

---

1   General information

The engine electrical systems include all ignition, charging and starting components. Because of their engine-related functions, these components are discussed separately from chassis electrical devices such as the lights, the instruments, etc. (which are included in Chapter 12).

Always observe the following precautions when working on the electrical systems:

a) *Be extremely careful when servicing engine electrical components. They are easily damaged if checked, connected or handled improperly.*

b) *Never leave the ignition switch on for long periods of time with the engine off.*

c) *Don't disconnect the battery cables while the engine is running.*

d) *Maintain correct polarity when connecting a battery cable from another vehicle during jump starting.*

e) *Always disconnect the negative cable first and hook it up last or the battery may be shorted by the tool being used to loosen the cable clamps.*

It's also a good idea to review the safety-related information regarding the engine electrical systems located in the *Safety first!* Section near the front of this manual before beginning any operation included in this Chapter.

2   Battery - removal and installation

Refer to illustrations 2.1a and 2.1b

1   **Caution:** *Always disconnect the negative cable first and hook it up last or the battery may be shorted by the tool being used to loosen the cable clamps.* Disconnect both cables from the battery terminals **(see illustrations)**.

2   Remove the battery hold-down clamp or strap.

3   Lift out the battery. Be careful - it's heavy.

4   While the battery is out, inspect the carrier (tray) for corrosion (see Chapter 1).

5   If you are replacing the battery, make sure that you get one that's identical, with the same dimensions, amperage rating, cold cranking rating, etc.

6   Installation is the reverse of removal.

3   Battery - emergency jump starting

Refer to the Booster battery (jump) starting procedure at the front of this manual.

4   Battery cables - check and replacement

Refer to illustrations 4.2a, 4.2b and 4.2c

1   Periodically inspect the entire length of each battery cable for damage, cracked or burned insulation and corrosion. Poor battery cable connections can cause starting problems and decreased engine performance.

2   Check the cable-to-terminal connections at the ends of the cables **(see illustrations)** for cracks, loose wire strands and corrosion. The presence of white, fluffy deposits under the insulation at the cable terminal connection is a sign that the cable is corroded and should be replaced. Check the terminals for distortion, missing mounting bolts and corrosion.

3   When removing the cables, always dis-

2.1a To remove the battery, disconnect both cables from the battery terminals (large arrows) - negative cable first, then the positive cable; then, on earlier models, remove the nuts from the battery hold-down plate (small arrows) and then remove the plate ...

2.1b ... or, on later models, remove the hold-down bolt (arrow) and clamp

4.2a The big red battery cable that's connected to the battery positive terminal at its upper end is connected to a terminal on the starter solenoid (arrow) at its lower end

connect the negative cable first and hook it up last or the battery may be shorted by the tool used to loosen the cable clamps. Even if only the positive cable is being replaced, be sure to disconnect the negative cable from the battery first (see Chapter 1 for further information regarding battery cable removal).

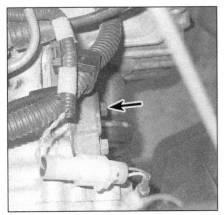

4.2b The larger battery negative cable is connected to the engine or transaxle

4.2c The smaller negative battery cable is connected to the body by a bolt (arrow)

4   Disconnect the old cables from the battery, then trace each of them to their opposite ends and detach them from the starter solenoid and ground terminals. Note the routing of each cable to ensure correct installation.

5   If you are replacing either or both of the old cables, take them with you when buying new cables. It is vitally important that you replace the cables with identical parts. Cables have characteristics that make them easy to identify: positive cables are usually red, larger in cross-section and have a larger diameter battery post clamp; ground cables are usually black, smaller in cross-section and have a slightly smaller diameter clamp for the negative post.

6   Clean the threads of the solenoid or ground connection with a wire brush to remove rust and corrosion. Apply a light coat of battery terminal corrosion inhibitor, or petroleum jelly, to the threads to prevent future corrosion.

7   Attach the cable to the solenoid or ground connection and tighten the mounting nut/bolt securely.

8   Before connecting a new cable to the battery, make sure that it reaches the battery post without having to be stretched.

9   Connect the positive cable first, followed by the negative cable.

---

5   Ignition system - general information and precautions

The ignition system consists of the following components:

> *Battery*
> *Ignition switch*
> *Ignition module (Chevrolet Sprint) or igniter(s) (Geo Metro)*
> *Ignition coil (Chevrolet Sprints, three-cylinder Geo Metros and 1995 through 1997 four-cylinder Geo Metros); two ignition coils (1998 and later Geo Metros)*
> *Primary (low voltage) and secondary (high voltage) wiring circuits*
> *Distributor (Chevrolet Sprints, three-cylinder Geo Metros and 1995 through 1997 four-cylinder Geo Metros)*
> *Spark plugs*

Chevrolet Sprints, all three-cylinder Geo Metro models and 1995 through 1997 Metro four-cylinder models are equipped with distributors. Sprint and 1989 through 1992 Geo Metro distributors are equipped with a pick-up coil. On these models, the pick-up coil is the timing device for the ignition module, which controls spark timing by turning the ignition coil on and off. All 1993 and later Metro models use a Camshaft Position (CMP) sensor instead of a pick-up coil. On these models, the CMP sensor is the timing device for the ECM (or, on 1996 and later models, the PCM), which controls spark timing by turning the igniter on and off.

There is no distributor in the ignition system used on 1998 and later Metro four-cylinder models. The CMP sensor is relocated into a housing that is bolted to the cylinder head at the same location as the distributor on earlier models. The ignition system is controlled by the Powertrain Control Module (PCM). Using data provided by the Camshaft Position (CMP) sensor and by the Crankshaft Position (CKP) sensor, the PCM ensures a perfectly timed spark under all conditions. For more information on the CMP and CKP sensors, refer to Chapter 6.

On Chevrolet Sprints and on 1989 through 1991 Geo Metro LSi models, the spark advance is controlled by vacuum and centrifugal advance units. On 1989 through 1991 Geo Metro Base and XFi models, and on all 1992 and later Geo Metro models, spark advance is controlled by the ECM or PCM.

Chevrolet Sprints, all three-cylinder Geo Metro models and 1995 through 1997 four-cylinder Geo Metro models are equipped with a conventional ignition coil, which is mounted on the firewall. On 1998 and later four-cylinder Geo Metro models, which use distributorless ignition systems, there are *two* ignition coils. Each coil fires two spark plugs. On 1998 models, the coils are bolted to brackets at the left (driver's) end of the cylinder head. The No. 1 coil (the coil that's closer to the front of the vehicle) fires the spark plugs in cylinders 1 and 4; the No. 2 coil (the coil closer to the firewall) fires the plugs in cylinders 2 and 3. On 1999 and later four-cylinder Metro models, the two ignition coils are mounted on top of the valve cover. The No. 1 coil (the coil that's nearer the left end of the cylinder head) fires the spark plugs in cylinders 1 and 4; the No. 2 coil (the coil that's mounted nearer the right side of the vehicle) fires the plugs in cylinders 2 and 3. The No. 1 coil is actually mounted right on top of the No. 4 spark plug, i.e. there is no spark plug wire - the coil fires that plug directly. This coil fires the No. 1 spark plug through a spark plug wire. The No. 2 coil fires the No. 2 spark plug directly, and fires the No. 3 spark plug via a plug wire.

When working on the ignition system, take the following precautions:

a) *Don't keep the ignition switch on for more than 10 seconds if the engine won't start.*

b) *Always connect a tachometer in accordance with the manufacturer's instructions. Some tachometers may be incompatible with this ignition system. Consult a dealer service department before buying a tachometer for use with this vehicle.*

c) *Never allow the ignition coil terminals to touch ground. Grounding the coil could result in damage to the module/igniter and/or the ignition coil.*

d) *Don't disconnect the battery when the engine is running.*

e) *Make sure that the module/igniter is properly grounded.*

6   Ignition system - check

Refer to illustration 6.2

Warning: *Because of the very high voltage generated by the ignition system, extreme care should be taken when this check is performed.*

1   If the engine turns over but won't start, disconnect the spark plug wire from any spark plug and attach it to a calibrated ignition tester (available at most auto parts stores).

2   Connect the clip of a spark tester to a bolt or metal bracket on the engine **(see illustration)**. If you're unable to obtain a calibrated ignition tester, remove the wire from one of the spark plugs and, using an insulated tool, hold the end of the wire about 1/4-inch from a good ground.

3   Crank the engine and watch the end of the tester or spark plug wire to see if bright blue, well-defined sparks occur. If you're not using a calibrated tester, have an assistant crank the engine for you. **Warning:** *Keep clear of drivebelts and other moving engine components that could injure you.*

4   If sparks occur, sufficient voltage is reaching the plug to fire it (repeat the check at the remaining plug wires to verify the wires, distributor cap and rotor are OK). However, the plugs themselves may be fouled, so remove them and check them as described in Chapter 1.

5   If no sparks or intermittent sparks occur, remove the distributor cap and check the cap and rotor as described in Chapter 1. If moisture is present, dry out the cap and rotor, then reinstall the cap.

6   If there's still no spark, detach the coil secondary wire from the distributor cap and hook it up to the tester (reattach the plug wire to the spark plug), then repeat the spark check. Again, if you don't have a tester, hold the end of the wire about 1/4-inch from a good ground. If sparks occur now, the distributor cap, rotor or plug wire(s) may be defective.

7   On all models, if no sparks occur, check the wire connections at the coil to make sure they're clean and tight. Check for voltage to the coil. Make any necessary repairs, then repeat the check again.

8   If there's still no spark, the coil-to-cap wire may be bad (check the resistance with an ohmmeter - it should be 7000 ohms per foot or less). If a known good wire doesn't make any difference in the test results, the ignition module may be defective.

7   Distributor air gap - check and adjustment

Refer to illustrations 7.2 and 7.3

1   Remove the distributor cap and rotor (see Chapter 1).

2   On a vacuum/centrifugal advance distributor, use a feeler gauge to measure the air

6.2 An inexpensive spark tester like this unit is the safest and most convenient way to find out if you've got spark - simply unplug each wire (one at a time to prevent mixing up the leads), plug in the tester, clip it to a good ground, like a valve cover bolt, crank the engine and see if a spark jumps across the gap

gap between a tooth of the pole piece and the pick-up coil **(see illustration)**. On an electronic spark control distributor, measure the gap between a ridge of the signal rotor and the pick-up coil/CMP sensor assembly. Compare your measurement to the air gap listed in this Chapter's Specifications. If the air gap is incorrect, adjust it.

3   To adjust the air gap on a vacuum/centrifugal advance distributor, first remove the ignition control module/igniter (see Section 11) and then loosen the two screws which attach the pick-up coil **(see illustration)**. There is no module or igniter inside the electronic spark control distributor - simply loosen the generator assembly retaining screws **(see illustration 7.2)** to adjust the gap.

4   Using a screwdriver, move the pick-up coil/CMP sensor until the gap is within specification, then retighten the pick-up coil/CMP sensor screws and recheck the gap. Install the module/igniter (vacuum/centrifugal advance distributor), the rotor and the distributor cap.

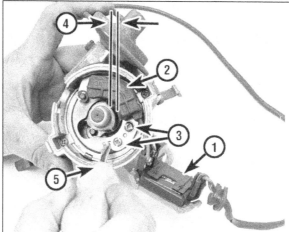

7.2 To measure the air gap, insert a feeler gauge between a tooth of the pole piece (lower arrow) and the pick-up coil (upper arrow) (vacuum/centrifugal advance distributor shown, electronic spark control distributor similar)

8   Ignition timing - check and adjustment

Vacuum/centrifugal advance distributor

Refer to illustration 8.6

1   Make sure the headlights, heater fan, engine cooling fan, rear defogger (if equipped) and air conditioner (if equipped) are off. If one of them is on, the idle up system operates and the engine idle speed will be higher than specified.

2   Start the engine and warm it up to normal operating temperature.

3   Check and, if necessary, adjust the idle speed (see Chapter 1).

4   If the vehicle is a Geo Metro LSi model, detach the vacuum hose at the intake manifold gas filter (it's screwed into the manifold near the timing belt cover). Plug the filter.

5   Hook up a timing light to the number one spark plug wire in accordance with the manufacturer's instructions.

7.3 To adjust the air gap, remove the module, loosen the pick-up coil retaining screws, move the pick-up coil assembly until the gap is correct, then retighten the screws (vacuum/centrifugal advance distributor shown, electronic spark control distributor similar)

1   Module
2   Pick-up coil
3   Pick-up coil retaining screws
4   Pole piece air gap
5   Screwdriver

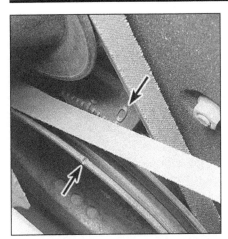

8.6 Typical stationary timing mark (upper arrow) and notch (lower arrow) in pulley

6   With the engine running at the specified idle speed (see Chapter 1), point the timing light at the stationary timing mark. The notch in the pulley should line up with the stationary mark **(see illustration)**. (The number of degrees of timing advance is listed in this Chapter's Specifications for Chevrolet Sprint models; refer to the VECI label in the engine compartment on Geo models.)

7   If the mark isn't aligned with the notch in the pulley, loosen the distributor hold-down bolts and turn the distributor housing to advance or retard the timing. Turning the distributor clockwise advances the timing; turning it counterclockwise retards the timing. When the timing is set, tighten the hold-down bolts. Recheck your work and repeat the procedure if necessary. Remove the timing light.

8   If the timing can't be adjusted properly, look for a vacuum leak, then check the vacuum advance unit (see Section 13) and the centrifugal advance unit (see Section 14).

9   If the vehicle is a Geo Metro LSi model, reattach the vacuum hose to the fuel filter at the intake manifold.

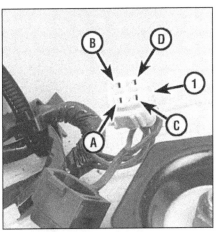

8.10 Diagnostic connector (models with electronic spark control distributor)

A   A/F duty check terminal
B   Diagnosis switch terminal
C   Ground
D   Test switch terminal
1   Diagnostic connector

Electronic spark control distributor

Refer to illustration 8.10

10   Remove the protective cap from the diagnostic connector **(see illustration)**, which is located next to the ignition coil.

11   Insert a jumper wire between terminals C and D of the diagnostic connector.

12   Hook up a timing light to the number one spark plug wire in accordance with the manufacturer's instructions.

13   Start the engine and aim the timing light at the stationary timing mark. The line on the pulley should line up with the timing mark. If it doesn't, loosen the distributor hold-down bolts and rotate the distributor - clockwise to retard, counterclockwise to advance - until the notch in the pulley is aligned with the tim-

ing mark. Tighten the hold-down bolts and recheck your work. (The specifications for ignition timing are on the VECI label in the engine compartment.)

14   Turn off the ignition switch, remove the jumper wire from the diagnostic connector and reinstall the protective cap.

Distributorless systems

15   The ignition timing is not adjustable on these models. If ignition timing problems occurs, have the vehicle checked by a dealer service department.

---

9   Ignition coil - check and replacement

Chevrolet Sprint, Geo Metro three-cylinder and 1995 through 1997 Metro four-cylinder models

Refer to illustrations 9.4, 9.5, 9.6a and 9.6b

1   Locate the ignition coil on the engine compartment firewall.

2   Using a voltmeter, check for voltage at the coil positive terminal. If there's no battery voltage, check the engine wire harness, the connector for the ignition coil and the underhood fuse/relay center for a blown fuse (see Chapter 12 for information about the fuses in your vehicle).

3   Detach the cable from the negative terminal of the battery.

4   Disconnect the coil high tension cable by gripping the boot firmly and turning it turn while pulling on the wire. Remove the protective cap from the coil **(see illustration)**. Disconnect the electrical connectors from the primary terminals of the coil.

5   Remove the coil from the firewall **(see illustration)**.

9.4 Remove the protective cap from the coil

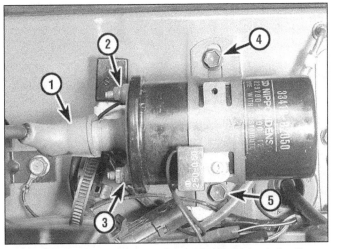

9.5 Disconnect the high tension cable (1), the leads for the positive (2) and negative (3) primary terminals, remove the bolts (4 and 5) from the coil mounting bracket and remove the coil

9.6a Measure the resistance between the primary terminals and compare your measurement with the resistance listed in this Chapter's Specifications - replace the coil if the indicated resistance is outside specification

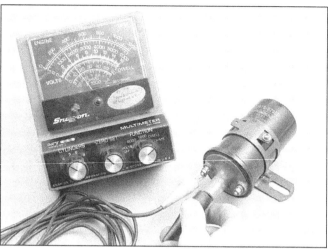

9.6b Measure the resistance between the positive terminal of the primary side of the coil and the high tension lead, then compare your measurement with the resistance listed in this Chapter's Specifications - replace the coil if the indicated resistance is outside specification

6   Using an ohmmeter, check the coil:
a) *Measure the resistance between the primary terminals* **(see illustration)**. *Compare your reading with the coil primary resistance listed in this Chapter's Specifications.*
b) *Measure the resistance between the positive terminal and high tension terminal* **(see illustration)**. *Compare your reading with the coil secondary resistance listed in this Chapter's Specifications.*

7   If either of the above tests yield resistance values outside the specified resistance, replace the coil. If the new coil doesn't come with a mounting bracket, loosen the clamp screw and slide the mounting bracket off the old coil and install it on the new one.

1998 and later Geo Metro four-cylinder models

Note 1: *There are two ignition coils on these models. The following procedure applies to either coil.*

Note 2: *Because the igniter is an integral part of the primary (low-voltage) circuit on these ignition coils, there is no way to measure the primary side of the coil. If the secondary (high-voltage) side of the coil is okay, and if the rest of the ignition circuit is in good condition, but the coil is putting out a weak spark, or no spark, take the coil to a dealer service department for further testing, or replace it.*

1998 models

8   Locate the ignition coils on the left (driver's) end of the cylinder head.
9   Unplug the electrical connector from the ignition coil.
10   Disconnect the two spark plug wire boots from the ignition coil high-tension towers.
11   Remove the two ignition coil mounting bolts and then remove the ignition coil unit.
12   Using an ohmmeter, measure the resistance between the two high tension towers and then compare your measurement to the coil secondary resistance listed in this Chapter's Specifications. If the secondary resis-

tance is incorrect, replace the ignition coil.
13   Installation is the reverse of removal.

1999 and later models

Refer to illustrations 9.15, 9.16, 9.17, 9.18, 9.19 and 9.20

14   Locate the ignition coils on top of the valve cover.
15   Unplug the electrical connector **(see illustration)** from the ignition coil.
16   Unplug the spark plug wire boot from the No. 1 or No. 4 spark plug **(see illustration)**.
17   Remove the two bolts **(see illustration)** that secure the ignition coil to the valve cover.
18   Pull straight up to disconnect the ignition coil assembly from the spark plug to which it's directly connected **(see illustration)**. Remove the ignition coil.
19   Remove the spark plug wire from the ignition coil **(see illustration)**.
20   Using an ohmmeter, measure the resistance between the two high tension towers **(see illustration)** and then compare your

9.15 Unplug the electrical connector from the ignition coil

9.16 Unplug the spark plug wire boot from the No. 1 or No. 4 spark plug

9.17 Remove the two bolts (arrows) that secure the ignition coil to the valve cover

9.18 To disconnect the ignition coil assembly from the spark plug to which it's directly connected, pull straight up

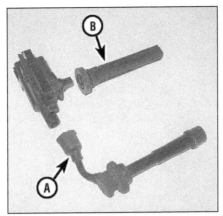

9.19 To remove the spark plug wire (A) and the spark plug boot (B) from the ignition coil, pull them straight off

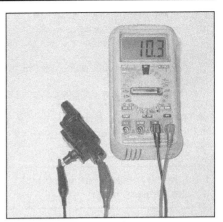

9.20 Using an ohmmeter, measure the resistance between the two high tension towers

10.2 Disconnect the electrical connector (arrow) for the module/igniter

10.5a Put an alignment mark (arrow) directly below the rotor on the distributor base

10.5b Put another alignment mark (arrows) on the distributor base and the cylinder head, then remove the hold-down bolt(s) (Chevrolet Sprint distributor shown, Geo Metro distributor similar, but has two hold-down bolts)

measurement to the coil secondary resistance listed in this Chapter's Specifications. If the secondary resistance is incorrect, replace the ignition coil.

21   Installation is the reverse of removal.

10   Distributor - removal and installation

Removal

Refer to illustration 10.2, 10.5a and 10.5b

1   Unplug the primary lead from the coil.
2   Unplug the electrical connector for the module/igniter **(see illustration)**. Follow the wires as they exit the distributor to find the connector.
3   Look for a raised "1" on the distributor cap. This marks the location for the number one cylinder spark plug wire terminal. If the cap does not have a mark for the number one terminal, locate the number one spark plug and trace the wire back to the terminal on the cap.
4   Remove the distributor cap (see Chapter 1) and turn the engine over until the rotor is pointing toward the number one spark plug terminal (see locating TDC procedure in Chapter 2).

5   Make a mark on the edge of the distributor base directly below the rotor tip and in line with it **(see illustration)**. Also, mark the distributor base and the cylinder head to ensure that the distributor is installed correctly **(see illustration)**.
6   Remove the distributor hold-down bolt(s) (Chevrolet Sprints have a single hold-down bolt; Geo Metros have two), then pull the distributor straight out to remove it. **Caution:** *DO NOT turn the crankshaft while the distributor is out of the engine, or the alignment marks will be useless.*

Installation

Refer to illustration 10.8
Note: *If you turned the crankshaft while the distributor was out, position the engine at Top Dead Center (TDC) compression for cylinder number one by following the procedure in Chapter 2 Part A.*
7   Insert the distributor into the cylinder head in exactly the same relationship to the head that it was in when removed.
8   The dogs on the end of the distributor coupling **(see illustration)** are offset. If they won't drop into the slot in the camshaft, you may have to turn the rotor slightly. If that

10.8 The dogs on the distributor coupling (arrow) are offset

doesn't work, recheck the alignment marks between the distributor base and the head to verify that the distributor is in the same position it was in before removal. Also check the rotor to see if it's aligned with the mark you made on the edge of the distributor base.

9   Loosely install the hold-down bolts.
10   Install the distributor cap.
11   Plug in the module/igniter electrical connector.
12   Reattach the spark plug wires to the plugs (if removed).
13   Connect the cable to the negative terminal of the battery.
14   Check the ignition timing (see Section 8) and tighten the distributor hold-down bolts securely.

11   Ignition module/igniter - check and replacement

Vacuum/centrifugal advance distributor

Refer to illustrations 11.3, 11.4, 11.5a, 11.5b, 11.6, 11.7 and 11.9

Note: *This component is referred to as a "module" on Chevrolet Sprints and an "igniter" on Geo Metros, but the procedures for checking and replacing it are the same.*

1   Remove the distributor cap and rotor (see Chapter 1).
2   Remove the distributor (see Section 10).
3   Remove the seal **(see illustration)**. If the seal is cracked or torn, discard it and get a new one.
4   Remove the module cover **(see illustration)**.
5   Remove the module screws **(see illustration)** and remove the module **(see illustration)**.
6   The module and the pick-up coil are connected by a pair of wires. One's red and the other's white. Disconnect both of them from the distributor base **(see illustration)** and unplug them from the module. Label each spade terminal on the module so you'll remember which one is for the red wire and which one is for the white wire.
7   Connect an ohmmeter, a test light and a 12-volt battery source to the module as

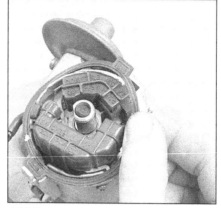

11.3 Remove the seal from the distributor base - if it's cracked or torn, replace it

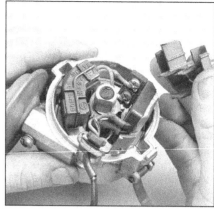

11.4 Remove the cover from the module

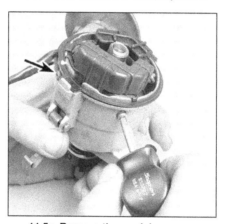

11.5a Remove the module screws (screwdriver and arrow) . . .

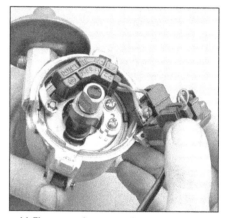

11.5b . . . and remove the module from the distributor

shown **(see illustration)**. Set the ohmmeter to the R x 1 range. Touch the negative probe to the module terminal for the red wire and the positive probe to the terminal for the white wire. **Caution:** *Don't reverse the leads of the ohmmeter or you will damage the module circuitry.*

8   If the bulb in the test light comes on, the module is working properly; if it doesn't come on, the module is faulty. Replace it.
9   Installation is the reverse of removal **(see illustration)**. Make sure you don't reverse the red and white leads from the pick-up coil.

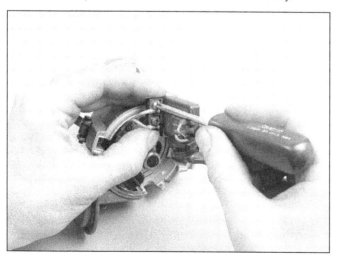

11.6 Detach the wires between the pickup coil and the module (one's red, the other is white) from the distributor base, then disconnect them from the module - be sure to label the terminals on the module so you don't reverse the wires during reassembly

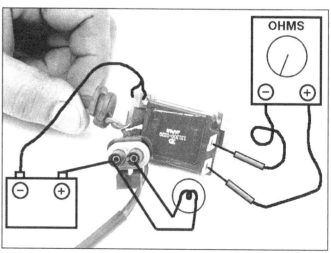

11.7 To check the module, hook it up as shown to the battery, an ohmmeter and a test light (1989 and earlier models only)

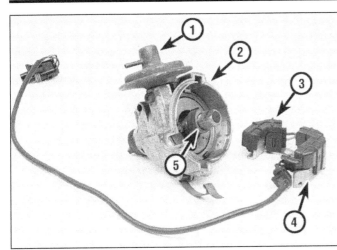

11.9 Disassembled view of a vacuum/centrifugal advance distributor

1 *Vacuum advance unit*
2 *Distributor housing*
3 *Pick-up coil*
4 *Module*
5 *Pole piece*

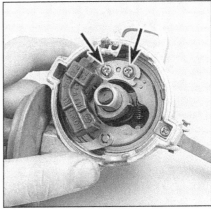

12.4 Remove the pick-up coil mounting screws (arrows) and detach the pick-up coil (vacuum/centrifugal advance distributor)

Electronic Spark Control distributor

Note: *In 1994, there were two basic emissions systems: The "upgraded emissions" package was used on base models with a manual transaxle and on base California models with an automatic transaxle. The "standard emissions" package was used on XFi and base Federal (49-state) models with an automatic transaxle. If you're not sure whether a 1994 vehicle with an automatic transaxle is a California or Federal model, look at the ignition coil. If it's a conventional, old-style coil (see illustration 9.5), the vehicle has the "standard emissions" package. If it's a new-style coil (one-piece ignition coil and bracket assembly), the vehicle has the "upgraded emissions" package.*

10   On 1990 through 1993 models with an Electronic Spark Control distributor, and on 1994 models with the "standard emissions" package, the igniter is located inside the Electronic Control Module (ECM). On these models, it's impossible to replace the igniter separately from the ECM.

11   On 1994 models with the "upgraded emissions" package, on all 1995 through 1997 models, and on 1998 and later three-cylinder models, the igniter is located on the firewall. To replace it, unplug the electrical connector and then remove the igniter mounting screws. Installation is the reverse of removal.

12   On 1998 and later four-cylinder models, the igniter is an integral part of each ignition coil. On these models, it's impossible to replace the igniter separately from the coil.

---

12   Pick-up coil - check and replacement

Vacuum/centrifugal advance distributor

Refer to illustrations 12.4 and 12.5

1   Remove the distributor cap and rotor (see Chapter 1).
2   Remove the distributor (see Section 10).
3   Remove the module, detach the dust cover from the module and disconnect the red and white wires from the module (see Section 11).

4   Remove the pick-up coil mounting screws **(see illustration)** and remove the pick-up coil **(see illustration)**.
5   Hook up an ohmmeter to the red and white wires as shown **(see illustration)** and measure the resistance of the pick-up coil. It should be within the range listed in this Chapter's Specifications. If it isn't, replace the pick-up coil.
6   Installation is the reverse of removal **(see illustration 11.9)**. Make sure you don't reverse the red and white leads or you may damage the module or pick-up coil.

Electronic spark control distributor

Refer to illustration 12.7

7   Unplug the lead to the pick-up coil **(see illustration)**, hook up an ohmmeter and measure the resistance. It should be within the range listed in this Chapter's Specifications. If it isn't, replace the pick-up coil assembly.
8   To replace the pick-up coil assembly, remove the distributor cap, the distributor

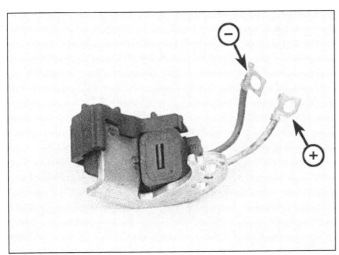

12.5 To check the pick-up coil on a vacuum/centrifugal advance distributor, touch the negative probe of an ohmmeter to the red wire and the positive probe to the white wire

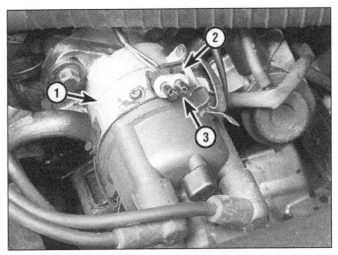

12.7 To check the pick-up coil on an electronic spark control distributor (1), simply unplug the connector (2) for the pick-up coil lead, hook up an ohmmeter to the terminals (3) and measure the resistance

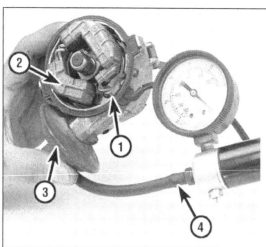

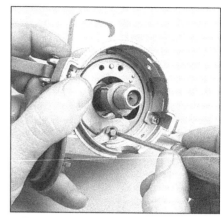

13.2 To test the vacuum advance unit, detach the vacuum hose, attach a hand-held vacuum tester in its place, apply about 15 inches (400 mm) Hg of vacuum and release it - the pick-up coil base plate should return smoothly to its normal (unadvanced) position - if it doesn't, either the base plate needs lubrication or the vacuum unit is faulty

1   *Pick-up coil base plate*
2   *Pick-up coil*
3   *Vacuum advance unit*
4   *Vacuum source (hand-operated vacuum pump)*

13.5 Pry off this C-clip from the post on the base plate

cap seal and the rotor (see Chapter 1).
9   Remove the dust cover from the pick-up coil assembly.
10   Remove the retaining screws from the pick-up coil assembly and remove the pick-up coil assembly.
11   Installation is the reverse of removal. Be sure to check and, if necessary, adjust the air gap (see Section 7) when you're done.

---

13   Vacuum advance unit - check and replacement

Refer to illustrations 13.2, 13.5, 13.6 and 13.7
1   Remove the distributor cap (see Chapter 1).
2   Detach the vacuum hose from the vacuum advance unit and attach a vacuum tester in its place **(see illustration)**.
3   Apply about 15 inches (400 mm) Hg of vacuum, then release it. The pick-up coil base plate should move smoothly. If it doesn't, try lubricating it (see Section 14). If that doesn't work, replace the vacuum advance unit.
4   To replace the vacuum advance unit, remove the distributor (see Section 10), the rotor (see Chapter 1), the module/igniter (see Section 11) and the pick-up coil (see Section 12).

13.7 Rotate the base plate clockwise to disengage the advance arm from the post, then remove the advance unit

5   Remove the C-clip that attaches the vacuum advance unit to the base plate **(see illustration)**.
6   Remove the vacuum advance unit retaining screw **(see illustration)**.
7   Rotate the base plate clockwise to disengage the advance arm from the post **(see illustration)** and remove the advance unit.
8   Installation is the reverse of removal **(see illustration 11.9)**. Be sure to check and, if necessary, adjust the air gap (see Section 7) and the ignition timing (see Section 8) when you're done.

---

14   Centrifugal advance - check and replacement

Refer to illustrations 14.4, 14.5 and 14.7
1   Remove the distributor cap (see Chapter 1).
2   Turn the rotor counterclockwise with your fingers, then release it. The rotor should return (rotate clockwise) smoothly by spring force. If it doesn't, remove and inspect the base plate.
3   To remove the base plate, remove the distributor (see Section 10), pull off the rotor, the module/igniter (see Section 11), the pick-up coil (see Section 12) and the vacuum

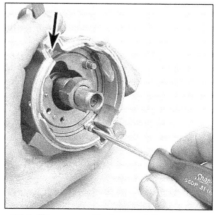

14.4 To remove the base plate, remove these two screws

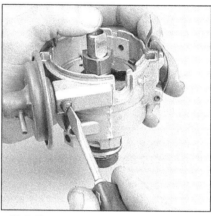

13.6 Remove the vacuum advance unit retaining screw from the distributor base

advance unit (see Section 13).
4   Remove the two base plate screws **(see illustration)** and remove the base plate. Do NOT attempt to disassemble the base plate.
5   Hold the outer ring of the base plate as shown, grasp the post for the centrifugal advance arm with your thumb and index finger and turn the inner plate back and forth **(see illustration)**. It should rotate smoothly. If

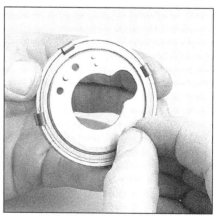

14.5 To check the base plate, hold it in one hand and slowly turn the inner plate with the other - if the inner plate doesn't rotate smoothly, the bearings or the races are worn - replace the plate (it can't be rebuilt)

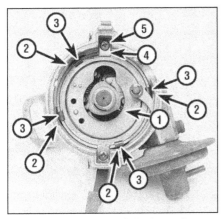

14.7 When you install the base plate, make sure the four clips are seated properly into their respective grooves in the distributor housing

1 Pick-up coil base plate
2 Groove
3 Clip
4 Base plate washer
5 Base plate screw

it doesn't, replace it.

6   Inspect the centrifugal advance springs and weights. If they're dirty or corroded, spray them with contact cleaner and lubricate them with light oil.

7   Install the base plate in the distributor housing. Fit the four clips on the base plate into the four grooves in the distributor housing as shown **(see illustration)**.

8   Installation is otherwise the reverse of removal. Be sure to check and, if necessary, adjust the air gap (see Section 7) and the ignition timing (see Section 8) when you're done.

9   If the distributor still fails to advance properly, the centrifugal advance assembly is sticking. This assembly isn't available individually; you'll have to replace the distributor base.

15   Charging system - general information and precautions

The charging system includes the alternator, an internal voltage regulator, a charge indicator, the battery, the fusible link and the wiring between all the components. The charging system supplies electrical power for the ignition system, the lights, the radio, etc. The alternator is driven by a drivebelt at the front of the engine.

The purpose of the voltage regulator is to limit the alternator's voltage to a preset value. This prevents power surges, circuit overloads, etc., during peak voltage output.

The fusible links are fuse-like units mounted on the relay panel in the engine compartment. See Chapter 12 for additional information regarding fusible links.

The charging system doesn't ordinarily require periodic maintenance. However, the

16.2 Check battery voltage with a voltmeter hooked up to the negative and positive terminals

drivebelt, battery and wires and connections should be inspected at the intervals outlined in Chapter 1.

The dashboard warning light should come on when the ignition key is turned to Start, then go off immediately. If it remains on, there is a malfunction in the charging system (see Section 16). Some vehicles are also equipped with a voltmeter. If the voltmeter indicates abnormally high or low voltage, check the charging system (see Section 16).

Be very careful when making electrical circuit connections to a vehicle equipped with an alternator and note the following:

a)   *When reconnecting wires to the alternator from the battery, be sure to note the polarity.*
b)   *Before using arc welding equipment to repair any part of the vehicle, disconnect the wires from the alternator and the battery terminals.*
c)   *Never start the engine with a battery charger connected.*
d)   *Always disconnect both battery leads before using a battery charger.*
e)   *The alternator is turned by an engine drivebelt which could cause serious injury if your hands, hair or clothes become entangled in it with the engine running.*
f)   *Because the alternator is connected directly to the battery, it could arc or cause a fire if overloaded or shorted out.*
g)   *Wrap a plastic bag over the alternator and secure it with rubber bands before steam cleaning the engine.*

16   Charging system - check

Refer to illustration 16.2

1   If a malfunction occurs in the charging circuit, don't automatically assume the alternator is causing the problem. First check the following items:

a)   *Check the drivebelt tension and condition (see Chapter 1). Replace it if it's worn or deteriorated.*

b)   *Make sure the alternator mounting and adjustment bolts are tight.*
c)   *Inspect the alternator wiring harness and the electrical connectors at the alternator and voltage regulator. They must be in good condition and tight.*
d)   *Check the fusible link (if equipped) (see Chapter 12).*
e)   *Start the engine and check the alternator for abnormal noises (a shrieking or squealing sound indicates a bad bearing).*
f)   *Check the specific gravity of the battery electrolyte. If it's low, charge the battery (doesn't apply to maintenance free batteries).*
g)   *Make sure the battery is fully charged (one bad cell in a battery can cause overcharging by the alternator).*
h)   *Disconnect the battery cables (negative first, then positive). Inspect the battery posts and the cable clamps for corrosion. Clean them thoroughly if necessary (see Chapter 1). Reconnect the cable to the positive terminal.*
i)   *With the key off, connect a test light between the negative battery post and the disconnected negative cable clamp.*

   1) *If the test light does not come on, reattach the clamp and proceed to the next Step.*
   2) *If the test light comes on, there is a short (drain) in the electrical system of the vehicle. The short must be repaired before the charging system can be checked.*
   3) *Disconnect the alternator wiring harness.*
      (a) *If the light goes out, the alternator is bad.*
      (b) *If the light stays on, pull each fuse until the light goes out (this will tell you which component is shorted).*

2   Using a voltmeter, check the battery voltage with the engine off **(see illustration)**. If should be approximately 12-volts.

3   Start the engine and check the battery voltage again. It should now be approximately 14 to 15-volts.

4   Turn on the headlights. The voltage should drop, and then come back up, if the charging system is working properly.

5   If the voltage reading is more than the specified charging voltage, replace the voltage regulator (see Section 18). If the voltage is less, the alternator diode(s), stator or rectifier may be bad or the voltage regulator may be malfunctioning.

17   Alternator - removal and installation

Refer to illustrations 17.2a, 17.2b, 17.3, 17.4a, 17.4b, 17.4c, 17.5a and 17.5b

Note: *Many new and rebuilt alternators DO NOT have a pulley installed, so you may have to switch the pulley from the old unit to the new/rebuilt one if you plan to buy an alterna-*

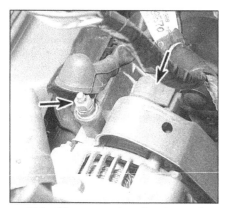

17.2a Disconnect the wires (arrows)
from the alternator

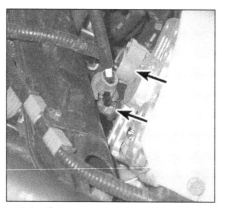

17.2b On some late-model alternators,
the connectors (arrows) are located on
the backside of the alternator

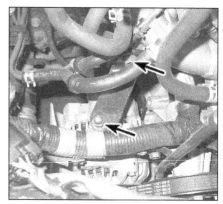

17.3 On 1998 and later four-cylinder
models, remove these bolts (arrows) and
then remove the support bracket between
the alternator and the intake manifold

17.4a Remove the adjustment bolt (arrow)
from the alternator

17.4b On 1998 and later four-cylinder
models, remove the adjustment bolt
(left arrow), remove the upper heat shield
bolt (right arrow) . . .

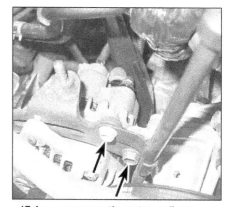

17.4c . . . remove these two adjustment
bracket bolts (arrows) and then remove
the adjustment bracket

*tor. Find out the shop's policy regarding pul-
leys - some shops will perform this service
free of charge, some won't. If you're going to
have to do it yourself, loosen the pulley bolts
now, before you remove the drivebelt.*
1   Detach the cable from the negative terminal of the battery.
2   Detach the electrical connectors from
the alternator **(see illustrations)**.
3   On 1998 and later four-cylinder models,

remove the support bracket between the
alternator and the intake manifold **(see illustration)**.
4   Loosen the alternator adjustment bolt
(see illustration). On 1998 and later four-
cylinder models, remove the adjustment bolt
and the three adjustment bracket bolts **(see
illustrations)** and then remove the adjust-

ment bracket.
5   Raise the vehicle and loosen the alternator pivot bolt **(see illustrations)**.
6   Remove the drivebelt.
7   Remove the adjustment and pivot bolts
and separate the alternator from the engine.
On 1998 and later four-cylinder models,
remove the alternator by pulling it out from
above.

17.5a Support the alternator and remove the pivot bolt (arrows)

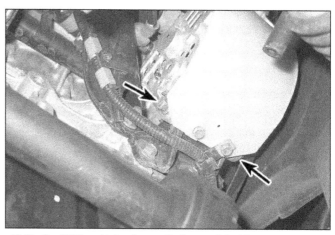

17.5b On 1998 and later four-cylinder models, it's not necessary
to remove the splash shield below the alternator; simply
remove the lower pivot bolt and nut (arrows)

18.2 To get the back cover off the alternator, loosen the insulator nut (upper right arrow), back it off far enough to slide the bushing clear of the cover, remove the three cover screws (other three arrows) and remove the cover

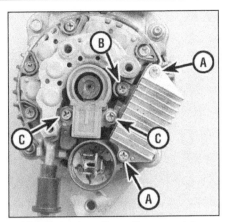

18.4 To replace the voltage regulator, remove the two mounting screws (A) and the screw between the brush holder and the regulator (B - secures the regulator lead to the brush holder) - to replace the brush holder, remove the three mounting screws (B and C) . . .

18.5 . . . and remove the brush holder - when you install the new holder, don't forget to attach the lead for the voltage regulator with the screw between the holder and the regulator

8   Remove the pulley, if necessary (see **Note** above).

9   If you're replacing the alternator, take the old one with you when purchasing a replacement unit. Make sure the new/rebuilt unit looks identical to the old alternator. Look at the terminals - they should be the same in number, size and location as the terminals on the old alternator. Finally, look at the identification numbers - they will be stamped into the housing or printed on a tag attached to the housing. Make sure the numbers are the same on both alternators.

10   Install the pulley on the new/rebuilt alternator, if necessary.

11   Installation is otherwise the reverse of removal.

12   After the alternator is installed, adjust the drivebelt tension (see Chapter 1).

13   Check the charging voltage to verify proper operation of the alternator (see Section 16).

18   Voltage regulator and alternator brushes - replacement

Refer to illustrations 18.2, 18.4 and 18.5

1   Remove the alternator (see Section 17).

2   Remove the B terminal insulator nut and bushing **(see illustration)**.

3   Remove the three rear end cover nuts and remove the rear end cover.

4   If you're replacing the voltage regulator, remove the two 4 mm regulator mounting screws **(see illustration)** and the screw between the regulator and the brush holder (this screw attaches the regulator lead to the brush holder). Remove the regulator.

5   If you're replacing the brush holder, remove the three brush holder screws **(see illustration 18.4)** and remove and brush holder **(see illustration)**.

6   Installation is the reverse of removal. Don't forget to reattach the regulator lead to the brush holder.

19   Starting system - general information and precautions

The sole function of the starting system is to turn over the engine quickly enough to allow it to start.

The starting system consists of the battery, the ignition switch, the clutch start switch (manual transaxle) or the neutral start switch (automatic transaxle), the starter solenoid, the starter motor and the wires connecting them. The solenoid is mounted directly on the starter motor.

The solenoid/starter motor assembly is installed on the lower part of the engine, next to the transaxle bellhousing.

When the ignition key is turned to the Start position, the starter solenoid is actuated through the starter control circuit. The starter solenoid then connects the battery to the starter. The battery supplies the electrical energy to the starter motor, which does the actual work of cranking the engine.

The starter motor on a vehicle equipped with a manual transaxle can only be operated when the clutch pedal is depressed; the starter on a vehicle equipped with an automatic transaxle can only be operated when the transaxle selector lever is in Park or Neutral.

Always observe the following precautions when working on the starting system:

a)   *Excessive cranking of the starter motor can overheat it and cause serious damage. Never operate the starter motor for more than 15 seconds at a time without pausing to allow it to cool for at least two minutes.*

b)   *The starter is connected directly to the battery and could arc or cause a fire if mishandled, overloaded or shorted out.*

c)   *Always detach the cable from the negative terminal of the battery before working on the starting system.*

20   Starter motor - testing in vehicle

Note: *Before diagnosing starter problems, make sure the battery is fully charged.*

1   If the starter motor does not turn at all when the switch is operated, make sure that the shift lever is in Neutral or Park (automatic transaxle) or that the clutch pedal is depressed (manual transaxle).

2   Make sure that the battery is charged and that all cables, both at the battery and starter solenoid terminals, are clean and secure.

3   If the starter motor spins but the engine is not cranking, the overrunning clutch in the starter motor is slipping and the starter motor must be replaced.

4   If, when the switch is actuated, the starter motor does not operate at all but the solenoid clicks, then the problem lies with either the battery, the main solenoid contacts or the starter motor itself (or the engine is seized).

5   If the solenoid plunger cannot be heard when the switch is actuated, the battery is bad, the fusible link is burned (the circuit is open) or the solenoid itself is defective.

6   To check the solenoid, connect a jumper lead between the battery (+) and the ignition switch wire terminal (the small terminal) on the solenoid. If the starter motor now operates, the solenoid is OK and the problem is in the ignition switch, neutral start switch or the wiring.

7   If the starter motor still does not operate, remove the starter/solenoid assembly for disassembly, testing and repair.

8   If the starter motor cranks the engine at an abnormally slow speed, first make sure the battery is fully charged and that all terminal connections are tight. If the engine is partially seized or has the wrong viscosity oil in it (in cold weather), it will crank slowly. Also, verify the battery's Cold Cranking Amp (CCA) rating is sufficient for the engine (an auto parts store can usually tell you what the minimum should be).

21.3a Disconnect the wires (upper right arrows) from the starter motor terminals. If you just want to replace the starter on earlier Nippondenso units like this one, remove the two long bolts (lower left arrows) that attach the starter motor to the drive housing

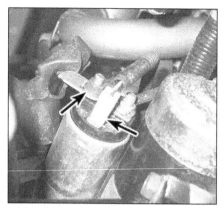

21.3b On most starter units, the electrical connections (arrows) are easier to disconnect from underneath the vehicle

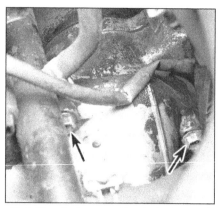

21.4 The starter mounting bolts (arrows) are located on the transaxle bellhousing, below the intake manifold

9   Run the engine until normal operating temperature is reached, then disconnect the coil wire from the distributor cap and ground it on the engine.

10   Connect a voltmeter positive lead to the positive battery post and connect the negative lead to the negative post. A fully charged battery should read about 12.6 volts. If the reading is lower, charge the battery before proceeding.

11   Crank the engine and take the voltmeter readings as soon as a steady figure is indicated. Do not allow the starter motor to turn for more than 15 seconds at a time. A reading of 9 volts or more, with the starter motor turning at normal cranking speed, is normal. If the reading is 9 volts or more but the cranking speed is slow, the solenoid contacts are burned, there is a bad connection or the starter motor is faulty. If the reading is less than 9 volts and the cranking speed is slow, the starter motor is bad or the battery is discharged.

21   Starter motor - removal and installation

Refer to illustrations 21.3a, 21.3b and 21.4

1   Detach the cable from the negative terminal of the battery.

2   Raise the front of the vehicle and place it securely on jackstands. **Note:** *It's not necessary to raise the vehicle to get to the starter motor bolts, which are located immediately behind the block and under the intake manifold. However, you will probably need to raise the vehicle to disconnect the electrical harness from the starter solenoid terminals.*

3   Clearly label, then disconnect the wires from the terminals on the starter motor solenoid **(see illustrations)**.

4   Working from above, locate the starter bolts **(see illustration)** (they face toward the driver's side of the vehicle) and then remove them from the transaxle bellhousing. [If you

just want to replace the starter on some earlier models with a Nippondenso starter unit like the one shown in illustration 21.2a, it's not necessary to remove the entire starter assembly. Simply remove the two long bolts that attach the starter motor to the drive housing **(see illustration 21.2a)**.]

5   Installation is the reverse of removal.

22   Starter solenoid - removal and installation

Refer to illustrations 22.3, 22.4 and 22.5

1   Disconnect the cable from the negative terminal of the battery.

2   Remove the starter motor (see Section 21).

3   Remove the retaining nut and disconnect the strap from the solenoid to the starter motor terminal **(see illustration)**. **Note:** *On vehicles with an automatic transaxle, the strap is attached between terminals on the sides of the starter motor and solenoid (as shown); on vehicles with a manual transaxle, you'll find the strap between terminals on the ends of the starter and solenoid (not shown).*

4   Remove the through-bolts which secure the solenoid to the starter drive housing **(see illustration)**.

22.3 Remove the retaining nut (arrow) and disconnect the strap between the solenoid and the starter motor (starter assembly for automatic model shown - strap and terminals are located on the ends of the starter and solenoid on models with manual transaxles)

5   Pull up and out to disengage the plunger hook from the shift lever **(see illustration)** and remove the solenoid.

6   Installation is the reverse of removal. Be sure to grease the plunger hook and install the solenoid with the plunger hook facing up.

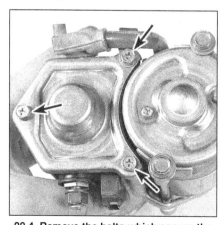

22.4 Remove the bolts which secure the solenoid to the starter motor

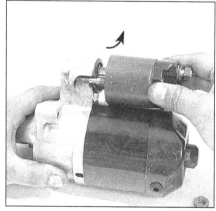

22.5 Pull up and out to disengage the plunger hook from the shift lever and remove the solenoid

Chapter 6
Emissions control systems

Contents

Specifications

Torque specifications

Ft-lbs (unless otherwise noted)

Camshaft Position (CMP) sensor bolt (1998 and later four-cylinder models)	89 in-lbs
Crankshaft Position (CKP) sensor bolt (1998 and later four-cylinder models)	89 in-lbs
Engine Coolant Temperature (ECT) sensor	
Sprint	114 to 144 in-lbs
Metro	
Three-cylinder	90 to 168 in-lbs
Four-cylinder	132 in-lbs
Idle Speed Control (ISC) motor mounting screws	24 to 36 in-lbs
Oxygen sensor	
Sprint	33 to 39.5
Metro (all engines, OBD-I and OBD-II, upstream and downstream)	29 to 36
Power Steering Pressure (PSP) switch (four-cylinder models)	20
Throttle Position (TP) sensor retaining screws	18
Transmission Range (TR) switch retaining bolt	17
Vehicle Speed Sensor (VSS) retaining bolt	72 in-lbs

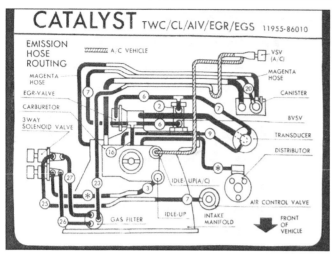

1.1a   Typical Sprint vacuum and emissions schematic

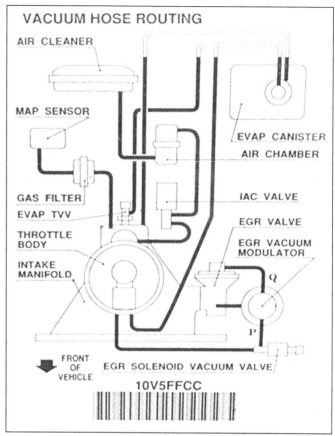

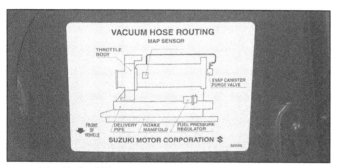

1.1c   Typical Metro vacuum routing diagram (Multiport Sequential Fuel Injection)

1.1b   Typical Metro vacuum and emissions schematic (Electronic Fuel Injection)

1.1d   Typical component locations on a three-cylinder engine with Electronic Fuel Injection (throttle body injection)

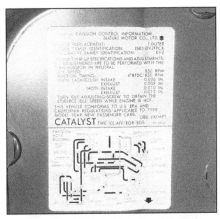

1.6a The Vehicle Emission Control Information (VECI) label contains tune-up specifications and vital information regarding the locations of the emission control devices and vacuum hose routing (Sprint model shown)

1 General information

Refer to illustrations 1.1a, 1.1b, 1.1c, 1.1d, 1.6a and 1.6b

Note: *Most of the components described in this Chapter are protected by a Federally-mandated extended warranty. Refer to your owner's manual or see your dealer for the details regarding your vehicle. It therefore makes little sense to either check or replace any of these parts yourself as long as they are still under warranty. However, once the warranty has expired, you may wish to perform some of the component checks and/or replacement procedures in this Chapter to save money.*

To minimize pollution of the atmosphere from incompletely burned and evaporating gases and to maintain good driveability and fuel economy, a number of emission control systems are used on these vehicles **(see illustrations)**. They include the:

Positive Crankcase Ventilation (PCV) system, which reduces hydrocarbons from crankcase blowby

Evaporative Emission Control (EVAP) system, which reduces evaporative hydrocarbons

Feedback system (Sprint models), which reduces hydrocarbons and carbon monoxide by regulating the operating conditions of the engine. For more information on the feedback system, refer to Chapter 4.

Exhaust Gas Recirculation (EGR) system, which reduces oxides of nitrogen emissions

Catalytic converter, which reduces hydrocarbons, carbon monoxide and oxides of nitrogen

Electronic Fuel Injection (EFI) system (Geo Metro three-cylinder and 1995 through 1997 four-cylinder models), or multiport **Sequential Fuel Injection (SFI) system** (1998 and later Metro four-cylinder models), which reduces all exhaust emissions by regu-

1.6b Typical VECI label on a late-model four-cylinder Metro with Multiport Sequential Fuel Injection (SFI)

lating the operating conditions of the engine. For more information on fuel injection systems, refer to Chapter 4.

The sections in this chapter include general descriptions, checking procedures within the scope of the home mechanic and component replacement procedures (when possible) for each of the systems listed above.

Before assuming an emissions control system is malfunctioning, check the fuel and ignition systems carefully (see Chapters 4 and 5). The diagnosis of some emission control devices requires specialized tools, equipment and training. If checking and servicing become too difficult or if a procedure is beyond the scope of your skills, consult your dealer service department.

This doesn't mean, however, that emission control systems are particularly difficult to maintain and repair. You can quickly and easily perform many checks and do most of the regular maintenance at home with common tune-up and hand tools. **Note 1:** *Because of a federally mandated extended warranty which covers the emission control system components, check with a dealer service department about warranty coverage before working on any emission related systems.* **Note 2:** *The most frequent cause of emissions problems is simply a loose or broken electrical connector or vacuum hose, so always check the electrical connectors and vacuum hoses first.*

Pay close attention to any special precautions outlined in this chapter. It should be noted that the illustrations of the various systems may not exactly match the system installed on your vehicle because of changes made by the manufacturer during production or from year to year.

The Vehicle Emissions Control Information (VECI) label is located in the engine compartment **(see illustrations)**. This label contains important emissions specifications and setting procedures, and a vacuum hose schematic with emissions components identified. When servicing the engine or emissions systems, the VECI label in your particular vehicle should always be checked for up-to-date information.

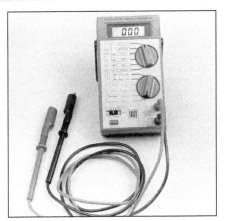

2.1 Digital multimeters can be used for testing all types of circuits; because of their high impedance, they are much more accurate than analog meters for measuring low-voltage computer circuits

Precautions

a) *Always disconnect the power by either turning off the ignition switch or disconnecting the battery terminals before removing any wiring connectors related to the computerized engine control system.*

b) *When installing a battery, be particularly careful to avoid reversing the positive and negative battery cables.*

c) *Do not subject carburetor components, fuel injection components, emissions related components or the PCM to severe impact during removal or installation.*

d) *Do not be careless during troubleshooting. Even slight terminal contact can invalidate a testing procedure and damage one of the numerous transistor circuits.*

e) *Never attempt to work on the PCM or open the PCM cover. The PCM is protected by a government mandated extended warranty that will be nullified if you tamper with or damage the PCM.*

f) *If you are inspecting electronic control system components during rainy weather, make sure that water does not enter any part. When washing the engine compartment, do not spray these parts or their connectors with water.*

2 On-Board Diagnostic (OBD) system and trouble codes

Diagnostic tool information

Refer to illustrations 2.1 and 2.2

1   Use a digital multimeter **(see illustration)** to check fuel injection and emission components. When working with electronic circuits, many of which are very low-voltage (5 volts or less), an accurate reading is essential. Most analog multimeters can't measure volts, ohms or amps to two or three decimal

places (and the few that can are difficult to read), yet even an inexpensive digital multimeter can easily make accurate measurements to two or three decimal places. That's because digital multimeters are equipped with internal circuitry of very high resistance (usually 10 million ohms). A voltmeter is hooked up in parallel with the circuit being tested, so it is critical that none of the voltage being measured be allowed to travel through the parallel path set up by the meter itself. This problem doesn't occur when measuring larger amounts of voltage (9 to 12-volt circuits), but if you are measuring a low-voltage circuit - oxygen sensor signal voltage, for example - a fraction of a volt is significant when diagnosing a problem.

2   Hand-held scanners **(see illustration)** are powerful and versatile tools for analyzing engine management systems used on later model vehicles. A scan tool must be compatible with (be able to communicate with) the PCM of the vehicle you are servicing. Some scan tools - the generic, usually less expensive units that most readers will buy - are designed to be compatible with a specific year, make and model. More expensive professional units can accept interchangeable cartridges, which allows one scan tool to interface with a manufacturer (Ford, GM, Chrysler, etc.). Some scan tools accept cartridges that enable the tool to be compatible with all vehicles manufactured in a country or geographic region (Germany, Japan or the USA, for example; or Asia, Europe, etc.).

3   A somewhat more sophisticated scan tool is needed to access a vehicle equipped with the Federally-mandated On-Board Diagnostics-II (OBD-II) system (which applies to all of the vehicles covered by this manual). Several tool manufacturers have already introduced relatively inexpensive generic OBD-II scan tools for the home mechanic. Ask the parts salesman at a local auto parts store for additional information concerning availability and cost.

2.2 Hand-held scan tools like these can extract trouble codes and also perform diagnostics

On-Board Diagnostics I (OBD-I) system

Note: *The following information applies to 1987 through 1995 models.*

General information

4   The PCM contains a built-in self-diagnosis system which detects and identifies malfunctions occurring in the network. When the PCM detects a problem, three things happen: the Check Engine Light (pre-1993 models)/ Malfunction Indicator Light (1993 and 1994 models) comes on, the trouble is identified and a diagnostic code is recorded and stored. The PCM stores the failure code assigned to the specific problem area until the diagnosis system is canceled by turning the diagnostic switch Off (1987 and 1988 models) or by removing the tail lamp fuse from the fuse block (1989 and later models).

5   The Check Engine warning light, which is located on the instrument panel, comes on when the ignition switch is turned to On and the engine is not running. When the engine is started, the warning light should go out. If the light remains on, the diagnosis system has detected a malfunction in the system.

Obtaining diagnosis code output

Refer to illustration 2.8

6   To obtain an output of diagnostic codes, verify first that the battery voltage is above 11 volts, the throttle is fully closed, the transaxle is in Neutral, the accessory switches are off and the engine is at normal operating temperature.

7   Turn the ignition switch to OFF. Do not start the engine. **Caution:** *The ignition key must be in the OFF position when disconnecting or reconnecting power to the PCM.*

8   On 1987 and 1988 models, turn the diagnosis switch located under the steering column area **(see illustration)**. On 1989 and later models, insert the spare fuse into the diagnostic terminal of the fuse block.

9   Turn the ignition key ON. Read the diagnosis code as indicated by the number of flashes of the "Check Engine/Malfunction Indicator Light" on the dash (see the accom-

panying chart). Normal system operation is indicated by Code 12 (no malfunctions) for all models. The "Check Engine" light displays a Code 12 by blinking the corresponding pattern one time only.

10   If there are any malfunctions in the system, their corresponding trouble codes are stored in computer memory and the light will blink the requisite number of times for the indicated trouble codes. If there's more than one trouble code in the memory, they'll be displayed in numerical order (from lowest to highest) with a pause interval between each one. After the code with the largest number flashes has been displayed, there will be another pause and then the sequence will begin all over again.

11   To ensure correct interpretation of the blinking "Check Engine/Malfunction Indicator Light," watch carefully for the interval between the end of one code and the beginning of the next (otherwise, you will become confused by the apparent number of blinks and misinterpret the display). The length of this interval varies with the model year.

Canceling a diagnostic code

12   After the malfunctioning component has been repaired/replaced, the trouble code(s) stored in computer memory must be canceled. To accomplish this, simply turn the diagnostic switch to OFF (1987 and 1988 models) or remove the tail lamp fuse (1988 and later models) from the fuse block with the ignition switch off. **Note:** *On 1988 and later models, the 30A main fuse will cancel the memory in the radio and the clock when removed. Use only the tail lamp fuse to cancel the codes in the PCM.*

13   Cancellation can also be affected by removing the cable from the battery negative terminal, but other memory systems (such as the clock) will also be canceled.

14   If the diagnosis code is not canceled it will be stored by the PCM and appear with any new codes in the event of future trouble.

15   Should it become necessary to work on engine components requiring removal of the battery terminal, first check to see if a diagnostic code has been recorded.

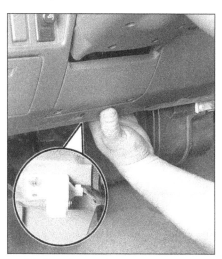

2.8 The diagnostic switch is located under the steering column on 1987 and 1988 models

1987 and 1988 Chevrolet Sprint trouble codes

Trouble codes	Circuit or system	Probable cause
Code 12 (1 flash, pause, 2 flashes)	Normal	This code will flash whenever the PCM is activated by turning the diagnostic switch ON, turning the ignition key ON and there are no codes stored in the PCM.
Code 13 (1 flash, pause, 3 flashes)	Oxygen sensor circuit (open circuit)	Check the wiring and connectors from the oxygen sensor Replace the oxygen sensor.*
Code 14 (1 flash, pause, 4 flashes)	Coolant temperature sensor (high temperature)	If the engine is experiencing overheating problems the problem must be rectified before continuing. Check all wiring and connectors associated with the coolant temperature sensor. Replace the coolant temperature sensor.*
Code 21 (2 flashes, pause, 1 flash)	Throttle position switches	Check for a faulty WOT or Idle switch. Check all wiring and connections between the switches and the PCM.
Code 23 (2 flashes, pause, 3 flashes)	Intake air temperature sensor	Check the Intake air temperature sensor and the circuit.
Code 32 (3 flashes, pause, 2 flashes)	Ambient pressure sensor	Ambient pressure sensor (located inside the ECM) is faulty
Code 51 (5 flashes, pause, 1 flash)	PCM	PCM faulty
Code 52 (5 flashes, pause, 2 flashes)	Fuel cut solenoid	Fuel cut solenoid or its circuit faulty. Also check ECM
Code 53 (5 flashes, pause, 3 flashes)	Second air solenoid	Second air three-way-solenoid or its circuit is faulty Also check ECM
Code 54 (5 flashes, pause, 4 flashes)	Mixture control solenoid	Mixture control solenoid or its circuit faulty. Also check ECM
Code 55 (5 flashes, pause, 5 flashes)	Bowl vent solenoid	Bowl vent solenoid or its circuit is faulty Also check ECM

1989 through 1995 Geo Metro trouble codes

Trouble codes	Circuit or system	Probable cause
Code 12 (1 flash, pause, 2 flashes)	Normal	Code 12 flashes when the diagnostic terminal is activated, the ignition key is switched to ON and there are no other codes stored in the PCM.
Code 13 (1 flash, pause, 3 flashes)	Oxygen sensor circuit (open circuit, or circuit voltage doesn't change)	Check the wiring and connectors from the oxygen sensor. Replace the oxygen sensor.*
Code 14 (1 flash, pause, 4 flashes)	Coolant sensor circuit (high-voltage input, low temperature indicated)	Check all wiring and connectors associated with the coolant temperature sensor. Replace the coolant temperature sensor.*
Code 15 (1 flash, pause, 5 flashes)	Coolant sensor circuit (low-voltage input, high temperature indicated)	If the engine is experiencing overheating problems the problem must be rectified before continuing. Then check the wiring connections at the PCM.
Code 21 (2 flashes, pause, 1 flash)	Throttle switch circuit (manual transmission only)	Check for a defective idle or WOT switch. Check all wiring and connections between the switch and the PCM. Adjust or replace the switch (see Section 12).*
Code 21 (2 flashes, pause, 1 flashes)	Throttle Position (TP) sensor circuit (high voltage input)	Check the TPS adjustment (Chapter 4). Check the PCM connector. Replace the TPS (Chapter 4).*
Code 22 (2 flashes, pause, 2 flashes)	Throttle Position (TP) sensor circuit (low voltage input)	Check the TPS adjustment (Chapter 4). Check the PCM connector. Replace the TPS (Chapter 4).*
Code 23 (2 flashes, pause, 3 flashes)	Manifold Air Temperature (MAT)/Intake Air Temperature (IAT) sensor circuit	Check the IAT/MAT sensor, wiring and connectors for an open circuit. Replace the IAT/MAT sensor.*
Code 24 (2 flashes, pause, 4 flashes)	Vehicle speed sensor circuit	A fault in this circuit should be indicated only when the vehicle is in motion. Check and repair the speedometer if it is not functioning.
Code 25 (2 flashes, pause, 5 flashes)	Manifold Air Temperature (MAT)/Intake Air Temperature (IAT) sensor circuit	Check the IAT/MAT sensor, wiring and connectors for an open circuit. Replace the IAT/MAT sensor.*
Code 31 (3 flashes, pause, 1 flash)	Manifold Absolute Pressure (MAP) sensor circuit	Check the circuit for stripped insulation or damaged electrical connectors. Replace the MAP sensor if necessary.
Code 32 (3 flashes, pause, 2 flashes)	Manifold Absolute Pressure (MAP) sensor circuit	Check the MAP sensor connector and wiring for an open circuit. Check the MAP sensor.
Code 41 (4 flashes, pause, 1 flash)	Ignition signal circuit (no signal)	Inspect and repair any damaged electrical connectors and wire in the harness. Note: The "Check Engine" light will not come on for this code.
Code 42 (4 flashes, pause, 2 flashes)	Crank angle sensor/Camshaft Position (CMP) sensor circuit (no signal for two seconds)	Check for a faulty connector or circuit. Also, on models with a distributor, measure and, if necessary, adjust the air gap (see Chapter 5).
Code 46 (4 flashes, pause, 6 flashes)	Idle Speed Control (ISC) motor circuit	Check the ISC motor connector and circuit.
Code 51 (5 flashes, pause, 1 flash)	Exhaust Gas Recirculation (EGR) system circuit	Check for a faulty connector or circuit. Also a high resistance in the solenoid coil.

Component replacement may not cure the problem in all cases. For this reason, you may want to seek professional advice before purchasing replacement parts.

On-Board Diagnostics II (OBD-II)

General description

16   All 1996 and later models described in this manual are equipped with the On-Board Diagnostic (OBD-II) system. The OBD-II system consists of an on-board computer, known as the Powertrain Control Module (PCM), information sensors and output actuators. The PCM is calibrated to optimize the emissions, fuel economy and driveability of the specific vehicle in which it's installed. The information sensors monitor various functions of the engine and send data to the PCM. Comparing this incoming data to its "map" (the program stored in the computer's memory), the PCM constantly alters the operating conditions of the engine (spark timing, fuel injection pulse width, etc.) through an array of relays, solenoids and other output actuators.

17   Because of a Federally mandated warranty that covers all emissions-related components and systems, and because owner-induced damage to the PCM, the sensors and/or the control devices could void the warranty, it's not a good idea to handle the PCM while the emissions warranty is still in effect. Take the vehicle to a dealer service department if the PCM or a system component malfunctions.

Information sensors

18   **Air conditioning (A/C) idle-up signal** - If the engine is idling when the A/C compressor clutch is energized, the extra load imposed upon the engine can cause the idle to drop and the engine to run roughly, or even stall. To prevent this from happening, the compressor control module sends an A/C idle-up signal to the PCM, which directs the idle air control valve to open the idle air passage, which speeds up the engine rpm slightly.

19   **Camshaft Position (CMP) sensor** - The CMP sensor is located inside the distributor on three-cylinder models and on 1996 and 1997 four-cylinder models, or inside a housing on the end of the cylinder head on 1998 and later four-cylinder models. The CMP sensor, which consists of a signal generator and a signal rotor, generates an alternating current (AC) signal as each pole piece of the signal rotor passes by the signal generator. There are four pole pieces, so the CMP sensor outputs four pulses per revolution. The CMP sensor provides information to the PCM regarding the position of the camshaft. The PCM uses this information to identify individual cylinders, to calculate engine speed, to trigger the ignition system, to help determine when to energize the fuel injectors and to monitor engine misfires.

20   **Crankshaft Position (CKP) Sensor** - The CKP sensor is located on the oil pan, behind the crank pulley, on three-cylinder engines, or on the engine block, behind the crank pulley, on four-cylinder engines. The CKP sensor, which consists of a magnet and coil, generates an alternating current (AC) each time the pole piece on the crankshaft

passes by. The PCM uses this information to initiate engine ignition and fuel delivery. Once the engine is running, the PCM uses the signal from the CKP sensor to calculate engine speed and to monitor misfires. The CKP sensor is not adjustable.

21   **Electrical load idle-up signal** - The diode module is located under the glovebox, near the cigarette lighter. When the heater blower motor, the rear window defogger and/or the headlights are turned on, the diode module sends a signal to the PCM, which responds by increasing the engine idle speed. On three-cylinder engines, the PCM increases the idle speed by activating the Idle Speed Control (ISC) motor, which opens the throttle plate slightly. On four-cylinder engines, the PCM increases idle speed by activating the Idle Air Control (IAC) valve, which opens the idle air passage slightly.

22   **Engine Coolant Temperature (ECT) sensor** - The ECT sensor is located in the cylinder head, near the distributor, on three-cylinder engines; it's located in the thermostat housing, near the ignition coil, on four-cylinder engines. The ECT sensor is a "thermistor" (a variable resistor that changes its resistance in proportion to the change in temperature). The ECT sensor is connected in series with a fixed resistor inside the PCM. The PCM supplies the ECT sensor with 5 volts, and then monitors the voltage drop across the ECT sensor as the temperature changes. The PCM converts the voltage reading into a temperature value and then uses this information to control fuel injection duration and ignition timing.

23   **Fuel level sensor** - The fuel level sensor, which is located inside the fuel tank, sends a signal to the PCM and to the fuel gauge in the instrument cluster. The PCM uses the signal from the fuel level sensor as one of several inputs for detecting EVAP system malfunctions. The PCM also uses the fuel level sensor signal to control the EVAP tank pressure control solenoid vacuum valve. If the fuel level exceeds a certain threshold, the PCM energizes the EVAP tank pressure control solenoid vacuum valve to prevent liquid fuel from entering the EVAP canister.

24   **Fuel tank pressure sensor** - The fuel tank pressure sensor, which is located on top of the fuel tank, measures the fuel vapor pressure in the fuel tank and compares it with the barometric pressure. The PCM supplies the fuel tank pressure sensor with 5 volts and then monitors the voltage signal from the sensor, which is also grounded through the PCM. The fuel tank pressure signal is one input used by the PCM to detect EVAP control system malfunctions.

25   **Heated Oxygen Sensor (HO2S)** - The oxygen sensors generate a voltage signal that varies with the difference between the oxygen content of the exhaust and the oxygen in the surrounding air. All 1996 and later vehicles covered by this manual are equipped with two oxygen sensors, an upstream HO2S and a downstream HO2S. The PCM uses the signal from the upstream

HO2S to determine whether the fuel system is running rich or lean; it uses the signal from the downstream HO2S to determine whether the catalytic converter is doing its job. Both HO2S units are heated, in order to reach their normal operating temperature as quickly as possible.

26   **Intake Air Temperature (IAT) sensor** - The IAT sensor is located in the air cleaner housing on three-cylinder engines, or in the intake duct on four-cylinder engines. The IAT sensor senses the temperature of the air entering the intake manifold. The IAT sensor is a "thermistor" (a variable resistor that changes its resistance in proportion to the change in temperature). The IAT sensor is connected in series with a fixed resistor inside the PCM. The PCM supplies the IAT sensor with 5 volts, and then monitors the voltage drop across the IAT sensor as the temperature changes. The PCM converts the voltage reading into a temperature value and then uses this information to control fuel injection duration. [On 1989 and 1990 models, this sensor is referred to as a *Manifold* Air Temperature (MAT) sensor.]

27   **Manifold Absolute Pressure (MAP) sensor** - As the engine load and rpm change, so does the pressure (or vacuum) inside the intake manifold. The MAP sensor measures these changes in pressure/vacuum. The MAP sensor is a piezoresistive crystal, a semi-conductor type of pressure-sensing element that converts any change in pressure/vacuum into an electrical signal. The PCM sends a 5-volt signal to the MAP sensor. The amount of voltage that returns to ground through the PCM is determined by, and is proportional to, the pressure/vacuum that's present inside the manifold. So the PCM is able to determine the pressure/vacuum in the manifold.

28   **Power Steering Pressure (PSP) switch** - The PSP switch is located on top of the power steering pump. More steering effort is required when the vehicle is turned at low speed because of the increased friction between the tires and the road. On vehicles with power steering, this higher effort is mitigated by the power steering system, but it imposes an additional load on the engine that is noticeable at low speed. When the steering wheel is turned enough to increase the power steering fluid pressure, the PSP switch closes, sending a signal to the PCM, which activates the Idle Air Control (IAC) valve, which increases the idle speed before the power steering load can cause a rough idle.

29   **Throttle Position (TP) sensor** - The TP sensor senses throttle movement and position. The TP sensor is a potentiometer mounted on the end of the throttle valve shaft. The PCM sends a 5-volt reference signal to the TP sensor, and the voltage also travels to ground through the PCM. As the throttle plate opens, the amount of voltage reaching the PCM changes. This analog voltage signal enables the PCM to determine when the throttle is closed, in cruise position or wide open. The PCM uses this information to control fuel injector(s), the Idle Speed Con-

trol (ISC) motor (three-cylinder engines) and the EGR solenoid vacuum valve. The PCM also converts the analog TP sensor voltage signal into a digital ON/OFF signal for use by the automatic transaxle.

30 Transaxle range switch - Used only on vehicles with an automatic transaxle, the transaxle range switch, also referred to as the Park-Neutral Position (PNP) switch, is located on the transaxle, at the manual selector lever. The range switch sends an ON/OFF signal to the PCM, which uses this input to control the fuel injectors, the Idle Air Control (IAC) valve and the performance of the automatic transaxle.

31 Vehicle Speed Sensor (VSS) - The VSS tells the PCM the vehicle's speed. On vehicles with a manual transaxle, the VSS is an integral part of the speedometer assembly consisting of a reed switch and a magnet. As the magnet turns with the speedometer cable, the reed switch turns ON and OFF. The ON/OFF frequency of the reed switch increases and decreases in proportion to vehicle speed. On vehicles with an automatic transaxle, the VSS is a magnetic pick-up coil, mounted on the transaxle case, that converts the transaxle countershaft revolutions into a pulsing AC voltage output to the PCM. When the countershaft rotates, the gear tooth cuts the magnetic field created by the VSS magnet, generating a voltage pulse in the VSS coil that's proportional to the vehicle speed.

Output actuators

32 A/C cutout signal - When you turn on the air conditioning system under normal driving conditions, the A/C compressor clutch engages. But during wide-open-throttle conditions, the PCM, which controls the ground side of the circuit between the A/C compressor control module and the compressor, can override your selection by keeping the control module circuit open.

33 Cooling fan control relay - The cooling fan control relay, which is located in the relay box in the engine compartment, controls the circuit for the engine cooling fan motor. When the ECT sensor indicates that the coolant is too hot, the PCM energizes the cooling fan control relay, which closes the circuit for the fan motor.

34 EVAP canister purge valve - The EVAP canister purge valve solenoid, which is located on the air intake plenum, is controlled by the PCM. The EVAP canister absorbs fuel vapors when the engine isn't running. When the engine is running and fully warmed up, the PCM energizes the canister purge valve solenoid, opening the canister purge valve and rerouting fuel vapors to the intake manifold.

35 EVAP canister vent valve - The EVAP canister vent valve, which is located near the EVAP canister, on top of the fuel tank, allows air to enter the EVAP canister. The canister vent valve is normally open, but is closed by the PCM when it checks for EVAP system leaks.

36 EVAP tank pressure control valve - The EVAP tank pressure control valve, which is located on top of the fuel tank, controls vapor flow into the EVAP canister. The EVAP tank pressure control valve is turned on (no vapor flow) by the PCM when the vehicle is stopped, the engine is running and the fuel level is lower than the specified value. It's also turned on when the vehicle is moving and the fuel level is higher than specified. The control valve is turned off (vapor flow) when the engine is turned off, and when the engine is running and the fuel level is higher than specified.

37 Fuel injectors - The PCM opens the fuel injectors individually in firing order sequence. The PCM also controls the time the injector is held open (pulse width). The pulse width of the injector (measured in milliseconds) determines the amount of fuel delivered. For more information on the fuel delivery system and the fuel injectors, including injector replacement, refer to Chapter 4.

38 Fuel pump control relay - The fuel pump control relay controls the fuel pump circuit. When the ignition switch is turned to the ON position, the relay is activated momentarily by the PCM, in order to energize the fuel pump so that it can pressurize the system fuel line. The fuel pump control relay is always activated by the PCM when the ignition switch is in the START or RUN position. For more information on fuel pump check and replacement, refer to Chapter 4.

39 Heated Oxygen Sensor (HO2S) heaters - All HO2S units are equipped with a heating element. Heating the HO2S allows it to reach operating temperature quickly. The PCM controls these heaters.

40 Idle Air Control (IAC) valve - The IAC valve, which is located on the throttle body, is controlled by the PCM. The IAC valve maintains the correct engine idle speed even when an additional load is placed on the engine. This load can be electrical or mechanical. Typical loads include turning on the headlights, shifting the (automatic) transaxle into gear and turning on the air conditioning. Engine idle is also affected by atmospheric pressure and by engine wear. The IAC valve also improves cold starts, improves driveability of the engine during warm-up and compensates for changes in the air-fuel mixture during deceleration.

41 Ignition coils - The PCM controls spark delivery and ignition timing depending on engine operation conditions. Refer to Chapter 5 for more information on the ignition system.

42 Main relay - The main relay, which is located in the relay box in the engine compartment, is controlled by the PCM. When the ignition key is turned to ON or when the engine is running, the main relay coil is grounded through the PCM. The main relay provides power to many PCM-controlled outputs, including the ignition system, the fuel injectors, the fuel pump, the EVAP control system solenoids and the IAC valve.

43 Malfunction Indicator Light (MIL) - The MIL, which is PCM-controlled, is located on the instrument cluster. The PCM illuminates the MIL if a malfunction occurs in the electronic engine control system.

44 Up-Shift Indicator Lamp (manual transaxle models) - The up-shift indicator lamp, which is PCM-controlled, is located on the instrument cluster. The PCM determines the optimal upshift point for best fuel economy, based on engine speed and load, and then illuminates the up-shift indicator lamp to tell the driver to upshift.

Obtaining diagnostic trouble codes

Refer to illustration 2.47

Note: *The diagnostic trouble codes on all models can only be extracted from the Engine Control Module (PCM) using a specialized scan tool. Have the vehicle diagnosed by a dealer service department or other qualified automotive repair facility if the proper scan tool is not available.*

45 The PCM monitors the circuits of most of the information sensors and actuators described above. When a malfunction is detected in one of these circuits, a trouble code is "set" (stored) in the PCM.

46 The PCM also illuminates the CHECK ENGINE light (also known as the Malfunction Indicator Lamp, or MIL) when it stores a trouble code. The MIL is located on the right side of the instrument cluster on 1996 through 2000 models; on 2001 models, the MIL is located in the left center part of the cluster, in the lower portion of the tachometer face. The MIL will remain illuminated until the problem is repaired and the code is cleared or the PCM does not detect any malfunction for several consecutive drive cycles.

47 The diagnostic trouble codes for the On-Board Diagnostic (OBD) system can only be extracted from the PCM using a scan tool, which interfaces with the OBD system by plugging into the diagnostic connector **(see illustration)**.

Clearing diagnostic trouble code

48 After the system has been repaired, the codes must be cleared from the PCM memory using a scan tool. Do not attempt to clear the codes by disconnecting battery power. If battery power is disconnected from the PCM,

2.47 The diagnostic connector (arrow) on OBD-II vehicles is located under the left end of the dash

the PCM will lose the current engine operating parameters and driveability will suffer until the PCM is programmed with a scan tool.

49   Always clear the codes from the PCM before starting the engine after a new electronic emission control component is installed onto the engine. The PCM stores the operating parameters of each sensor. The PCM may set a trouble code if a new sensor is allowed to operate before the parameters from the old sensor have been erased.

Diagnostic trouble code identification

50   The accompanying list of diagnostic trouble codes is a compilation of all the codes that may be encountered using a generic scan tool. Additional trouble codes may be available with the use of the manufacturer specific scan tool. Not all codes pertain to all models and not all codes will illuminate the Check Engine light when set. All models require a scan tool to access the diagnostic trouble codes.

Trouble codes

Code	Code Identification
P0106	Manifold Absolute Pressure (MAP) sensor, system performance
P0107	Manifold Absolute Pressure (MAP) sensor circuit, low voltage
P0108	Manifold Absolute Pressure (MAP) sensor circuit, high voltage
P0111	Intake Air Temperature (IAT) sensor performance
P0112	Intake Air Temperature (IAT) sensor circuit, low voltage
P0113	Intake Air Temperature (IAT) sensor circuit, high voltage
P0116	Engine Coolant Temperature (ECT) sensor circuit performance
P0117	Engine Coolant Temperature (ECT) sensor circuit, low voltage
P0118	Engine Coolant Temperature (ECT) sensor circuit, high voltage
P0121	Throttle Position (TP) sensor system performance
P0122	Throttle Position (TP) sensor circuit, low voltage
P0123	Throttle Position (TP) sensor circuit, high voltage
P0125	Engine Coolant Temperature (ECT), taking too long to go into closed loop fuel control
P0131	Heated Oxygen Sensor (HO2S) circuit, low voltage (pre-converter sensor)
P0132	Heated Oxygen Sensor (HO2S) circuit, high voltage (pre-converter sensor)
P0133	Heated Oxygen Sensor (HO2S), slow response (pre-converter sensor)
P0134	Heated Oxygen Sensor (HO2S) circuit, insufficient activity detected (pre-converter sensor)
P0135	Heated Oxygen Sensor (HO2S) heater circuit (pre-converter sensor)
P0136	Heated Oxygen Sensor (HO2S) circuit (post-converter sensor)
P0141	Heated Oxygen Sensor (HO2S) heater circuit (post-converter sensor)
P0171	Fuel trim system lean
P0172	Fuel trim system rich
P0300	Engine misfire detected
P0301	Cylinder no. 1 misfire detected
P0302	Cylinder no. 2 misfire detected
P0303	Cylinder no. 3 misfire detected
P0304	Cylinder no. 4 misfire detected
P0335	Crankshaft Position (CKP) sensor circuit
P0340	Camshaft Position (CMP) sensor circuit
P0420	Three Way Catalyst (TWC) system, low efficiency
P0440	EVAP system
P0450	EVAP system pressure sensor circuit
P0451	EVAP system pressure sensor performance
P0455	EVAP system leak detected
P0461	Fuel level sensor circuit performance
P0463	Fuel level sensor circuit high voltage
P0480	Cooling fan control circuit
P0500	Vehicle Speed Sensor (VSS) circuit
P0505	Idle control system
P0506	Idle speed low
P0507	Idle speed high
P0601	Powertrain Control Module (PCM) memory
P0603	Powertrain Control Module (PCM) memory reset

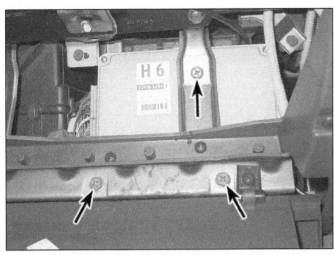

3.3 To remove the Powertrain Control Module (PCM), remove these three mounting bracket bolts (arrows), pull the PCM out of the dash and then disconnect the electrical connectors from the left end of the PCM

4.4 To detach the Camshaft Position (CMP) sensor from the cylinder head, disconnect the electrical connector from the sensor and then remove the sensor retaining bolt (arrows)

3   Powertrain Control Module (PCM) - removal and installation

Refer to illustration 3.3

Note: *On 1993 and later models the PCM is located under the driver's side of the dash.*

1   Disconnect the negative battery cable.

2   Remove the glove box (see Chapter 11).

3   Remove the three PCM mounting bracket bolts **(see illustration)** and then disconnect the three electrical connectors from the PCM.

4   Remove the PCM.

5   Installation is the reverse of removal.

4   Camshaft Position (CMP) sensor - replacement

1   Disconnect the negative battery cable.

1993 and later Metro three-cylinder models

2   On these models, the ignition system pick-up coil inside the distributor functions as the CMP sensor. To replace the CMP sensor, refer to Section 12 in Chapter 5.

1998 and later Metro four-cylinder models

Refer to illustration 4.4

3   Disconnect the negative battery cable.

4   Locate the CMP sensor at the left end of the cylinder head. Disconnect the electrical connector **(see illustration)** from the CMP sensor.

5   Remove the CMP sensor retaining bolt and then remove the sensor.

6   While the CMP sensor is removed, inspect the condition of the sensor signal rotor teeth with a flashlight and a small mirror.

7   Installation is the reverse of removal. Be

sure to tighten the CMP sensor retaining bolt to the torque listed in this Chapter's Specifications.

5   Crankshaft Position (CKP) sensor (1996 and later models) - replacement

Refer to illustration 5.2

1   Raise the vehicle and place it securely on jackstands.

2   Disconnect the electrical connector from the CKP sensor **(see illustration)**.

3   Remove the sensor retaining bolt and then remove the sensor.

4   While the sensor is removed, inspect the condition of the sensor signal rotor teeth with a flashlight and a small mirror.

5   Installation is the reverse of removal. Be

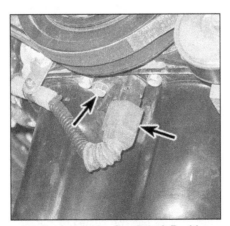

5.2 To detach the Crankshaft Position (CKP) sensor from the block, disconnect the electrical connector from the sensor and then remove the sensor retaining bolt (arrows)

sure to tighten the CKP sensor retaining bolt to the torque listed in this Chapter's Specifications.

6   Engine Coolant Temperature (ECT) sensor - replacement

Refer to illustrations 6.1, 6.2a and 6.2b

1   The ECT sensor is located in the intake manifold, near the EGR valve, on Sprint models. On 1989 through 1994 Metro three-cylinder models, the ECT sensor is on the side of the TBI unit. On all 1995 and later models, it's located at the left end of the cylinder head **(see illustration)**.

2   To check the coolant temperature sensor on 1995 and earlier models, unplug the ECT sensor electrical connector and then use an ohmmeter to measure the resistance between the two terminals **(see illustra-**

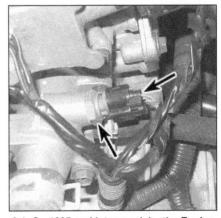

6.1 On 1995 and later models, the Engine Coolant Temperature (ECT) sensor (upper arrow) is located at the left end of the cylinder head; to replace an ECT sensor, simply unplug the electrical connector and then unscrew the sensor at the base (lower arrow)

6.2a Use an ohmmeter to observe the change in resistance of the coolant temperature sensor as the engine is warmed-up

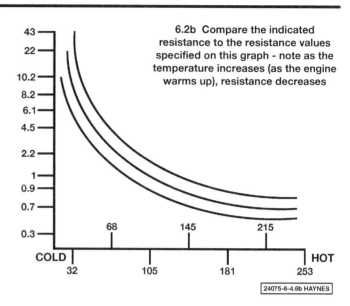

6.2b Compare the indicated resistance to the resistance values specified on this graph - note as the temperature increases (as the engine warms up), resistance decreases

24075-6-4.6b HAYNES

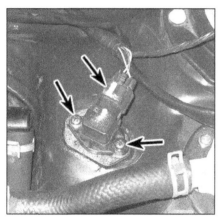

7.2 To detach the Fuel Tank Pressure (FTP) sensor from the fuel tank, unplug the electrical connector (upper arrow) and remove these two screws (lower arrows)

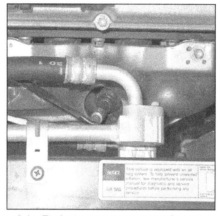

8.4a Typical upstream Heated Oxygen Sensor (HO2S) (1995 and later models)

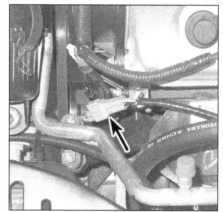

8.4b To locate the electrical connector (arrow) for the upstream HO2S, trace the electrical lead from the HO2S to the connector

tions). (Check the ECT sensor on 1996 and later vehicles, all of which are OBD-II models, with an aftermarket scan tool, or have a dealer check it for you.)

3   If the indicated resistance is out of range, replace the sensor.

4   To replace the ECT sensor, simply unplug the electrical connector and then unscrew the sensor.

5   Installation is the reverse of removal. Be sure to use Teflon tape or thread sealant on the threads of the new sensor to prevent leaks. Tighten the ECT sensor to the torque listed in this Chapter's Specifications.

7   Fuel Tank Pressure (FTP) sensor - replacement

Refer to illustration 7.2

1   Remove the fuel tank (see Chapter 4).

2   Disconnect the electrical connector from the FTP sensor **(see illustration)**.

3   Remove the two FTP sensor mounting screws and then remove the FTP sensor from

the fuel tank.

4   Installation is the reverse of removal. Be sure to tighten the FTP sensor bolts to the torque listed in this Chapter's Specifications.

8   Oxygen Sensor (O2S)/Heated Oxygen Sensor (HO2S) - replacement

Refer to illustrations 8.4a, 8.4b, 8.4c, 8.4d, 8.5a and 8.5b

1   All 1995 and earlier vehicles use a single oxygen sensor (O2S), which is located in the exhaust manifold or in the exhaust pipe below the manifold. 1995 models use a Heated Oxygen Sensor (HO2S). 1996 and later vehicles use two heated oxygen sensors, one upstream (ahead of) the catalytic converter and one downstream (behind the converter). However, the replacement procedure for all sensors is the same.

2   Disconnect the negative battery cable.

3   If you're replacing a downstream HO2S on a late-model vehicle, raise the front of the

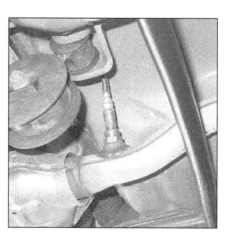

8.4c Typical downstream Heated Oxygen Sensor (HO2S) (1996 and later models)

vehicle and place it securely on jackstands.

4   Locate the O2S, or the upstream or downstream HO2S and then disconnect the O2S or HO2S electrical connector **(see illustrations)**.

8.4d To locate the electrical connector (arrow) for the downstream HO2S, trace the electrical lead from the HO2S to the connector

8.5a Using a special oxygen sensor socket to remove the upstream HO2S

8.5b Using a special oxygen sensor socket to remove the downstream HO2S

5   Unscrew and remove the O2S or HO2S. If you're planning to reuse the sensor, be careful not to damage the electrical lead. There are special oxygen sensor sockets, available from automotive tool companies, which protect the sensor and the electrical leads during removal and installation **(see illustrations)**.
6   Installation is the reverse of removal. Be sure to tighten the sensor to the torque listed in this Chapter's Specifications.

9   Intake Air Temperature (IAT) sensor - replacement

Refer to illustrations 9.2 and 9.3

1   The Manifold Air Temperature (MAT) sensor (1989 and 1990 models) or Intake Air Temperature (IAT) sensor (1991 and later models), is located in the side of the air cleaner housing (three-cylinder models) or in the air intake duct (four-cylinder models).
2   Unplug the MAT/IAT sensor electrical connector **(see illustration)**.
3   Pull the MAT/IAT sensor out of its grommet **(see illustration)**.
4   Inspect the sensor grommet. Make sure that it's not cracked or torn. If it is, replace it.
5   Installation is the reverse of removal.

10   Manifold Absolute Pressure (MAP) sensor - replacement

Refer to illustrations 10.1a and 10.1b

1   The Manifold Absolute Pressure (MAP) sensor is located on the firewall on three-cylinder models **(see illustration)** or on the air intake plenum on four-cylinder models **(see illustration)**.
2   Make sure that the MAP sensor vacuum hose connections are tight and that the hoses are in good condition. Look for damage to the MAP sensor electrical connector. If the MAP sensor assembly appears to be in good condition, but a trouble code still indicates a problem, have the MAP sensor checked by a dealer service department.

9.2 Unplug the IAT sensor electrical connector (arrow)

Three-cylinder models

3   Disconnect the MAP sensor electrical connector **(see illustration 10.1a)**.
4   Loosen the hose clamp and disconnect the vacuum hose from the MAP sensor.
5   Remove the MAP sensor filter from the

10.1a On three-cylinder models, the Manifold Absolute Pressure (MAP) sensor (1) is mounted on the firewall; to detach the MAP sensor from the firewall, unplug the electrical connector (2), disconnect the vacuum line (3) and then remove the sensor retaining bolt (near the connector, not visible in this photo)

9.3 To replace the IAT sensor, simply pull it out of its grommet

vacuum hose.
6   Disconnect the MAP sensor retaining bolt and then remove the MAP sensor and mounting bracket from the firewall.
7   Separate the MAP sensor from the sensor mounting bracket.
8   Installation is the reverse of removal.

10.1b On four-cylinder models, the Manifold Absolute Pressure (MAP) sensor (1) is mounted on the air intake plenum; to detach the MAP sensor from the plenum, unplug the electrical connector (2) and then remove the sensor bracket bolt (3)

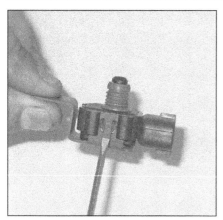

10.11 To separate the MAP sensor from its mounting bracket on four-cylinder models, pry open this locking tab and pull the MAP sensor straight up

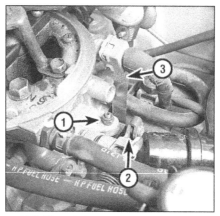

12.2 Throttle lever stop screw details

1   Throttle stop screw
2   Throttle lever
3   Feeler gauge

13.1 The Throttle Position (TP) sensor (arrow) is located on the throttle body unit on 1989 through 1991 models and on 1992 and later three-cylinder models

Four-cylinder models

Refer to illustration 10.11

9   Disconnect the MAP sensor electrical connector **(see illustration 10.1b)**.
10   Remove the MAP sensor mounting bolt.
11   Remove the MAP sensor and bracket and then separate the sensor from the bracket **(see illustration)**.
12   Installation is the reverse of removal.

11   Power Steering Pressure (PSP) switch - replacement

1   Remove the right engine mount (see Chapter 2A).
2   Locate the PSP switch on top of the power steering pump.
3   Disconnect the PSP switch electrical connector.
4   Unscrew and remove the PSP switch from the power steering pump.
5   Installation is the reverse of removal. Be sure to use a new O-ring and tighten the switch to the torque listed in this Chapter's Specifications.

12   Throttle Switch (TS) (1989 through 1991 manual transaxle models) - check and adjustment

Check

Refer to illustration 12.2

1   The Throttle Switch (TS) is located on the throttle body. By monitoring the output voltage from the TS, the PCM can determine fuel delivery based on throttle valve angle (driver demand). The Throttle Switch consists of two contact points: one contact closes at idle and the other closes during a wide-open-throttle condition.
2   To check the throttle switch, disconnect the TS electrical connector and insert a 0.012 inch (0.3 mm) feeler gauge between the throt-tle lever stop screw and throttle lever **(see illustration)**.

3   There should be continuity between terminal A and terminal B **(see illustration 13.2)**. If there is no continuity, adjust the switch.

Adjustment

4   To adjust the TS, insert a 0.024 inch (0.6 mm) feeler gauge between the throttle lever stop screw and the throttle lever.
5   Connect an ohmmeter between terminal A and terminal B **(see illustration 13.2)**.
6   Loosen the TS bolts and move the switch clockwise until the ohmmeter reads open (no continuity). Tighten the screws to 18 in-lbs.
7   Check that there is no continuity with a 0.35 inch (0.9 mm) feeler gauge between the throttle lever stop screw and throttle lever.
8   Check that there is continuity with a 0.12 inch (0.3 mm) feeler gauge between the throttle lever stop screw and throttle lever.
9   If the readings are correct, reinstall the electrical connector and the air cleaner assembly. If the throttle switch requires diag-

nosis or replacement, it should be done by a dealer service department (because of the need for special tools and test equipment).

13   Throttle Position (TP) sensor - replacement

1989 through 1992 automatic transaxle models, 1993 and later three-cylinder models and 1995 through 1997 four-cylinder models

Check

Refer to illustrations 13.1 and 13.2
Note: *The following procedure applies to 1989 through 1991 manual transaxle models and to all 1992 and later three-cylinder models.*

TEST	TPS Terminal	Throttle Position	Valves
1	Terminal A - use ⊖ probe Terminal D - use ⊕ probe	Closed or in the idle position	Between 4.37 to 8.13k ohms
2	Terminal A - use ⊖ probe Terminal C - use ⊕ probe	Closed or in the idle position	Between 240 to 1140 ohms
3	Terminal A - use ⊖ probe Terminal C - use ⊕ probe	Wide open position	Between 3.17 to 6.6k ohms
4	Between A and B terminals (Idle switch)	When throttle lever-to-stop screw clearance is 0.3 mm (0.012 in.)	Continuity
5	Between A and B terminals (Idle switch)	When throttle lever-to-stop screw clearance is 0.9 mm (0.035 in.)	No Continuity

24075-4-15.31 HAYNES

13.2 Throttle Position (TP) sensor resistance (1989 through 1991 models with automatic transaxle and all 1992 and later three-cylinder models)

13.11 Disconnect the Throttle Position (TP) sensor electrical connector (arrow) (1998 and later four-cylinder models)

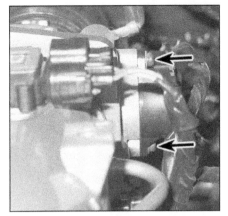

13.12 To detach the TP sensor from the throttle body on a 1998 or later four-cylinder model, remove these two screws (arrows)

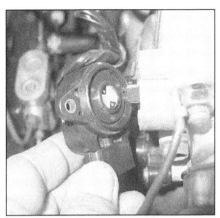

13.14 When installing the TP sensor on a 1998 or later four-cylinder model, position the sensor on the throttle body so that the upper sensor screw hole is slightly to the right (passenger's side) of the screw hole in the throttle body; this aligns the tab on the throttle shaft with the notch in the TP sensor

1   The TP sensor (see illustration) is located on the throttle body.
2   Unplug the TPS electrical connector and check the resistance between the terminals when the throttle is fully open and fully closed (see illustration). If the readings aren't as specified, adjust the throttle position sensor. If the correct readings cannot be obtained even after adjustment, then replace the sensor.

Replacement

3   Unplug the TP sensor electrical connector.
4   Remove the TP sensor retaining screws.
5   Remove the TP sensor.
6   Installation is the reverse of removal. Be sure to adjust the TP sensor before tightening the sensor retaining screws to the torque listed in this Chapter's Specifications.

Adjustment

7   To adjust the sensor, insert a 0.012 inch (0.3 mm) feeler gauge between the throttle lever stop screw and the throttle lever (see illustration 12.2).
8   Connect an ohmmeter between terminal A and terminal B (see illustration 13.2).
9   Loosen the TP sensor retaining screws and then move the switch counter-clockwise until the ohmmeter reads 0 ohms (continuity). Tighten the screws to the torque listed in this Chapter's Specifications.
10   Using a 0.35 inch (0.9 mm) feeler gauge

between the throttle lever stop screw and throttle lever, verify that there is no continuity.

1998 and later four-cylinder models

Refer to illustrations 13.11, 13.12 and 13.14
Note: There is no adjustment procedure for these models.
11   Disconnect the TP sensor electrical connector (see illustration).
12   Remove the TP sensor retaining screws (see illustration).
13   Remove the TP sensor.
14   When installing the TP sensor, place the sensor in position on the throttle body with the sensor screw hole slightly to the right (toward the passenger side); this correctly engages the throttle lever tab with the groove in the TP sensor (see illustration).

14   Transmission Range (TR) switch (automatic transaxle) - replacement

Refer to illustrations 14.1 and 14.4
Note: On 1994 and earlier models, the TR switch is referred to as a Park/Neutral Position (PNP) switch, but it looks virtually identical to a TR switch, and is removed, installed

and adjusted the same way.
1   Disconnect the TR switch electrical connector (see illustration), detach the wiring harness from the clip on the transaxle and set the harness aside.
2   Remove the TR switch retaining bolt (see illustration 14.1) and then remove the TR switch.
3   Move the transaxle shift lever to the Neutral position (at the transaxle end, the manual selector lever is all the way forward when it's in Park, so Neutral is two detent positions - two clicks - to the rear).
4   If the shift lever was in Neutral before you removed the TR switch, the switch should still be in Neutral. If the shift lever wasn't in Neutral, or if you don't know whether the TR switch has been turned since it was removed, use a flat-bladed screwdriver to turn the TR switch counterclockwise until it stops; this is the Park position. Then turn it clockwise until you hear two clicks (see illustration); this is Neutral.

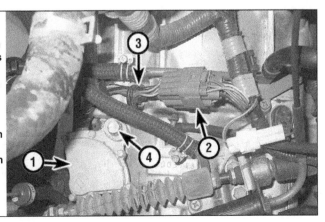

14.1 The Transmission Range (TR) switch (1) is located on top of the transaxle; to remove it, disconnect the TR switch electrical connector (2), detach the wiring harness from the clip (3), set the harness aside, and then remove the TR switch retaining bolt (4)

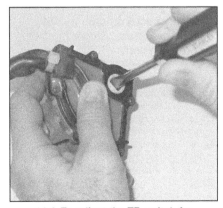

14.4 To adjust the TR switch for installation, use a flat-bladed screwdriver to turn the TR switch counterclockwise until it stops (this is Park) and then turn it clockwise two clicks (the Neutral position)

5   Install the TR switch on the transaxle, install the switch retaining bolt and tighten it to the torque listed in this Chapter's Specifications.

6   Plug in the electrical connector and reattach the harness to the transaxle with the harness clamp.

7   To verify that the TR switch is correctly installed:

a) *Apply the parking brake and then block the wheels.*

b) *With the manual selector lever in the Park position, turn the ignition switch key to the START position and verify that the starter motor operates.*

c) *Stop the engine and then turn the ignition switch key to the ON position.*

d) *Move the manual selector lever from the Park position to the Neutral position. Turn the ignition switch to the START position and then verify that that the starter motor operates.*

e) *Verify that the starter motor does NOT operate in Drive, 2nd, Low or Reverse.*

---

15   Vehicle Speed Sensor (VSS) - replacement

Manual transaxle models

1   The Vehicle Speed Sensor (VSS) is built into the speedometer. If the VSS is defective, replace the instrument cluster (see Chapter 12).

Automatic transaxle models

Refer to illustration 15.2

2   The VSS **(see illustration)** is located on top of the transaxle.

3   Disconnect the VSS electrical connector.

4   Remove the VSS retaining bolt and then remove the VSS.

5   Installation is the reverse of removal. Be sure to tighten the VSS retaining bolt to the torque listed in this Chapter's Specifications.

---

16   Early Fuel Evaporation (EFE) heater (three-cylinder models) - replacement

1   The Early Fuel Evaporation (EFE) heater, which is located between the throttle body and the intake manifold, heats up the air/fuel mixture during cold starts. The EFE heater is controlled by the PCM, which uses the ECT and IAT sensors to determine when to turn on the EFE heater and when to turn it off.

2   Remove the throttle body (see Chapter 4).

3   Disconnect the EFE heater electrical connector.

4   Remove the EFE heater and gasket from the intake manifold.

5   Installation is the reverse of removal. Be sure to use a new EFE heater gasket.

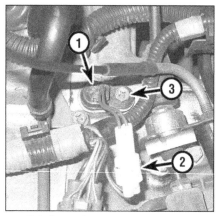

15.2 On vehicles with an automatic transaxle, the VSS (1) is located on top of the transaxle; to remove the VSS, simply unplug the electrical connector (2) and then remove the retaining bolt (3)

---

17   Idle Speed Control (ISC) motor (1994 and later three-cylinder and 1995 through 1997 four-cylinder models) - replacement

Note: *The throttle body on pre-1994 models is equipped with an Idle Air Control (IAC) valve (see Section 18). In 1994, 49-state models with the "standard emissions" package also used an IAC valve; California models with the "upgraded emissions" package are equipped with an Idle Speed Control (ISC) motor. All 1995 and later three-cylinder models, and 1995 through 1997 four-cylinder models, are equipped with an ISC motor.*

Check for tampering

1   Before replacing the ISC motor, inspect the throttle lever screw and the throttle stop screw for tampering:

a) *Open the throttle lever to the wide open position and verify that the yellow paint hasn't come off the throttle lever screw.*

b) *Turn the resin cap on the throttle stop screw (throttle body side) with your fingers. Inspect the cap for damage.*

2   If any yellow paint has come off the throttle lever screw, or if the resin cap is damaged, do not replace the ISC motor - replace the entire throttle body.

Replacement

3   Remove the air cleaner (see Chapter 4).

4   Disconnect the ISC motor electrical connector.

5   Remove the three ISC motor mounting screws and then remove the ISC motor from the throttle body.

6   Installation is the reverse of removal. Be sure to tighten the ISC motor mounting screws to the torque listed in this Chapter's Specifications.

7   When you're done, start the engine, warm it up and verify that the idle speed is within the normal range (see Chapter 1).

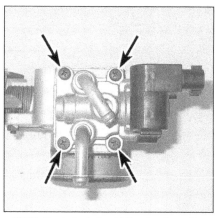

18.12 To detach the IAC valve from the throttle body, remove these four retaining screws (arrows), separate the IAC valve from the throttle body and then remove and discard the old rubber gasket

---

18   Idle Air Control (IAC) valve - replacement

1989 through 1994 models

Fast-idle air valve

1   The fast-idle air valve, also referred to more simply as the air valve, raises the idle speed by allowing additional air to bypass the throttle plate when the engine is cold. The fast-idle air valve consists of a thermo-wax pellet, a piston and a spring-loaded valve. When the cold engine is started, the thermo-wax pellet is contracted and the spring-loaded valve is open, allowing some extra air to enter the intake manifold, which raises the engine rpm to a fast idle. As the engine and throttle body warm up, the thermo-wax pellet begins to expand, pressing against the piston, which starts to close the valve, which lowers the idle rpm. By the time the temperature reaches about 140 degrees to 176 F (60 to 80 degrees C), the pellet is fully expanded and the piston has closed the valve, cutting off the fast-idle air passage and lowering the engine idle to its normal speed. If the engine doesn't run at a fast idle when it is cold, or if it doesn't idle down once it's warmed up, check the operation of the air valve as follows.

2   Remove the throttle body (see Chapter 4), remove the components from the throttle body, remove the three screws that attach the upper and lower halves of the throttle body unit and then separate the throttle body halves.

3   Place a pot of water on the stove and immerse the thermo-wax pellet part of the fast-idle air valve in the water. **Caution:** *Do NOT submerge the throttle body in the water. Doing so could cause damaging corrosion to the throttle body.*

4   Heat the water and then verify that the fast-idle air valve closes as the water temperature increases. If the valve is not fully closed by the time the water temperature reaches 176 degrees F (80 degrees C), replace the fast-idle air valve.

19.1 A typical late-model downstream catalytic converter

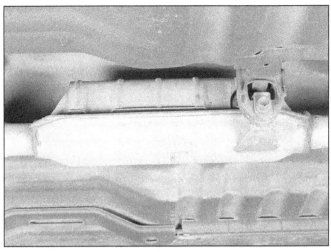

19.3 If the catalytic converter is mounted under the vehicle, periodically inspect the shield for dents and other damage - if a dent is deep enough to touch the surface of the converter, replace the shield

5   Reassemble and install the throttle body (see Chapter 4).

Idle Air Control (IAC) valve

Note: *On 1989 through 1992 models, this system is referred to as the Idle Speed Control (ISC) solenoid valve. But it's identical to the Idle Air Control (IAC) valve used on 1993 and 1994 models.*

6   The Idle Air Control (IAC) valve opens and closes an air bypass passage to maintain a smooth engine idle during all operating conditions. For example, it maintains the specified idle speed when an additional load, such as the air conditioning compressor or power steering pump, is imposed upon the engine. Unlike the fast-idle air valve, the IAC valve is controlled by the Powertrain Control Module (PCM). When the PCM determines that the engine idle is being dragged down by a load, it energizes the solenoid valve inside the IAC valve, which opens a passage through the valve and increases the idle rpm.

7   Disconnect the IAC valve electrical connector.

8   Remove the two IAC valve retaining screws and then remove the IAC valve and bracket from the throttle body.

9   Separate the IAC valve from the valve mounting bracket.

10   Installation is the reverse of removal.

1998 and later four-cylinder models

Refer to illustration 18.12

11   Remove the throttle body (see Chapter 4).

12   Remove the four IAC valve retaining screws **(see illustration)** and then remove the IAC valve from the throttle body.

13   Remove and discard the old rubber gasket.

14   Installation is the reverse of removal. Be sure to use a new rubber gasket and tighten

the four IAC valve retaining screws to the torque listed in this Chapter's Specifications.

---

19   Catalytic converter - removal and installation

Note: *Because of a federally mandated extended warranty which covers emissions-related components such as the catalytic converter, check with a dealer service department before replacing the converter at your own expense.*

General description

Refer to illustration 19.1

1   To reduce hydrocarbon, carbon monoxide and oxides of nitrogen emissions, all vehicles are equipped with a three-way catalytic converter **(see illustration)** which oxidizes and reduces these chemicals, converting them into harmless nitrogen, carbon dioxide and water. On 1995 and later models, there are *two* catalytic converters: The upstream converter is bolted to the exhaust manifold, and the downstream converter is located under the engine.

Checking

Refer to illustrations 19.3 and 19.4

2   Periodically inspect the catalytic converter-to-exhaust pipe mating flanges and bolts. Make sure that there are no loose bolts and no leaks between the flanges.

3   Look for dents in or damage to the catalytic converter protector **(see illustration)**. If any part of the protector is damaged or dented enough to touch the converter, repair or replace it.

4   Inspect the heat insulator for damage. Make sure that there is adequate clearance between the heat insulator and the catalytic converter **(see illustration)**.

Replacement

Refer to illustrations 19.6a, 19.6b and 19.7

Warning: *Make sure that the engine exhaust system is cool before beginning this procedure.*

5   Raise the front of the vehicle and place it securely on jackstands.

6   To replace an upstream catalytic converter, remove the lower flange bolts and nuts and then remove the upper flange bolts

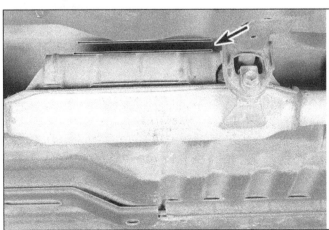

19.4 If the catalytic converter is mounted under the vehicle, periodically inspect the heat insulator to make sure there is adequate clearance between it and the converter

and nuts **(see illustrations)**. You might have to apply penetrant to the threads of the bolts to loosen the nuts.

7   To replace a downstream converter, remove the upper flange bolts and nuts **(see illustration 19.6b)** and then remove the rear flange nuts and bolts **(see illustration)**. You might have to apply penetrant to the threads of the bolts to loosen the nuts.

8   Installation is the reverse of removal. Be sure to use new gaskets where applicable and new fasteners. Tighten all nuts and bolts securely.

20   Evaporative Emission (EVAP) system - component replacement

General description

Refer to illustration 20.1

1   The Evaporative Emission Control (EVAP) system **(see illustration)** is designed to capture and store fuel vapors (raw, unburned hydrocarbons) - from the fuel tank, the carburetor (Sprint models) and the intake manifold - that would otherwise escape into the atmosphere. The EVAP system consists of a charcoal-filled canister, the lines connecting the canister to the fuel tank and various devices that control the flow of vapors between the fuel tank, the canister and the intake manifold. Fuel vapors are routed from the fuel tank, carburetor, intake manifold, etc. to the canister, where they're stored when the engine isn't running. When the engine is running and fully warmed up, the fuel vapors are purged from the canister to the intake manifold, from which they're drawn into the

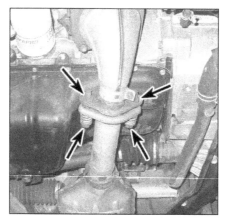

19.6a To remove an upstream catalytic converter, remove the lower flange nuts and bolts (arrows) . . .

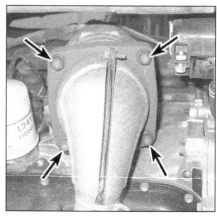

19.6b . . . and then remove the upper flange nuts and bolts (arrows)

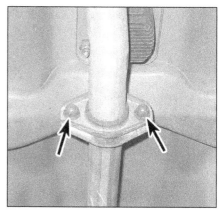

19.7 To remove a downstream catalytic converter, remove the front flange nuts and bolts (see illustration 19.6b) and then remove the rear flange nuts and bolts (arrows)

combustion chambers with the air/fuel mixture and then burned in the normal combustion process. All EVAP systems are similar, but the means by which vapors are routed from the fuel tank to the canister, and from the canister to the intake manifold, vary from year to year and from model to model. The EVAP system is also protected by a Federally-mandated extended warranty, so we recommend having it serviced by a dealer while it's under warranty. Once it's out of warranty, you can easily replace the canister, and some of the other components, on earlier models. However, EVAP system troubleshooting, particularly on OBD-II (1996 and later) models, is complicated. These models have highly complex EVAP systems that are capable of detecting leaks in the system and capturing vapors released during refueling. If the integrity of an OBD-II EVAP system is vio-

20.1 Schematic of the EVAP system

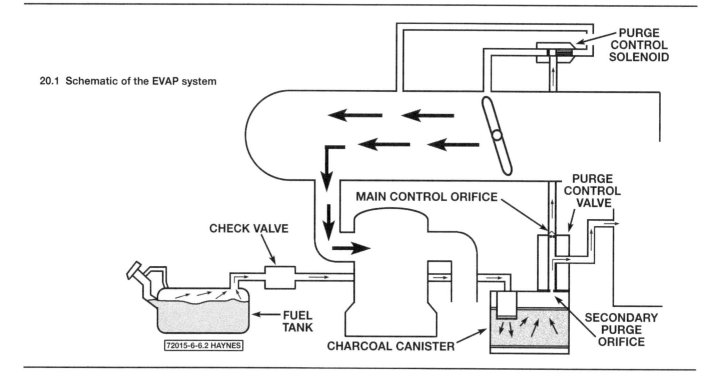

PURGE CONTROL SOLENOID

MAIN CONTROL ORIFICE

PURGE CONTROL VALVE

CHECK VALVE

FUEL TANK

CHARCOAL CANISTER

SECONDARY PURGE ORIFICE

72015-6-6.2 HAYNES

20.12 Typical early charcoal canister mounting details (Sprint model shown, early Geo Metro models similar)

A   Carburetor vent line
B   Purge control valve

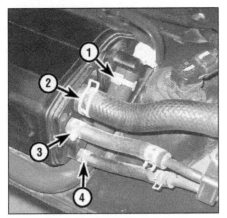

20.16a Before removing the EVAP canister on OBD-II models, disconnect these hoses (arrows)

1   To EVAP canister vent solenoid
2   To Fuel limiter vent (shut-off) valve
3   To EVAP canister purge valve
4   To fuel vapor separator

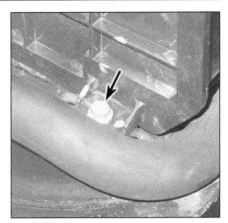

20.17 To detach the EVAP canister from the fuel tank, remove this bolt (arrow)

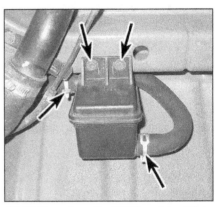

20.20 To replace the EVAP canister filter, loosen these two hose clamps (arrows), disconnect the two hoses, and then remove the two filter mounting bolts (arrows)

lated, it can trigger a host of diagnostic trouble codes. Finally, OBD-II EVAP systems are located on top of the fuel tank, so no diagnosis is even possible without dropping the fuel tank.

2   When fuel evaporates, it emits raw hydrocarbons in a gaseous state. These vapors increase the pressure inside the fuel tank. They escape by pushing open a one-way check valve known as a tank pressure control (TPC) valve (1998 and later models have two TPC valves: one for the tank, and another for the filler neck and the shut-off valve). From the TPC valve, the vapors travel to the canister, where they're absorbed by the activated charcoal inside the canister. When the engine is started, any vapors stored in the canister are "purged," i.e. drawn out of the charcoal by intake vacuum and routed to the intake manifold. A canister purge valve, also known as a solenoid purge valve, which is turned on and off by the PCM, either prevents fuel vapors from escaping the canister, or allows them to travel to the intake manifold. But the canister purge valve only determines whether the vapors can leave the canister or not; it doesn't control the actual ported vacuum source that pulls the vapors into the intake manifold. This job is handled by another device.

3   On 1992 and earlier models, a bi-metal vacuum switching valve (BVSV) controls the ported vacuum signal from the intake manifold to the canister. When the coolant temperature is low, the BVSV closes and does not allow ported vacuum to the canister purge control valve. When the coolant has reached operating temperature, the BVSV allows vacuum to the canister purge control valve which in turn allows the vapor from the canister to be sucked through the purge control valve and then through the hose between the valve and the intake manifold. On 1993 and 1994 models, a thermal vacuum valve (TVV) replaces the BVSV, but the system controls the ported vacuum source to the canister purge valve the same way as a BVSV.

On 1995 and later models, the PCM determines when the purge valve is on or off. On these models, there is no BVSV or TVV.

4   On 1998 and later EVAP systems, an On-Board Refueling Vapor Recovery (ORVR) system prevents any fuel vapors from escaping during refueling. These models are also equipped with a self-testing capability that allows the PCM to check the system for leaks. The EVAP systems on these models are complex, and troubleshooting should be left to a dealer service department.

Checking

5   Poor idle, stalling and poor driveability can be caused by an inoperative purge valve, a damaged canister, split or cracked hoses or hoses connected to the wrong fittings. Check the fuel filler cap for a damaged or deformed gasket (see Chapter 1).

6   Evidence of fuel loss or fuel odor can be caused by liquid fuel leaking from fuel lines, a cracked or damaged canister, an inoperative purge valve, disconnected, misrouted, kinked, deteriorated or damaged vapor or control hoses.

7   Inspect each hose attached to the canister for kinks, leaks and cracks along its entire length. Repair or replace as necessary.

8   Inspect the canister. If it's cracked or damaged, replace it.

9   Look for fuel leaking from the bottom of the canister. If fuel is leaking, replace the canister and check the hoses and hose routing.

10   Further testing should be left to a dealer service department.

Component replacement
Charcoal canister

1994 and earlier models
Refer to illustration 20.12

11   Detach the negative cable from the battery.

12   Clearly label, then detach the vacuum hoses from the canister (see illustration).

13   Remove the canister mounting bolts and lift it out of the vehicle.

14   Installation is the reverse of removal.

1995 and later models
Refer to illustrations 20.16 and 20.17

15   Remove the fuel tank (see Chapter 4).

16   Disconnect the four vapor hoses from the canister (see illustration).

17   Remove the canister retaining bolt (see illustration) and then remove the canister.

18   Installation is the reverse of removal.

Other EVAP components (1996 and later vehicles)

EVAP canister filter
Refer to illustration 20.20

19   Raise the rear of the vehicle and place it securely on jackstands.

20   Locate the EVAP canister filter (see illustration), near the fuel filler neck pipe, at the left rear corner of the vehicle.

21   To replace the EVAP canister filter, loosen the hose clamps, disconnect the two hoses and then remove the two filter mounting bolts.

22   Installation is the reverse of removal.

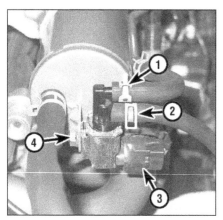

20.23 To replace the EVAP canister purge valve, disconnect the vapor hoses (1 and 2), unplug the electrical connector (3) and then remove the purge valve mounting nut (4)

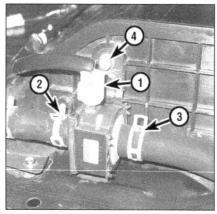

20.27 Locate the EVAP canister vent valve next to the EVAP canister; to replace the canister vent valve, unplug the electrical connector (1), loosen the hose clamps (2 and 3), disconnect the vapor hoses and then remove the vent valve mounting bracket bolt (4)

20.29 When installing the canister vent valve, make sure that the alignment tab on the valve is aligned with its corresponding hole in the EVAP canister before installing the valve mounting bracket bolt.

EVAP canister purge valve

Refer to illustration 20.23

23   Open the hood and locate the EVAP canister purge valve **(see illustration)** on the end of the air intake plenum.

24   To replace the EVAP canister purge valve, clearly label and then disconnect the vapor hoses, unplug the electrical connector and then remove the purge valve mounting nut.

25   Installation is the reverse of removal.

EVAP canister vent valve

Refer to illustrations 20.27 and 20.29

26   Remove the fuel tank (see Chapter 4).

27   Locate the EVAP canister vent valve **(see illustration)** next to the EVAP canister.

28   To replace the EVAP canister vent valve, unplug the electrical connector, loosen the hose clamps, disconnect the vapor hoses and then remove the vent valve mounting

bracket bolt.

29   When installing the canister vent valve, make sure that the alignment tab on the valve is aligned with its corresponding hole in the EVAP canister **(see illustration)** before installing the valve mounting bracket bolt.

30   Installation is otherwise the reverse of removal.

EVAP tank pressure control valve

Refer to illustration 20.32

31   Remove the fuel tank (see Chapter 4).

32   Locate the EVAP tank pressure control valve **(see illustration)** near the front edge of the fuel tank.

33   To replace the tank pressure control valve, disconnect the vapor hoses, unplug the electrical connector and then remove the valve mounting bracket bolt.

34   Installation is the reverse of removal.

Fuel limiter vent (shut-off) valve

Refer to illustration 20.36

35   Remove the fuel tank (see Chapter 4).

36   Locate the fuel limiter vent (shut-off) valve **(see illustration)** on top of the fuel tank, just ahead of the fuel pump/fuel gauge sending unit assembly.

37   To replace the fuel limiter vent valve, loosen the hose clamps, disconnect the two vapor hoses and then remove the vent valve mounting screws.

38   Installation is the reverse of removal.

Fuel vapor separator

Refer to illustration 20.40

39   Remove the fuel tank (see Chapter 4).

40   Locate the fuel vapor separator **(see illustration)** on the right rear edge of the fuel tank.

41   To replace the fuel vapor separator, disconnect the two vapor hoses and then remove the vapor separator retaining screw.

42   Installation is the reverse of removal.

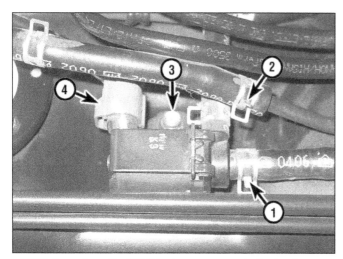

20.32 To replace the EVAP tank pressure control valve, loosen the hose clamps (1 and 2), disconnect the vapor hoses, unplug the electrical connector (3) and then remove the valve bracket mounting bolt (4)

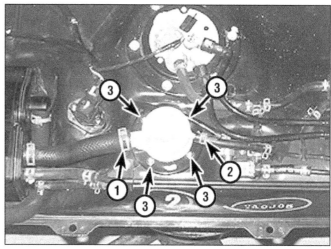

20.36 To replace the fuel limiter vent valve, loosen the hose clamps (1 and 2), disconnect the two vapor hoses and then remove the vent valve mounting screws (3)

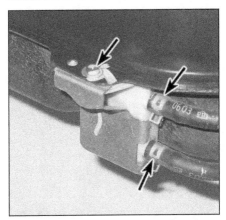

20.40 To replace the fuel vapor separator, loosen the hose clamps (right arrows), disconnect the two vapor hoses and then remove the vapor separator retaining screw (left arrow)

21   Exhaust Gas Recirculation (EGR) system (three-cylinder models) - component replacement

General description

Refer to illustration 21.3

1   To reduce oxides of nitrogen emissions, some of the exhaust gases are recirculated through the EGR valve to the intake manifold to lower combustion temperatures.

2   On carburetor-equipped models, the EGR system consists of the EGR valve, EGR vacuum modulator and a bi-metal vacuum switching valve (BVSV). The EGR valve, which is operated by ported vacuum recirculates gases in accordance with engine load (intake air volume). To eliminate recirculation at idle, the vacuum signal is ported above the idle throttle position. During cold engine operation, the thermovalve opens, bleeding off ported vacuum and keeping the EGR valve closed. When the engine coolant temperature exceeds the set temperature of the BVSV, it closes and ported vacuum is applied to the EGR valve. The EGR vacuum modula-

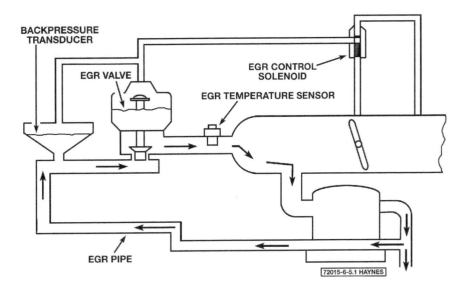

21.3 Typical EGR system

tor controls the EGR valve by controlling the vacuum signal to the EGR valve with an atmospheric bleed. This bleed is controlled by the amount of exhaust pressure which acts on the bottom of the EGR vacuum modulator.

3   On fuel-injected models, the EGR system **(see illustration)** consists of the EGR valve, the EGR modulator, vacuum switching valve (VSV), the Electronic Control Module (ECM) and various sensors. The PCM memory is programmed to produce the ideal EGR valve lift for each operating condition.

Check

EGR valve

Refer to illustration 21.4

Warning: *The EGR valve becomes very hot during engine operation - wear gloves when checking the valves to avoid burning your fingers.*

4   Check the EGR valve diaphragm by pushing on it slightly with your finger **(see**

illustration). The diaphragm should move and not be frozen in one spot.

5   Start the engine and allow it to idle. Detach the vacuum hose from the EGR valve and attach a hand vacuum pump in its place.

6   Apply vacuum to the EGR valve. Vacuum should remain steady and the engine should run poorly.

   a) *If the vacuum doesn't remain steady and the engine doesn't run poorly, replace the EGR valve and recheck it.*

   b) *If the vacuum remains steady but the engine doesn't run poorly, remove the EGR valve and check the valve and the intake manifold for blockage. Clean or replace parts as necessary and recheck.*

EGR vacuum modulator valve

Refer to illustrations 21.7a, 21.7b and 21.9

7   Remove the valve (see Step 13 below). Pull the cover off and check the filters **(see illustrations)**. Clean them with compressed air, reinstall the cover and the modulator.

21.4 Gently push the EGR diaphragm with your finger

21.7a To remove the EGR vacuum modulator filter for cleaning, remove this cap . . .

21.7b . . . then pull out the filter and blow it out with compressed air

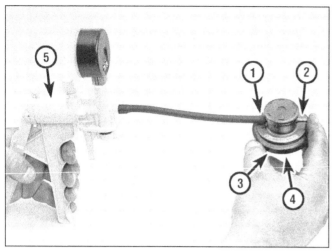

21.9 The EGR modulator should hold vacuum when it is applied to port P, port Q is plugged and the air is blown into port A

1	Port P	4	Air
2	Port Q	5	Vacuum pump gauge
3	Port A		

21.11 The EGR valve is located at the base of the intake manifold (carburetor removed for clarity)

21.13 Label each vacuum hose before removing them from the EGR modulator

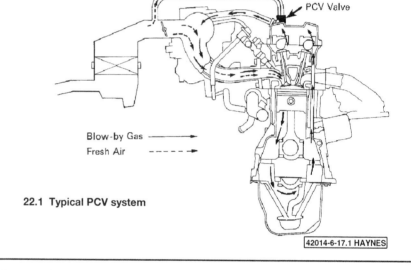

22.1 Typical PCV system

8   Plug one of the vacuum ports on the side of the valve and blow into the other port. Air should pass through the valve and come out the filter at the top of the valve.

9   Connect a vacuum pump to the modulator **(see illustration)**, plug the other vacuum port with your finger then blow into the port on the bottom. Operate the pump - the modulator should hold vacuum as long as air is being blown into the bottom port.

10   Any further checking of the EGR systems requires special tools and test equipment. Take the vehicle to a dealer service department for checking.

Component replacement

EGR valve

Refer to illustration 21.11

11   Disconnect the threaded fitting that attaches the EGR pipe to the EGR valve, remove the two EGR valve mounting bolts **(see illustration)**, remove the EGR valve from the intake manifold and check it for sticking

and heavy carbon deposits. If the valve is sticking or clogged with deposits, clean or replace it.

12   Installation is the reverse of removal.

EGR vacuum modulator valve

Refer to illustration 21.13

13   Label and disconnect the vacuum hoses **(see illustration)** and remove the EGR vacuum modulator from its bracket.

14   Installation is the reverse of removal.

22   Positive Crankcase Ventilation (PCV) system

Refer to illustration 22.1

1   To reduce hydrocarbon (HC) emissions, crankcase blow-by gas is routed to the intake manifold for combustion in the cylinders **(see illustration)**.

2   The main components of the PCV system are the PCV valve, a fresh air filtered inlet

and the vacuum hoses connecting these components with the engine.

3   To maintain idle quality, the PCV valve restricts the flow when the intake manifold vacuum is high. If abnormal operating conditions arise, the system is designed to allow excessive amounts of blow-by gases to flow back through the crankcase vent tube into the air cleaner to be consumed by normal combustion.

4   Checking and replacement of the PCV valve is covered in Chapter 1.

23   Pulse Air Control system - check and component replacement

General information

1   Additional oxygen is needed to aid the oxidation of HC and CO in the catalytic converter. The Pulse Air Control system supplies secondary air into the exhaust manifold when the engine coolant temperature is low or dur-

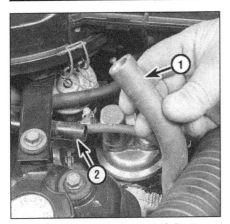

23.2 Remove the second air hose from the air cleaner

1   *Second air hose*
2   *Air cleaner*

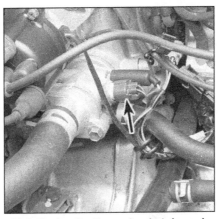

23.6 Unplug the thermal switch (arrow), then use a jumper wire to connect the terminals in the connector

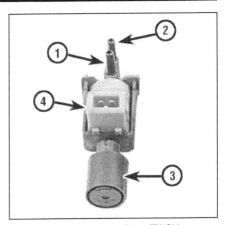

23.9 Diagram of the TWSV

1   *Nozzle*
2   *Nozzle*
3   *Filter*
4   *TWSV (Blue)*

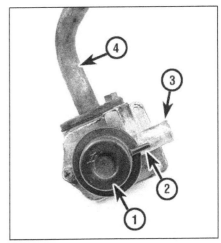

23.11 Secondary air valve

1   *Air valve*
2   *Vacuum hose*
3   *Air hose*
4   *Air pipe*

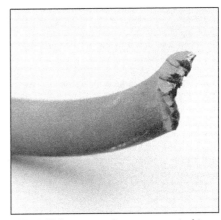

23.14 This vacuum hose was routed too close to an exhaust manifold - after being overheated repeatedly, it finally cracked and broke

ing deceleration. The air control valve is a simple reed-type valve that opens when exhaust gas pressure is low. When open, oxygen-rich air is drawn into the exhaust from the air cleaner to aid the converter with the oxidation process. When the exhaust gas pressure is high, the reed valve is closed (because of the high exhaust pressure) and prevents backflow of the exhaust gases. The alternating conditions cause the system to "pulse."

Checking

Non-PCM controlled (1985 and 1986 models only)

Refer to illustrations 23.2, 23.6, 23.9 and 23.11

2   Disconnect the second air hose from the air cleaner when the engine is cold **(see illustration)**.

3   Start the engine and allow it to idle. Air should be drawn into the hose and a bubbling sound should be heard from the inside of the hose.

4   Allow the engine to warm up to operating temperature. There should be no air drawn into the hose (no bubbling sound heard).

5   Accelerate the engine to 3,000 rpm and release the pedal quickly. Air should be drawn into the hose (bubbling sound is heard).

6   Check the operation of the thermal switch. Disconnect the coupler of the thermal switch and jump the terminals with a wire lead **(see illustration)**. Air should be drawn into the hose during idle (bubbling sound should be heard).

7   If the thermal switch is defective, replace it with a new unit. Reconnect the air hose and the thermal switch connector.

8   Check the operation of the Three Way Solenoid Valve. Disconnect the two vacuum hoses from the TWSV (Blue).

9   Blow air into nozzle no. 1. Air should come out of no. 3 but not out of no. 2 **(see illustration)**.

10   Use an external source of power or con-

nect leads from the battery and apply 12 volts to the TWSV. Blow air into no. 1. Air should come out of no. 2. Replace the valve with a new unit if it is faulty.

11   Check the operation of the Secondary Air Valve (SAV). Remove the SAV with the air pipe **(see illustration)**.

12   While applying vacuum to the diaphragm of the SAV, blow into the air hose. Air should come out of the pipe. Blow air into the pipe. There should not be any air coming out of the hose.

13   If the SAV is faulty, replace it with a new unit.

PCM controlled (1987 and 1988 models only)

Refer to illustration 23.14

14   Check for cracked hoses, pipes, damaged electrical connectors or any exposed wires in the electrical harness **(see illustration)**. Have the system checked by a dealer service department or other repair shop.

Component replacement

Non-PCM controlled (1985 and 1986 models only)

15   To replace the thermal switch, use an open end wrench or a box end wrench to remove the switch **(see illustration 23.6)**. **Warning:** *Wait until the engine is completely cool before removing the thermal switch. Install the new switch as quickly as possible (to minimize coolant loss). Check the coolant level and add some, if necessary (see Chapter 1).*

16   To replace the Three Way Solenoid Valve (TWSV), disconnect the vacuum lines and electrical connector from the valve **(see illustration 23.9)**.

17   Remove the bolts that hold the TWSV to the firewall. Replace it with a new unit.

18   To replace the Secondary Air Valve, use an open end wrench and separate the flanged nut from the exhaust manifold. Spray penetrating lubricant onto the pipe threads. Remove the bracket that holds the air valve to the engine and remove the SAV assembly.

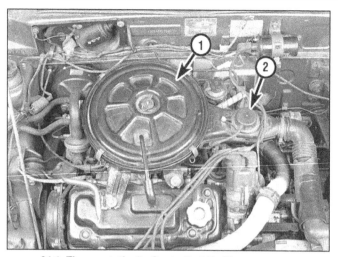

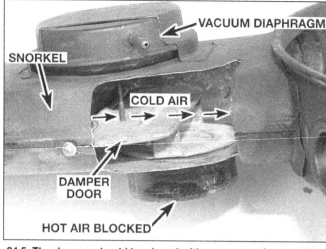

24.1 Thermostatically Controlled Air Cleaner system

1   *Air cleaner*   2   *Air control actuator*

24.5 The damper should be closed with a warm engine condition

PCM controlled (1987 and 1988 models only)

19   If the Three Way Solenoid Valve (TWSV) is diagnosed as faulty, remove the bolts that retain it to the firewall. Replace the valve with a new unit.
20   Replace any damaged hoses or electrical connectors.

---

24   Intake air temperature control system - check and component replacement

---

General description

Refer to illustration 24.1

1   The Intake Air Temperature Control system consists of a thermo wax and damper mounted in the snorkel of the air cleaner **(see illustration)**. **Note:** *On early models (1985 and 1986), the system is slightly different and is referred to as the Thermostatically Controlled Air Cleaner (TAC) system.*
2   This system controls the intake air temperature by mixing the warm air (pre-heated air) and cold air (non-preheated air) to improve fuel vaporization and engine warm-up characteristics.
3   With the engine temperature cold, the thermo-wax contracts and closes the air duct that allows cold air into the air cleaner (cold air duct) and opens the duct that allows warm air from near the exhaust manifold into the air cleaner (warm air duct).
4   As the engine is warmed up, the temperature of the incoming air causes the

thermo-wax to expand. Gradually, the cold air duct is opened to allow the cool air to mix with the warm air. The temperature of the intake air is controlled to maximize the efficiency of the air/fuel mixture. An improperly functioning ITAC system can result in poor cold or hot driveability and excessive emissions.

Check

Air Control Actuator (ACA) and damper (1985 and 1986 models only)

Refer to illustration 24.5

5   With the engine off, disconnect the warm air hose from the air intake housing and feel the damper **(see illustration)** - it should be shut (covering the warm air inlet duct).
6   Start the engine. If the air cleaner is cold, the damper should now open up. Allow the engine to warm up to normal operating temperature - the door should begin to close, allowing cool air to pass into the air intake housing.
7   If the damper doesn't move as described above, disconnect the vacuum hose to the ACA and connect a hand-held vacuum pump to it. Apply vacuum and feel for the damper to open - if it doesn't, replace the ACA.

Thermo sensor

8   If the damper valve does open, check the thermo sensor. Disconnect the two hoses from the sensor, which is located on the air cleaner lid **(see illustration 24.5)**.
9   Measure the air temperature around the

sensor and write it down. Attach a length of hose to one side of the sensor, cover the other port with your finger and blow into the hose. If the recorded temperature is above 104-degrees F (45-degrees C), air should bleed out of the valve, indicating the valve is open.
10   If the temperature is below 77-degrees F (25-degrees C), no air should come out. If the thermo sensor doesn't work as described, replace it.

Component replacement

Air control actuator (ACA)

11   Pull the warm air duct out of the outlet on the air intake housing. Disconnect the vacuum hose from the ACA.
12   Remove the mounting bolt from the ACA, then pull the ACA out. Lift the ACA to disconnect the rod from the damper door.
13   When installing the ACA, position the rod with the bent end pointing down, push in on the damper door, then engage the rod with the hole in the damper door. A small mirror and flashlight will be helpful when doing this.
14   Install the ACA into the hole.
15   Reconnect the vacuum hose and the warm air duct.

Thermo sensor

16   Unplug the vacuum hoses from the sensor. Remove the top of the air cleaner, along with the air intake case, from the carburetor.
17   Pry the spring clip off the ports on the thermo sensor and remove the sensor from the under side of the air cleaner lid.
18   Installation is the reverse of the removal procedure.

Chapter 7 Part A
Manual transaxle

Contents

Specifications

Lubricant type .... See Chapter 1

Torque specifications

	Ft-lbs (unless otherwise indicated)	Nm
Transaxle-to-engine bolts	44	60
Shift lever nuts	53 in-lb	6
Shift lever-to-guide plate nuts	53 in-lb	6
Shift lever guide plate bolts	89 in-lb	10
Shift rod bolts and nuts	15	20
Shift lever guide case bolts	106 in-lb	12

1 General information

The vehicles covered by this manual are equipped with either a five-speed manual or a three-speed automatic transaxle. Information on the manual transaxle is in this Part of Chapter 7. Service procedures for the automatic transaxle are in Chapter 7, Part B.

The manual transaxle is a compact, two-piece, lightweight aluminum alloy housing containing both the transmission and differential assemblies. Because of the complexity, unavailability of new parts and special tools necessary, internal repair procedures for the manual transaxle are not recommended for the home mechanic. For readers who wish to tackle a transaxle rebuild, exploded views are provided. The bulk of information in this Chapter is devoted to removal and installation procedures.

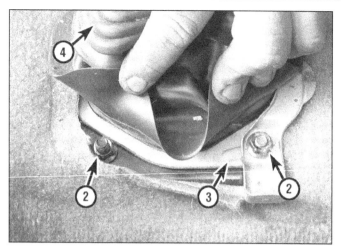 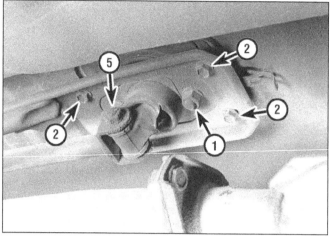

2.2 Shift lever adjustment details

| 1 | *Shift lever housing nut* | 2 | *Guide plate bolt* | 3 | *Guide plate* | 4 | *Shift lever* | 5 | *Shift rod bolt/nut* |

2   Shift lever - adjustment

Refer to illustration 2.2

1   The shift lever is properly adjusted if, when in Neutral, it rests in the center of the shift pattern. If it doesn't, the shift lever position must be adjusted to center it.

2   Loosen the shift assembly housing nuts and guide plate bolts, then move the guide plate forward or back so the shift lever in the middle of the plate is at a right angle **(see illustration)**.

3   Once the shift lever is centered, tighten the guide plate bolts, followed by the housing nuts.

3   Shift lever assembly - removal and installation

Refer to illustration 3.3

1   Remove the shifter boot and center console (Chapter 12).

2   Raise the vehicle and support it securely on jackstands.

3   From under the vehicle, remove the nut and bolt, detach the control rod from the shift lever, then remove the four guide plate bolts **(see illustration)**.

4   From inside the vehicle, remove the two lever cover screws (if equipped).

5   Scribe around the shift lever housing with a marking pen so it can be reinstalled to the same position, remove the four housing nuts, then push the control lever boots in the hole in the floor.

6   Move the control rod out of the way and remove the shift lever assembly by lowering it through the hole in the floor.

7   Installation is the reverse of removal.

4   Manual transaxle - removal and installation

Removal

1   Disconnect the negative cable from the battery.

2   Raise the vehicle and support it securely on jackstands.

3   Drain the transaxle lubricant (see Chapter 1).

4   Disconnect the shift and clutch linkage from the transaxle.

5   Remove the starter motor and, on Geo models, the starter plate.

6   Detach the speedometer cable and electrical connectors from the transaxle.

7   Remove the exhaust system components as necessary for clearance.

8   Support the engine. This can be done from above with an engine hoist, or by placing a jack (with a block of wood as an insulator) under the engine oil pan. The engine must remain supported at all times while the transaxle is out of the vehicle!

9   Remove any chassis or suspension components that will interfere with transaxle removal (see Chapter 10).

10   Disconnect the driveaxles from the transaxle (see Chapter 8).

11   Support the transaxle with a jack, then remove the bolts securing the transaxle to the engine.

12   Remove the transaxle mount nuts and bolts.

13   Make a final check that all wires and hoses have been disconnected from the transaxle, then carefully pull the transaxle and jack away from the engine.

14   Once the input shaft is clear, lower the transaxle and remove it from under the vehicle. **Caution:** *Do not depress the clutch pedal while the transaxle is out of the vehicle.*

15   With the transaxle removed, the clutch components are now accessible and can be inspected. In most cases, new clutch components should be routinely installed when the transaxle is removed.

Installation

16   If removed, install the clutch components (see Chapter 8).

17   With the transaxle secured to the jack with a chain, raise it into position behind the engine, then carefully slide it forward, engag-

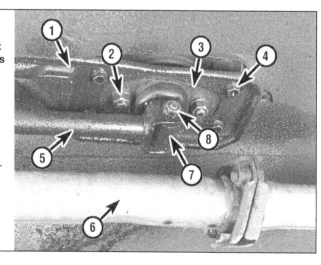

3.3 Shift lever to shift rod and chassis details

1   *Shift assembly extension rod*
2   *Housing nut*
3   *Guide plate*
4   *Guide plate bolt*
5   *Shift Rod*
6   *Exhaust pipe*
7   *Shift control lever*
8   *Shift rod nut and bolt*

ing the input shaft with the clutch plate hub splines. Do not use excessive force to install the transaxle - if the input shaft does not slide into place, readjust the angle of the transaxle so it is level and/or turn the input shaft so the splines engage properly with the clutch plate hub.

18   Install the transaxle-to-engine bolts. Tighten the bolts securely.

19   Install the transaxle mount nuts or bolts.

20   Install the chassis and suspension components which were removed. Tighten all nuts and bolts securely.

21   Remove the jacks supporting the transaxle and engine.

22   Install the various items removed previously, referring to Chapter 8 for installation of the driveaxles and Chapter 4 for information regarding the exhaust system components.

23   Make a final check that all wires, hoses, linkages and the speedometer cable have been connected and that the transaxle has been filled with lubricant to the proper level (see Chapter 1).

24   Connect the negative battery cable.

Road test the vehicle for proper operation and check for leaks.

5   Manual transaxle overhaul - general information

Overhauling a manual transaxle is a difficult job for the do-it-yourselfer. It involves the disassembly and reassembly of many small parts. Numerous clearances must be precisely measured and, if necessary, changed with select fit spacers and snap-rings. As a result, if transaxle problems arise, it can be removed and installed by a competent do-it-yourselfer, but overhaul should be left to a transmission repair shop. Rebuilt transaxles may be available - check with a dealer parts department and auto parts stores. At any rate, the time and money involved in an overhaul is almost sure to exceed the cost of a rebuilt unit.

Nevertheless, it's not impossible for an inexperienced mechanic to rebuild a transaxle if the special tools are available and the job is done in a deliberate step-by-step manner so nothing is overlooked.

The tools necessary for an overhaul include internal and external snap-ring pliers, a bearing puller, a slide hammer, a set of pin punches, a dial indicator and possibly a hydraulic press. In addition, a large, sturdy workbench and a vice or transaxle stand will be required.

During disassembly of the transaxle, make careful notes of how each piece comes off, where it fits in relation to other pieces and what holds it in place. Noting how they're installed when you remove the parts will make it much easier to get the transaxle back together.

Before taking the transaxle apart for repair, it will help if you have some idea what area of the transaxle is malfunctioning. Certain problems can be closely tied to specific areas in the transaxle, which can make component examination and replacement easier. Refer to the *Troubleshooting* Section at the front of this manual for information regarding possible sources of trouble.

Notes

Chapter 7 Part B
Automatic transaxle

Contents

Specifications

General

Fluid type	See Chapter 1
Throttle valve (TV) cable boot-to-inner cable stopper clearance	0 to 0.02 in (0 to 0.5 mm)
Torque converter flange nut-to-transaxle face clearance	27/32 in (21.4 mm)

Torque specifications

	Ft-lbs	Nm
Transaxle-to-engine bolts	40	55
Torque converter-to-driveplate bolts	14	19

1   General information

All vehicles covered in this manual come equipped with either a five-speed manual or a three-speed automatic transaxle. All information on the automatic transaxle is included in this Part of Chapter 7. Information for the manual transaxle can be found in Part A of this Chapter.

Due to the complexity of the automatic transaxles covered in this manual and the need for special equipment to perform most service operations, this Chapter contains only general diagnosis, routine maintenance, adjustment and removal and installation procedures.

If the transaxle requires major repair work, it should be left to a dealer service department or an automotive or transmission repair shop. You can, however, remove and install the transaxle yourself and save the expense, even if the repair work is done by a transmission shop.

2   Diagnosis - general

Note: *Automatic transaxle malfunctions may be caused by five general conditions: poor engine performance, improper adjustments, hydraulic malfunctions, mechanical malfunctions or malfunctions in the computer or its signal network. Diagnosis of these problems should always begin with a check of the easily repaired items: fluid level and condition (see Chapter 1), shift linkage adjustment and* *throttle linkage adjustment. Next, perform a road test to determine if the problem has been corrected or if more diagnosis is necessary. If the problem persists after the preliminary tests and corrections are completed, additional diagnosis should be done by a dealer service department or transmission repair shop. Refer to the Troubleshooting Section at the front of this manual for information on symptoms of transaxle problems.*

Preliminary checks

1   Drive the vehicle to warm the transaxle to normal operating temperature.
2   Check the fluid level as described in Chapter 1:
  a) *If the fluid level is unusually low, add enough fluid to bring the level within the designated area of the dipstick, then check for external leaks (see below).*
  b) *If the fluid level is abnormally high, drain off the excess, then check the drained fluid for contamination by coolant. The presence of engine coolant in the automatic transmission fluid indicates that a failure has occurred in the internal radiator walls that separate the coolant from the transmission fluid (see Chapter 3).*
  c) *If the fluid is foaming, drain it and refill the transaxle, then check for coolant in the fluid, or a high fluid level.*
3   Check the engine idle speed. **Note:** *If the engine is malfunctioning, do not proceed with the preliminary checks until it has been repaired and runs normally.*
4   Check the throttle valve cable for freedom of movement. Adjust it if necessary (see Section 5). **Note:** *The throttle cable may function properly when the engine is shut off and cold, but it may malfunction once the engine is hot. Check it cold and at normal engine operating temperature.*
5   Inspect the shift cable (see Section 3). Make sure that it's properly adjusted and that the linkage operates smoothly.

Fluid leak diagnosis

6   Most fluid leaks are easy to locate visually. Repair usually consists of replacing a seal or gasket. If a leak is difficult to find, the following procedure may help.
7   Identify the fluid. Make sure it's transmission fluid and not engine oil or brake fluid (automatic transmission fluid is a deep red color).
8   Try to pinpoint the source of the leak. Drive the vehicle several miles, then park it over a large sheet of cardboard. After a minute or two, you should be able to locate the leak by determining the source of the fluid dripping onto the cardboard.
9   Make a careful visual inspection of the suspected component and the area immediately around it. Pay particular attention to gasket mating surfaces. A mirror is often helpful for finding leaks in areas that are hard to see.
10   If the leak still cannot be found, clean the suspected area thoroughly with a

3.3 Loosen the cable nut then turn the adjusting nut until it contacts the shift cable joint

1   *Adjusting nut*
2   *Shift cable joint*
3   *Cable nut*

degreaser or solvent, then dry it.
11   Drive the vehicle for several kilometers at normal operating temperature and varying speeds. After driving the vehicle, visually inspect the suspected component again.
12   Once the leak has been located, the cause must be determined before it can be properly repaired. If a gasket is replaced but the sealing flange is bent, the new gasket will not stop the leak. The bent flange must be straightened.
13   Before attempting to repair a leak, check to make sure that the following conditions are corrected or they may cause another leak. **Note:** *Some of the following conditions cannot be fixed without highly specialized tools and expertise. Such problems must be referred to a transmission shop or a dealer service department.*

Gasket leaks

14   Check the pan periodically. Make sure the bolts are tight, no bolts are missing, the gasket is in good condition and the pan is flat (dents in the pan may indicate damage to the valve body inside).
15   If the pan gasket is leaking, the fluid level or the fluid pressure may be too high, the vent may be plugged, the pan bolts may be too tight, the pan sealing flange may be warped, the sealing surface of the transaxle housing may be damaged, the gasket may be damaged or the transaxle casting may be cracked or porous. If sealant instead of gasket material has been used to form a seal between the pan and the transaxle housing, it may be the wrong sealant.

Seal leaks

16   If a transaxle seal is leaking, the fluid level or pressure may be too high, the vent may be plugged, the seal bore may be damaged, the seal itself may be damaged or improperly installed, the surface of the shaft

protruding through the seal may be damaged or a loose bearing may be causing excessive shaft movement.
17   Make sure the dipstick tube seal is in good condition and the tube is properly seated. Periodically check the area around the speedometer gear or sensor for leakage. If transmission fluid is evident, check the O-ring for damage.

Case leaks

18   If the case itself appears to be leaking, the casting is porous and will have to be repaired or replaced.
19   Make sure the oil cooler hose fittings are tight and in good condition.

Fluid comes out vent pipe or fill tube

20   If this condition occurs, the transaxle is overfilled, there is coolant in the fluid, the case is porous, the dipstick is incorrect, the vent is plugged or the drain back holes are plugged.

---

3   Shift cable - check and adjustment

Refer to illustration 3.3

Check

1   With the engine running and your foot firmly on the brake, place the shift lever in each detent and make sure the transaxle shifts into the corresponding gear as the indicator is moved.

Adjustment

2   Raise the vehicle and support it securely on jackstands.
3   Loosen the cable nut on the manual shift lever at the transaxle **(see illustration)**.
4   Move the transaxle shift lever to the Neutral position.
5   Move the shift lever inside the vehicle to the Neutral position.
6   Turn the adjusting nut until it contacts the shift cable joint, then tighten the cable nut securely.
7   Check the operation of the transaxle in each shift lever position (try to start the engine in each gear - the starter should operate in the Park and Neutral positions only).

---

4   Backdrive system - description and adjustment

Refer to illustration 4.3

Description

1   The backdrive system used on early models ensures that the shift lever in Park whenever the engine is started and won't allow the ignition key to be removed except in Park. The system is operated by the backdrive cable which connects the ignition

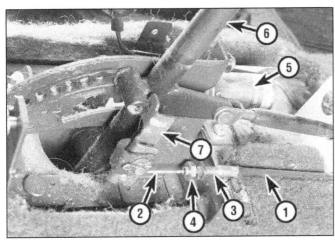

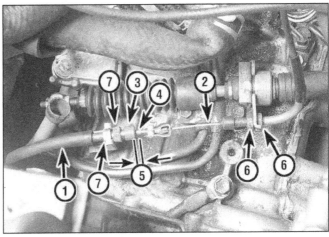

4.3 Backdrive cable adjustment details

1	Backdrive cable casing	5	Solenoid
2	Backdrive cable	6	Shift lever
3	Nut	7	Key release plate
4	Nut		

5.3 TV cable details

1	TV cable	5	Boot-to-inner cable
2	Accelerator cable		clearance
3	Boot	6	Upper cable adjusting nut
4	Inner cable stopper	7	Lower cable adjusting nut

switch and the shifter and a solenoid. Should it be necessary to shift from Park to another detent position with the key removed, such as when the battery is dead or the key is unavailable, a manual release knob is provided to release the backdrive mechanism.

Adjustment

2   Remove the console (see Chapter 11).
3   With the shift lever in Park, completely loosen nuts 3 and 4 on the backdrive cable **(see illustration)**.
4   Pull the cable casing forward until there is no deflection on the inner cable, hand tighten nut 3 first, then nut 4. Tighten the nuts securely.
5   After adjustment, make sure that when the shift lever is in Park the ignition key can be turned from the Accessory to the Lock position, then removed. When the shift lever is in any other position than Park, it should not be possible to turn the key from Accessory to Lock.

5   Throttle Valve (TV) cable - check and adjustment

Refer to illustration 5.3
1   Warm up the engine to normal operating temperature. Shut off the engine and remove the TV cable cover.
2   Have an assistant hold the accelerator pedal down while you watch the TV cable link in the engine compartment to make sure it opens fully.
3   If the link doesn't open all the way, use a feeler gauge to check the TV cable boot-to-inner cable clearance **(see illustration)**. Compare this clearance measurement to the Specifications at the beginning of this Chapter.
4   If the clearance is not as specified, loosen adjusting nuts (6 in **illustration 5.3**) and adjust the cable clearance. If this doesn't achieve the specified clearance, turn upper

adjusting nuts (7 in **illustration 5.3**). After adjustment, tighten the nuts.
5   Install the cable cover.

6   Neutral start switch - replacement and adjustment

Refer to illustrations 6.2, 6.6 and 6.7
1   Disconnect the negative cable from the battery.
2   Shift the transaxle into Neutral **(see illustration)**.
3   Remove the nut and lift off the shift lever.
4   Unplug the electrical connector.
5   Remove the bolt and detach the switch from the shift shaft.
6   Before installing the switch, insert a screwdriver into the switch slot and turn the screwdriver until you hear a click **(see illustration)**.

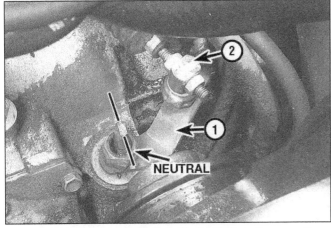

6.2 Make sure the transaxle is in Neutral before disconnecting the shift lever

1	Shift lever	2 Shift cable

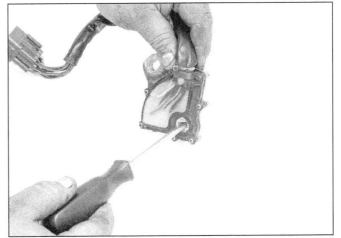

6.6 Insert a screwdriver into the slot in the switch and turn it until the switch clicks into Neutral

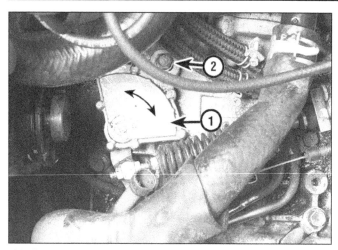

6.7 Rotate the switch until it clicks into Neutral before tightening the bolt

    1   Neutral start switch   2   Bolt

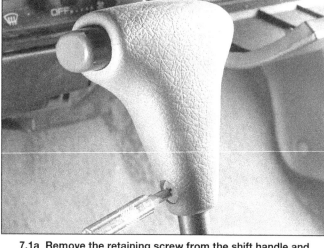

7.1a Remove the retaining screw from the shift handle and remove the knob

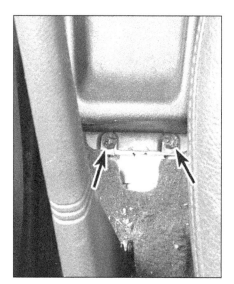

7.1b Remove the two rear mounting screws and . . .

7   Push the switch onto the shaft and rotate it until the switch clicks, indicating that it is in Neutral, then install the bolt **(see illustration)**. Tighten the bolt securely.

8   Connect the negative battery cable and verify that the engine will start only in Neutral or Park.

7   Shift control assembly - removal and installation

Refer to illustration 7.1a, 7.1b, 7.1c and 7.1d

1   Remove the shift control knob, the center console, the shift indicator and the housing **(see illustrations)**. **Warning:** *Some models covered by this manual are equipped with an airbag. Always disable the airbag system before working in the vicinity of the steering column, impact sensors or dashboard to avoid the accidental deployment of the* airbag, which could cause personal injury. *See Chapter 12 for the airbag disarming procedure.*

2   Remove the split pin, pull out the pin and detach the cable from the shift control assembly.

3   Detach the backdrive cable.

4   Unplug the electrical connectors.

5   From under the vehicle, remove the four shift control assembly mounting nuts.

6   Remove the shift control assembly.

7   Installation is the reverse of removal.

8   Shift cable - replacement

Refer to illustrations 8.3 and 8.4

1   Detach the cable from the negative battery terminal.

2   Remove the center console (see Section 7).

3   Detach the shift cable from the shift

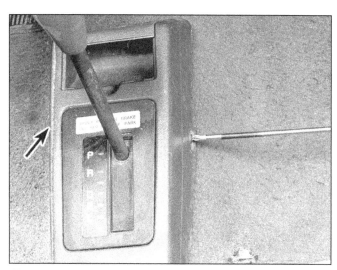

7.1c . . . then remove the retaining screws from each side of the console and lift the console out of the way

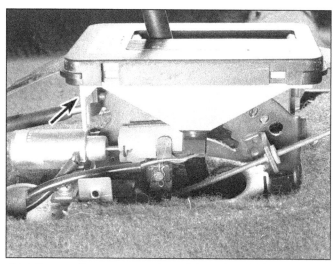

7.1d Remove the cap screws (arrow) from the shift indicator and remove the indicator (early model shown, later models similar)

8.3 Remove the horse shoe clip (arrow) and pull the cable housing from the bracket

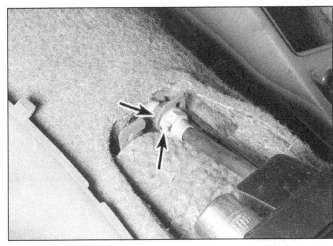

8.4 Loosen the nuts (arrows) and detach the cable from the floor bracket

control assembly **(see illustration)**.
4   Loosen the nuts and detach the cable from the floor bracket **(see illustration)**.
5   Raise the vehicle and place it securely on jackstands.
6   Remove the nut and detach the shift cable from the shift lever.
7   Detach the shift cable from the bracket on the transaxle.
8   Detach the shift cable from the cable clips on the crossmember.
9   Remove the cable from under the vehicle.
10   Installation is the reverse of removal. After installation, adjust the cable as described in Section 3.

9   Oil seals - replacement

Refer to illustrations 9.10a and 9.10b
1   Oil leaks frequently occur due to wear of the driveaxle oil seals, and/or the speedometer cable or drive gear O-ring. Replacement of these seals is relatively easy, since the

repairs can usually be performed without removing the transaxle from the vehicle.
2   The driveaxle oil seals are located at the sides of the transaxle, where the driveaxles are attached. If leakage at the seal is suspected, raise the vehicle and support it securely on jackstands. If the seal is leaking, lubricant will be found on the sides of the transaxle.
3   Refer to Chapter 8 and remove the driveaxles.
4   Using a screwdriver or pry bar, carefully pry the oil seal out of the transaxle bore.
5   If the oil seal cannot be removed with a screwdriver or pry bar, a special oil seal removal tool (available at auto parts stores) will be required.
6   Using a large section of pipe or a large deep socket as a drift, install the new oil seal. Drive it into the bore squarely and make sure it's completely seated.
7   Install the driveaxle(s). Be careful not to damage the lip of the new seal.
8   The speedometer cable and driven gear housing or speed sensor connection is located on the transaxle housing. Look for

lubricant around the cable housing to determine if the grommet is leaking.
9   Disconnect the speedometer cable from the transaxle. On Sprint models, grasp the speedometer cable securely near the connection to the transaxle and pull out sharply to disconnect it. On Geo models, remove the bolt and separate the driven gear from the transaxle.
10   On Sprint models, remove the bolt in the transaxle, withdraw the cable end, and use a small screwdriver to replace the O-ring. On Geo models, use a hooked tool to remove the O-ring from the driven gear housing, press the new one in evenly and reinstall the speedometer cable assembly **(see illustrations)**.

10   Transaxle mounts - check and replacement

Refer to illustration 10.1
1   Insert a large screwdriver or pry bar between each mount and the transaxle and

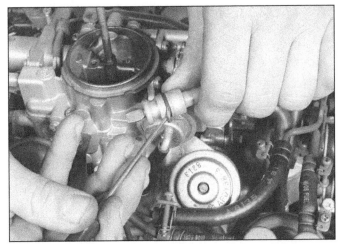

9.10a On Sprint models, use a small screwdriver to remove the O-ring from the end of the speedometer cable

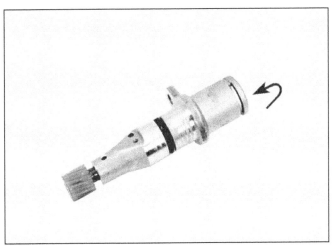

9.10b Use a hooked piece of wire to remove the grommet (arrow) from the speedometer gear (Metro)

pry up while watching the mount **(see illustration)**.

2   If the rubber separates from the metal plate on the mount, or the case moves up, but not down (mount bottomed out), replace the mount.

3   To replace a mount, support the transaxle with a jack, remove the nuts and bolts and detach the mount. It may be necessary to raise the transaxle slightly to provide enough clearance to remove the mount.

4   Installation is the reverse of removal.

11   Automatic transaxle - removal and installation

Refer to illustration 11.6

Removal

1   Disconnect the negative cable from the battery. Remove the battery and tray.

2   Raise the vehicle and support it securely on jackstands.

3   Drain the transaxle fluid (see Chapter 1).

4   Remove the torque converter cover.

5   Mark the relationship of the torque converter and driveplate with white paint so they can be installed in the same position.

6   Remove the torque converter-to-driveplate bolts **(see illustration)**. Turn the crankshaft pulley bolt for access to each bolt.

7   Remove the starter motor (see Chapter 5).

8   Disconnect the driveaxles from the transaxle (see Chapter 8).

9   Disconnect the speedometer cable.

10   Disconnect the wire harness from the transaxle.

11   Remove any exhaust components which will interfere with transaxle removal (see Chapter 4).

12   Disconnect the TV cable.

13   Disconnect the shift linkage.

14   Support the engine using a hoist from above or a jack and a block of wood under the sump to spread the load.

15   Support the transaxle with a jack - preferably a special jack made for this purpose. Safety chains will help steady the transaxle on the jack.

16   Remove any chassis or suspension components which will interfere with transaxle removal.

17   Remove the bolts securing the transaxle to the engine.

18   Remove the transaxle mount nuts and bolts.

19   Lower the transaxle slightly and disconnect and plug the transaxle cooler lines.

20   Move the transaxle back to disengage it from the engine block dowel pins and make sure the torque converter is detached from the driveplate. Secure the torque converter to the transaxle so it will not fall out during removal. Lower the transaxle from the vehicle.

Installation

21   Prior to installation, make sure the torque converter hub is securely engaged in the pump. Measure the distance from the face of the torque converter flange nut to the face of the transaxle housing. If the measurement is not as specified at the beginning of this Chapter, it isn't seated properly. In necessary, remove the torque converter and reinstall it securely. After installation, lubricate the cup in the center of the torque cover

with multi-purpose grease.

22   With the transaxle secured to the jack, raise it into position. Be sure to keep it level so the torque converter does not slide out. Connect the fluid cooler lines.

23   Turn the torque converter to line up the drive studs with the holes in the driveplate. The white paint mark on the torque converter and the stud made in Step 5 must line up.

24   Move the transaxle forward carefully until the dowel pins and the torque converter are engaged.

25   Install the transaxle housing-to-engine bolts. Tighten them to the torque listed in this Chapter's Specifications.

26   Install the torque converter-to-driveplate bolts. Tighten the bolts to the specified torque.

27   Install any suspension and chassis components which were removed. Tighten the bolts and nuts to the torque listed in Chapter 10 Specifications.

28   Remove the jacks supporting the transaxle and the engine.

29   Install the starter motor (see Chapter 5).

30   Connect the vacuum hose(s) (if equipped).

31   Connect the shift and TV linkage.

32   Plug in the transaxle electrical connectors.

33   Install the torque converter cover.

34   Connect the driveaxles (see Chapter 8).

35   Connect the speedometer cable.

36   Adjust the shift cable (see Section 3).

37   Install any exhaust system components that were removed or disconnected.

38   Lower the vehicle.

39   Fill the transaxle (see Chapter 1), run the vehicle and check for fluid leaks.

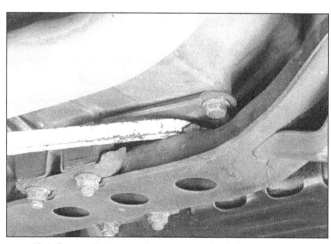

10.1 Pry on the transaxle mount to check for movement

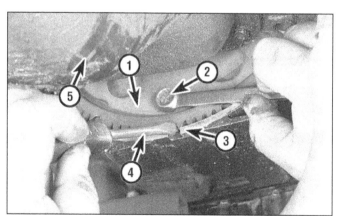

11.6 Prevent the driveplate from turning by inserting a screwdriver through the notch in the transaxle case, then remove the torque converter bolts

1	*Driveplate*	3	*Notch*	5	*Oil pan*
2	*Bolt*	4	*Screwdriver*		

Chapter 8
Clutch and driveaxles

Contents

Specifications

Clutch release arm freeplay	5/64 to 11/64 in (2 to 4 mm)

Torque specifications
Ft-lbs

Drive axle to hub nut	
Castellated nut	108.5 to 195.0
Staked nut	129
Pressure plate-to-flywheel bolts	17

---

1   General information

The information in this Chapter deals with the components from the rear of the engine to the front wheels, except for the transaxle, which is dealt with in the previous Chapter. For the purposes of this Chapter, these components are grouped into two categories - clutch and driveaxles. Separate Sections within this Chapter offer general descriptions and checking procedures for both groups.

Since nearly all the procedures covered in this Chapter involve working under the vehicle, make sure it's securely supported on sturdy jackstands or a hoist where the vehicle can be easily raised and lowered.

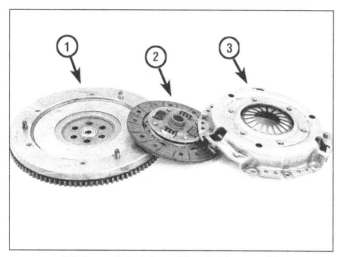

2.1 An exploded view of the clutch assembly

1 Flywheel	2 Clutch disc	3 Clutch cover

3.2 Remove the nut (arrow) and detach the clutch cable from the release arm

2   Clutch - description and check

Refer to illustration 2.1

1   All vehicles with a manual transaxle use a single dry plate, diaphragm spring type clutch **(see illustration)**. The clutch disc has a splined hub which allows it to slide along the splines of the transaxle input shaft. The clutch and pressure plate are held in contact by spring pressure exerted by the diaphragm in the pressure plate.

2   The clutch release system is cable-operated. It includes the clutch pedal with adjuster mechanism, a clutch cable which actuates the clutch release lever and the release bearing.

3   When pressure is applied to the clutch pedal to release the clutch, mechanical pressure is exerted against the outer end of the clutch release lever. As the lever pivots, the shaft fingers push against the release bearing. The bearing pushes against the fingers of the diaphragm spring of the pressure plate assembly, which in turn releases the clutch plate.

4   Terminology can be a problem when discussing the clutch components because common names are in some cases different from those used by the manufacturer. For example, the driven plate is also called the clutch plate or disc and the clutch release bearing is sometimes called a throwout bearing.

5   Other than replacement of components with obvious damage, some preliminary checks should be performed to diagnose clutch problems.

a) *To check "clutch spin down time," run the engine at normal idle speed with the transaxle in Neutral (clutch pedal up - engaged). Disengage the clutch (pedal down), wait several seconds and shift the transaxle into Reverse. No grinding noise should be heard. A grinding noise would most likely indicate a problem in*

the pressure plate or the clutch disc.

b) *To verify full clutch release, run the engine (with the parking brake applied to prevent movement) and hold the clutch pedal approximately 1/2-inch from the floor. Shift the transaxle between 1st gear and Reverse several times. If the shift is hard or the transaxle grinds, component failure is indicated.*

c) *Visually inspect the pivot bushing at the top of the clutch pedal to make sure there is no binding or excessive play.*

d) *A clutch pedal that's difficult to operate is most likely caused by a faulty clutch cable. Check the cable where it enters the housing for frayed wires, rust and other signs of corrosion. If it looks good, lubricate the cable with penetrating oil. If pedal operation improves, the cable is worn out. Replace it.*

3   Clutch cable - removal, installation and adjustment

Removal and installation

Refer to illustration 3.2

1   Detach the cable from the negative battery terminal.

2   Remove the clutch cable retaining nut from the release arm **(see illustration)** and disconnect the cable from the release arm.

3   Remove the retaining bolts from the clutch cable bracket and detach the bracket from the cable.

4   Remove the retaining pin from the upper end of the clutch pedal and detach the cable clevis from the pedal.

5   Pull the cable through the firewall and remove it from the vehicle.

6   Inspect the cable for frayed areas, kinks, worn ends, excessive friction and broken boots. If there are any signs of damage or wear, replace the cable.

7   Grease the clevis and pin of the clutch

cable and attach the cable to the clutch pedal. Installation is otherwise the reverse of removal. Adjust the cable when you're done.

Adjustment

8   Loosen the locknut on the clutch pedal stop bolt. Turn the stop bolt until the clutch pedal is the same height as the brake pedal. Tighten the locknut.

9   Measure the freeplay at the clutch release arm. Compare your measurement to the freeplay listed in this Chapter's Specifications. If the freeplay at the release arm is out of specification, loosen or tighten the clutch cable joint nut **(see illustration 3.2)** to bring it within spec.

4   Clutch components - removal, inspection and installation

Warning: *Dust produced by clutch wear and deposited on clutch components may contain asbestos, which is hazardous to your health. DO NOT blow it out with compressed air and DO NOT inhale it. DO NOT use gasoline or petroleum-based solvents to remove the dust. Brake system cleaner should be used to flush the dust into a drain pan. After the clutch components are wiped clean with a rag, dispose of the contaminated rags and cleaner in a covered, marked container.*

Removal

Refer to illustration 4.5

1   Access to the clutch components is normally accomplished by removing the transaxle, leaving the engine in the vehicle. If, of course, the engine is being removed for major overhaul, then check the clutch for wear and replace worn components as necessary. However, the relatively low cost of the clutch components compared to the time and trouble spent gaining access to them warrants their replacement anytime the engine or transaxle is removed, unless they

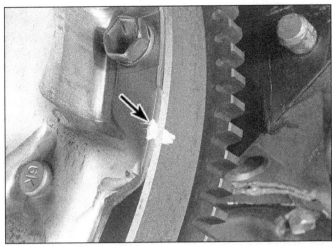

4.5 Mark the relationship of the pressure plate to the flywheel (if you're planning to re-use the old pressure plate)

4.10 Inspect the clutch disc for excessive wear and damage such as distorted lining material, chewed-up rivets, worn hub splines and distorted damper cushions or springs

NORMAL FINGER WEAR

EXCESSIVE WEAR

EXCESSIVE FINGER WEAR

BROKEN OR BENT FINGERS

4.12a Replace the pressure plate friction surface if excessive wear is noted

are new or in near perfect condition. The following procedures are based on the assumption the engine will stay in place.

2   Referring to Chapter 7 Part A, remove the transaxle from the vehicle. Support the engine while the transaxle is out. Preferably, an engine hoist should be used to support it from above. However, if a jack is used underneath the engine, make sure a piece of wood is positioned between the jack and oil pan to spread the load. **Caution:** *The pickup for the oil pump is very close to the bottom of the oil pan. If the pan is bent or distorted in any way, engine oil starvation could occur.*

3   The release fork, shaft and bearing can remain attached to the transaxle housing for the time being.

4   To support the clutch disc during removal, install a clutch alignment tool through the clutch disc hub.

5   Carefully inspect the flywheel and pressure plate for indexing marks. The marks are usually an X, an O or a white letter. If you can't find a mark, scribe your own so the pressure plate and the flywheel will be in the same alignment during installation **(see illustration)**.

6   Turning each bolt 1/4-turn at a time, loosen the pressure plate-to-flywheel bolts. Work in a criss-cross pattern until all spring pressure is relieved. Then hold the pressure

plate securely and completely remove the bolts, followed by the pressure plate and clutch disc.

Inspection

Refer to illustrations 4.10, 4.12a and 4.12b

7   Ordinarily, when a problem occurs in the clutch, it can be attributed to wear of the clutch driven plate assembly (clutch disc). However, all components should be inspected at this time.

8   Inspect the flywheel for cracks, heat checking, grooves and other obvious defects. If the imperfections are slight, a machine shop can machine the surface flat and smooth, which is highly recommended regardless of the surface appearance. Refer to Chapter 2 for the flywheel removal and installation procedure.

9   Inspect the pilot bearing (see Section 6).

10   Inspect the lining on the clutch disc. There should be at least 1/16-inch of lining above the rivet heads. Check for loose rivets, distortion, cracks, broken springs and other obvious damage **(see illustration)**. As mentioned above, ordinarily the clutch disc is routinely replaced, so if in doubt about the condition, replace it with a new one.

11   The release bearing should also be replaced along with the clutch disc (see Section 5).

4.12b Examine the pressure plate friction surface for score marks, cracks and evidence of overheating

12   Check the machined surfaces and the diaphragm spring fingers of the pressure plate **(see illustrations)**. If the surface is grooved or otherwise damaged, replace the pressure plate. Also check for obvious damage, distortion, cracking, etc. Light glazing can be removed with medium grit emery cloth. If a new pressure plate is required, new and factory-rebuilt units are available.

4.14 Center the clutch disc in the pressure plate with a clutch alignment tool

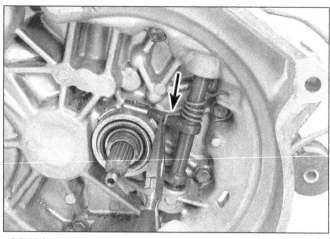

5.3 Using a pair of pliers, unhook the return spring (arrow) from the clutch release shaft fork

Installation

Refer to illustration 4.14

13   Before installation, clean the flywheel and pressure plate machined surfaces with lacquer thinner or acetone. It's important that no oil or grease is on these surfaces or the lining of the clutch disc. Handle the parts only with clean hands.

14   Position the clutch disc and pressure plate against the flywheel with the clutch held in place with an alignment tool **(see illustration)**. Make sure it's installed properly (most replacement clutch plates will be marked "fly-wheel side" or something similar - if not marked, install the clutch disc with the damper springs toward the transaxle).

15   Tighten the pressure plate-to-flywheel bolts only finger tight, working around the pressure plate.

16   Center the clutch disc by ensuring the alignment tool extends through the splined hub and into the pilot bearing in the crankshaft. Wiggle the tool up, down or side-to-side as needed to bottom the tool in the pilot bearing. Tighten the pressure plate-to-flywheel bolts a little at a time, working in a criss-cross pattern to prevent distorting the cover. After all of the bolts are snug, tighten them to the torque listed in this Chapter's Specifications. Remove the alignment tool.

17   Using high-temperature grease, lubricate the inner groove of the release bearing (see Section 5). Also place grease on the release lever contact areas and the transaxle input shaft bearing retainer.

18   Install the clutch release bearing as described in Section 5.

19   Install the transaxle and all components removed previously. Tighten all fasteners to the proper torque specifications.

---

5   Clutch release bearing and fork - removal, inspection and installation

Warning: *Dust produced by clutch wear and*

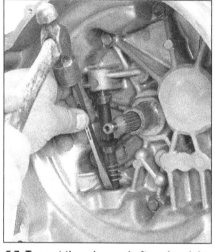

5.7 Tap out the release shaft seal and the No. 1 bushing with a hammer and chisel

deposited on clutch components may contain asbestos, which is hazardous to your health. DO NOT blow it out with compressed air and DO NOT inhale it. DO NOT use gasoline or petroleum-based solvents to remove the dust. Brake system cleaner should be used to flush the dust into a drain pan. After the clutch components are wiped clean with a rag, dispose of the contaminated rags and cleaner in a covered, marked container.

Removal

Refer to illustrations 5.3, 5.7 and 5.8

1   Disconnect the negative cable from the battery.

2   Remove the transaxle (see Chapter 7).

3   Note how the return spring is hooked onto the release shaft fork, then detach it from the shaft fork **(see illustration)**.

4   Slide the clutch release bearing off the transaxle input shaft.

5   Remove the nut and pinch bolt from the clutch release arm and remove the arm from the release shaft.

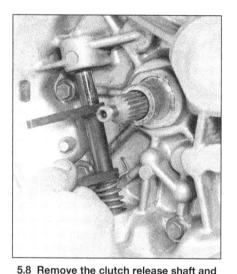

5.8 Remove the clutch release shaft and the return spring from the transaxle

6   Remove the cover from the release shaft seal.

7   Remove the release shaft seal and No. 1 bushing **(see illustration)** with a bushing remover or a chisel.

8   Remove the release shaft and return spring from the transaxle **(see illustration)**.

9   Remove the No. 2 bushing from the transaxle.

Inspection

Refer to illustration 5.10

10   Hold the center of the bearing and rotate the outer portion while applying pressure. If the bearing doesn't turn smoothly or if it's noisy, replace it. Wipe the bearing with a clean rag and inspect it for damage, wear and cracks. Don't immerse the bearing in solvent - it's sealed for life and to do so would ruin it. Also check the release lever and fork for cracks and other damage. Finally, inspect the bearing retainer **(see illustration)** for wear and damage. If it's damaged, replace the right transaxle case (see Chapter 7, Part A).

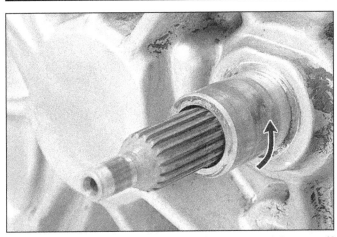

5.10 Inspect the bearing retainer (arrow) for wear and damage - if it's damaged or seriously worn, the right transaxle case must be replaced

5.14 After the new seal is installed, stake the transaxle case against the sides of the seal at two points

Installation

Refer to illustrations 5.14 and 5.17

11   Tap the No. 2 bushing into the transaxle with a hammer and a large socket with an outside diameter slightly smaller than the outside diameter of the bushing. Grease the inside of the bushing.

12   Install the release shaft and spring.

13   Grease the inside of the No. 1 bushing, slide it onto the end of the release shaft and, using a large socket with an outside diameter slightly smaller than the outside diameter of the bushing, tap the bushing into place.

14   Grease the lip of the new seal and drive the seal into place using a large socket with an outside diameter slightly smaller than the outside diameter of the seal. Stake the transaxle case against the seal at two points **(see illustration)**.

15   Slide the shaft seal cover onto the release shaft.

16   Install the clutch release arm. Be sure to align the punch mark on the release arm with the punch mark on the release shaft **(see illustration 3.2)**.

17   Grease the clutch release bearing **(see illustration)**, then slide it onto the input shaft.

18   Hook the return spring over the release shaft fork.

19   Install the transaxle (see Chapter 7, Part A).

20   Adjust the clutch cable (see Section 3).

6   Pilot bearing - inspection and replacement

Refer to illustrations 6.5, 6.8, 6.9 and 6.10

1   The clutch pilot bearing is a needle roller type bearing which is pressed into the rear of the crankshaft. It's greased at the factory and does not require additional lubrication. Its primary purpose is to support the front of the transaxle input shaft. Because of its inaccessibility, you should always inspect the pilot bearing any time the clutch components are removed. If you're in doubt about the condition of the pilot, replace it.

2   Remove the transaxle (see Chapter 7, Part A). **Note:** *If you have already removed*

the engine from the vehicle, disregard this Step.

3   Remove the clutch components (see Section 4).

4   Inspect the pilot bearing for excessive wear, scoring, lack of grease, dryness or obvious damage. You'll need a flashlight to illuminate the recess. If the pilot bearing is damaged or worn, replace it.

5   The pilot bearing can be removed with a slide hammer and puller attachment **(see illustration)**, but an alternative method may work if it's done carefully.

6   Find a solid steel bar which is slightly smaller in diameter than the bearing. Alternatives to a solid bar would be a wood dowel or a socket with a bolt fixed in place to make it solid.

7   Check the bar for fit - it should just slip into the bearing with very little clearance.

8   Pack the bearing and the area behind it (in the crankshaft recess) with heavy grease **(see illustration)**. Pack it tightly to eliminate as much air as possible.

9   Insert the bar into the bearing bore and

5.17 Lubricate the new release bearing at the indicated points (arrows)

6.5 Use a slide hammer for removing the pilot bearing

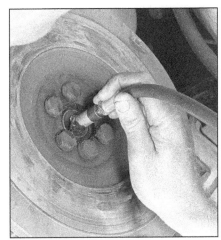

6.8 Here's an alternative to a slide hammer: fill the cavity behind the pilot bushing with grease . . .

6.9 . . . then force the bushing out hydraulically with a steel rod slightly smaller than the bore in the bushing - when the hammer strikes the rod, the grease will transmit the force to the backside of the bushing and push it out

6.10 Install the new pilot bearing with a bushing driver or a socket with the outside diameter slightly smaller than the outside diameter of the pilot bearing

strike the bar sharply with a hammer. This will force the grease to the back side of the bearing and push it out **(see illustration)**. Remove the bearing and clean all grease from the crankshaft recess.

10   To install the new bearing, lightly lubricate the outside surface with lithium-based grease, then drive it into the recess with a soft-face hammer **(see illustration)**. Make sure the seal faces out (the brand, size, etc. faces toward you).

11   Install the clutch components, transaxle and all other components removed previously, tightening all fasteners properly.

7   Clutch start switch - check and replacement

Refer to illustrations 7.3 and 7.4

1   Firmly apply the parking brake and put the gear shift lever in Neutral.

2   Reach under the dash and locate the clutch start switch at the upper end of the

clutch pedal. Trace the switch leads back to their connector and unplug it.

3   Loosen the locknut and screw out the switch **(see illustration)**.

4   Push down the clutch pedal as far as it will go, then let it come up about 0.6 to 1.2 inches (15 to 30 mm) **(see illustration)**.

5   Hook up an ohmmeter to the switch and slowly screw in the switch until the indicated resistance goes to zero - the switch is now on.

6   Hold the switch at this position and tighten the locknut securely. Recheck your resistance reading and make sure it occurs when the clutch pedal is 0.6 to 1.2 inches above the bottom of its travel.

7   Reconnect the lead wire.

8   Driveaxles - general information and inspection

Power is transmitted from the transaxle to the wheels through a pair of driveaxles.

The inner end of each driveaxle is connected to its respective differential side gear in the transaxle by a splined stub axle which is an integral part of the inner CV joint. The splined stub axle at the outer end of each driveaxle is inserted through the steering knuckle and axle hub and secured by a large nut.

The inner CV joints are the sliding tripot type, which are capable of both angular and axial motion. Each inner joint assembly consists of a "spider" (tripot and bearings) and a CV joint housing. The tripot can slide freely in and out as the driveaxle moves up and down with the wheel. The inner CV joints are rebuildable (see Section 10).

The outer CV joints are the ball-and-cage type. Ball bearings run between an inner race and an outer race (the joint housing), and are held in proper relationship to each other by a cage. These CV joints are capable of angular - but not axial - movement. The outer joints are neither rebuildable nor removable. If one of them fails, a new driveaxle/outer joint assembly must be installed.

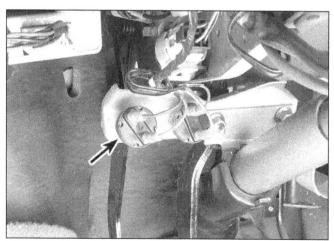

7.3 Once you find the clutch start switch (arrow) at the upper end of the clutch pedal, disconnect the electrical connector and loosen the locknut

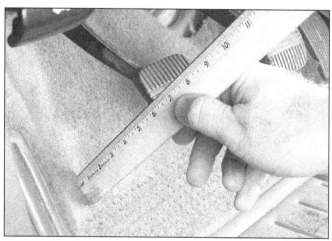

7.4 To check the clutch start switch, you must first position the clutch pedal at the proper height - push the pedal all the way to the floor, then let it back up about 0.6 to 1.2 inches (15 to 30 mm)

9.3 Loosen the driveaxle/hub nut with a long breaker bar - depending on the design of the wheel, it may be necessary to install the spare wheel and tire to access the hub nut

9.6 Use a large screwdriver or a prybar to pop the inner CV joint out of the differential side gear

The boots should be inspected periodically for damage and leaking lubricant. Damaged CV joint boots must be replaced immediately or the joints can be damaged. Boot replacement involves the removal of the driveaxle (see Section 9). **Note:** *Some auto parts stores carry "split" type replacement boots, which can be installed without removing the driveaxle from the vehicle. This is a convenient alternative; however, the driveaxle should be removed and the CV joint disassembled and cleaned to ensure the joint is free from contaminants such as moisture and dirt, which will accelerate CV joint wear. The most common symptom of worn out or damaged CV joints, besides lubricant leaks, is a clicking noise in turns, a clunk when accelerating after coasting and vibration at highway speeds.*

To check for wear in the CV joints and driveaxle shafts, grasp each axle (one at a time) and rotate it in both directions while holding the CV joint housings and feeling for play (indicating worn splines or sloppy CV joints). Also check the driveaxle shafts for cracks, dents and distortion.

9 Driveaxles - removal and installation

Note: *Some later models are equipped with an intermediate shaft assembly on the right side. If it's necessary to remove the intermediate shaft, first remove the right driveaxle, then unbolt and remove the intermediate shaft from the transaxle. Servicing of the intermediate shaft bearings should be left to a dealer service department or automotive machine shop.*

Removal
Refer to illustrations 9.3 and 9.6

1   Remove the wheel cover and break loose the hub nut. Loosen the wheel lug nuts, raise the front of the vehicle and support it securely on jackstands. Remove the front wheel.

2   Drain the oil from the transaxle (see Chapter 1).
3   Remove the cotter pin from the driveaxle hub nut and remove the nut. If equipped with a staked driveaxle hubnut, unstake the nut before loosening it. To prevent the hub from turning, place a pry bar between two of the wheel studs, then loosen the nut **(see illustration)**.
4   Remove both bolts from the stabilizer bar mounting bracket (see Chapter 10).
5   Remove the balljoint stud pinch bolt and detach the balljoint stud from the steering knuckle by pulling down on the stabilizer bar (see Chapter 10).
6   Using a large pry bar positioned between the transaxle housing and the CV joint housing, carefully pry the splined inner end of the driveaxle out of the differential side gear in the transaxle **(see illustration)**. A circlip in the end of the splined stub axle "locks" the driveaxle into the differential side gear. You'll feel this circlip pop loose when you pry on the CV joint. Once you've got the driveaxle disconnected from the transaxle, support the inner end with a piece of wire while you work on the outer end. **Note:** *On models equipped with center-bearing assemblies on the right side, the procedure for disconnecting the driveaxle is the same, except pry between the driveaxle and center bearing. If prying is difficult, tap the joint out of the bearing with a plastic mallet.*
7   Push the outer end of the driveaxle out of the hub and steering knuckle. Use a puller if necessary.
8   Support the CV joints and carefully remove the driveaxle from the vehicle.

Installation
9   Pry the old circlip from the inner end of the driveaxle and install a new one. Lubricate the differential seal with multi-purpose grease, raise the driveaxle into position while supporting the CV joints and insert the splined end of the inner CV joint into the differential side gear. Make sure the stub axle is

fully seated.
10   Apply a light coat of multi-purpose grease to the outer CV joint splines, pull out on the strut/steering knuckle assembly and install the stub axle into the hub.
11   Insert the balljoint stud into the steering knuckle and install the pinch bolt (see Chapter 10).
12   Install the hub nut. Lock the disc so it cannot turn, as described in Step 3, and tighten - but don't torque - the hub nut.
13   Grasp the inner CV joint housing (not the driveaxle) and try to pull it out to make sure the driveaxle is securely locked into place.
14   Install the wheel, install the wheel lug nuts, hand tighten them and lower the vehicle.
15   Tighten the wheel lug nuts to the torque listed in the Chapter 1 Specifications. Tighten the hub nut to the torque listed in this Chapter's Specifications. Install a new cotter pin, or if equipped with a staked nut, re-stake the nut into the groove. Install the wheel cover.

10 Driveaxle boot replacement and CV joint overhaul

Note: *If the CV joints are worn, indicating the need for an overhaul (usually due to torn boots), explore all options before beginning the job. Complete rebuilt driveaxles are available on an exchange basis, which eliminates much time and work. If you decide to rebuild a CV joint, check on the cost and availability of parts **before** disassembling the driveaxle.*

1   Remove the driveaxle from the vehicle (see Section 9).
2   Mount the driveaxle in a vise. The jaws of the vise should be lined with wood or rags to prevent damage to the driveaxle.

Inner CV joint
Disassembly
Refer to illustrations 10.3, 10.4, 10.5 and 10.6
3   Cut off both boot clamps **(see illustra-**

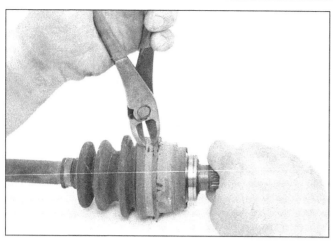

10.3 Cut off the boot clamps and discard them - don't try to reuse old clamps

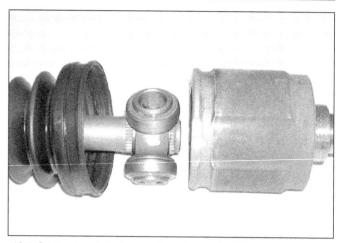

10.4 Scribe or paint alignment marks on the tripot assembly and the outer race, then slide the outer race off

10.5 Remove the snap-ring from the end of the axleshaft, then mark the relationship of the tripot to the axleshaft

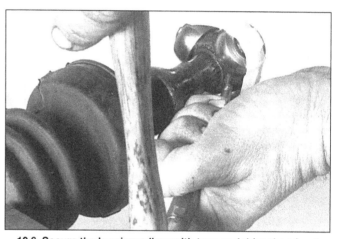

10.6 Secure the bearing rollers with tape and drive the tri-pot joint from the axleshaft with a brass punch and hammer - make sure you don't damage the bearing surfaces or the splines on the shaft, then remove the stop-ring

tion) and slide the boot towards the center of the driveaxle.

4   Mark or paint alignment marks on the outer race and the tripot bearing assembly **(see illustration)** so they can be returned to their original position, then slide the outer race off the tripot bearing assembly.

5   Remove the snap-ring from the end of the axleshaft, then mark the relationship of the tripot bearing assembly to the axleshaft **(see illustration).**

6   Secure the bearing rollers with tape, then remove the tripot bearing assembly from the axleshaft with a brass drift and a hammer **(see illustration).** Remove the tape, but don't let the rollers fall off and get mixed up.

7   Remove the stop-ring, slide the old boot off the driveaxle and discard it.

Inspection

8   Clean the old grease from the outer race and the tripot bearing assembly. Paint or scribe marks on each bearing roller and its respective shaft to ensure proper reassembly, then carefully disassemble each section

of the tripot assembly, one at a time, and clean the needle bearings with solvent.

9   Inspect the rollers, tripot, bearings and outer race for scoring, pitting or other signs of abnormal wear, which will warrant the replacement of the inner CV joint.

10.10 Wrap the splined area of the axleshaft with tape to prevent damage to the boot(s) when installing it

Reassembly

Refer to illustrations 10.10, 10.16, 10.17a and 10.17b

10   Wrap the splines of the axleshaft with tape to avoid damaging the new boot, then slide the boot onto the axleshaft **(see illus-**

10.16 Equalize the pressure inside the boot by inserting a small, dull screwdriver between the boot and the outer race

10.17a To install new fold-over type clamps, bend the tang down . . .

10.17b . . . and flatten the tabs to hold it in place

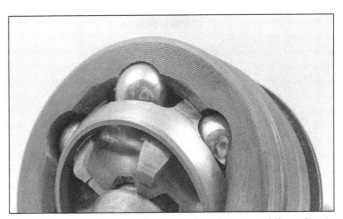

10.22 After the old grease has been rinsed away and the solvent has been blown out with compressed air, rotate the outer joint assembly through its full range of motion and inspect the bearing surfaces for wear and damage - if any of the ball bearings, the race or the cage look damaged, replace the driveaxle and outer joint assembly

tration). Remove the tape.

11   Align the match marks you made before disassembly and tap the tripot assembly onto the axleshaft with a hammer and brass drift.

12   Install the outer snap-ring.

13   Apply a coat of CV joint grease to the inner bearing surfaces to hold the needle bearings in place when reassembling the tripot assembly. Make sure each roller is installed on the same post as before.

14   Pack the outer race with half of the grease furnished with the new boot and place the remainder in the boot. Install the outer race. Make sure the marks you made on the tripot assembly and the outer race are aligned.

15   Seat the boot in the grooves in the outer race and the axleshaft.

16   With the CV joint positioned mid-way within its travel, equalize the pressure in the boot by inserting a blunt screwdriver between the boot and the outer race **(see illustration)**. Don't damage the boot with the tool.

17   Install and tighten the new boot clamps **(see illustrations)**.

18   Install the driveaxle assembly (see Section 9).

Outer CV joint and boot

Disassembly

19   Following Steps 3 through 7, remove the inner CV joint from the driveaxle and disassemble it.

20   Cut the boot clamps from the outer CV joint. Slide the boot off the shaft.

Inspection

Refer to illustration 10.22

21   Thoroughly wash the inner and outer CV joints in clean solvent and blow them dry with compressed air, if available. **Note:** *Because the outer joint can't be disassembled, it is difficult to wash away all the old grease and to rid the bearing of solvent once it's clean. But it is imperative that the job be done thoroughly, so take your time and do it right. On some later models, it may be possible to disassemble the CV joint for cleaning. Be sure to mark the relationship of the inner and outer*

race assemblies and housing so the joint can be reassembled in the same relationship.

22   Bend the outer CV joint housing at an angle to the axleshaft to expose the bearings, inner race and cage **(see illustration)**. Inspect the bearing surfaces for signs of wear. If the bearings are damaged or worn, replace the driveaxle.

Reassembly

23   Slide the new outer boot onto the axleshaft. It's a good idea to wrap tape around the splines of the shaft to prevent damage to the boot **(see illustration 10.10)**. When the boot is in position, add the specified amount of grease (included in the boot replacement kit) to the outer joint and the boot (pack the joint with as much grease as it will hold and put the rest into the boot). Slide the boot on the rest of the way and install the new clamps **(see illustrations 10.17a and 10.17b)**.

24   Clean and reassemble the inner CV joint by following Steps 8 through 17.

25   Install the driveaxle as outlined in Section 9.

Notes

Chapter 9 Brakes

Contents

Specifications

General

Brake fluid type	See Chapter 1
Booster mounting surface-to-clevis pin hole	
1985 to 1988	4-15/16 to 4-31/32 inches
1989 to 1994	4-1/2 to 4-35/64 inches
1995 to 1997	3-59/64 to 3-31/32 inches
1998	4-31/32 to 5 inches
1999 and later	3-59/64 to 3-31/32 inches
Power brake booster pushrod-to-master cylinder piston clearance	
1991 and earlier models	0.004-0.020 in
1992 and later models	zero

Disc brakes

Minimum brake pad thickness	See Chapter 1
Brake disc	
Standard thickness	
1994 and earlier	0.394 in (10.0 mm)
1995 and later	0.67 in (17.0 mm)
Minimum thickness*	
1994 and earlier	0.315 in (8.0 mm)
1995 and later	0.59 in (15.0 mm)
Runout limit	
Metro	0.004 in (0.1 mm)
Sprint	0.003 in (0.07 mm)

Drum brakes

Drum inside diameter	
Standard	7.09 in (180 mm)
Maximum*	7.16 in (182 mm)

*Refer to marks cast or stamped into the disc or drum (they supersede information printed here)

Torque specifications

	Ft-lbs (unless otherwise indicated)	Nm
Caliper mounting bolts		
Metro	22	30
Sprint	17.5 to 26	24 to 35
Brake hose-to-caliper banjo fitting bolt	14.5 to 18	20 to 24
Brake disc-to-hub bolts	37	50
Front hub nut	See Chapter 8	
Drum spindle nut		
Castellated nut	58 to 86	80 to 110
Staked nut	129	174
Master cylinder mounting nuts	120 in-lbs	13
Power booster mounting nuts	120 in-lbs	13
Wheel cylinder mounting bolts	90 to 108 in-lbs	122 to 146

1   General information

The vehicles covered by this manual are equipped with hydraulically operated front and rear brake systems. The front brakes are disc type and the rear brakes are drum type. Both the front and rear brakes are self adjusting. The front disc brakes automatically compensate for pad wear, while the rear drum brakes incorporate an adjustment mechanism.

Hydraulic system

The hydraulic system consists of two separate circuits. The master cylinder has separate reservoirs for the two circuits and in the event of a leak or failure in one hydraulic circuit, the other circuit will remain operative. A visual warning of circuit failure or air in the system is given by a warning light on the dashboard.

Proportioning valve

A proportioning valve, located on the engine compartment on the firewall, consists of sections providing the following functions. The metering section limits pressure to the front brakes until a predetermined front input pressure is reached and until the rear brakes are activated. There is no restriction at inlet pressures below 3 psi, allowing pressure equalization during non-braking periods. The proportioning section proportions outlet pressure to the rear brakes after a predetermined rear input pressure has been reached, preventing early rear wheel lock-up under heavy brake loads. The valve is also designed to assure full pressure to one brake system should the other system fail.

Anti-lock Brake System (ABS)

Refer to illustration 1.5

Some 1995 and later models are equipped with an Anti-lock Brake System (ABS). ABS systems maintain vehicle maneuverability, directional stability, and optimum deceleration under severe braking conditions on most road surfaces. It does so by monitoring the rotational speed of the wheels and controlling the brake line pressure to the wheels during braking. This prevents the wheels from locking up on slippery roads or during hard braking.

Hydraulic modulator/motor pack assembly

The hydraulic modulator/motor pack assembly, mounted on the side of the master cylinder, controls hydraulic pressure to the front calipers and rear wheel cylinders or calipers by modulating hydraulic pressure to prevent wheel lock-up **(see illustration)**.

The Electronic Brake Control Module (EBCM) is located under the left side of the dash. The EBCM monitors the ABS system and controls the anti-lock valve solenoids. It accepts and processes information received from the brake switch and wheel speed sen-

sors to control the hydraulic line pressure and avoid wheel lock up. It also monitors the system and stores fault codes which indicate specific problems.

Each sensor assembly consists of a variable reluctance sensor mounted adjacent to a "toothed ring" with an air gap between them. A wheel speed sensor and toothed ring are mounted in the hub/bearing unit of each front wheel and each rear wheel. The air gap between the sensors and the rings is not adjustable, and the sensors themselves are not rebuildable. If a sensor malfunctions, the sensor must be replaced. If a ring malfunctions, it must be driven off the hub/bearing unit and a new one pressed on.

A wheel speed sensor measures wheel speed by monitoring the rotation of the toothed ring. As the teeth of the ring move through the magnetic field of the sensor, an AC voltage signal is generated. This signal frequency increases or decreases in proportion to the speed of the wheel. The EBCM monitors these three signals for changes in wheel speed; if it detects the sudden deceleration of a wheel, i.e. wheel lockup, the EBCM activates the ABS system.

Warning lights

The ABS system has self-diagnostic capabilities. Each time the vehicle is started, the EBCM runs a self-test. There are two warning lights on the instrument panel, a red BRAKE light and an amber ABS light, each with their own functions. During starting, the red BRAKE warning light should come on briefly then go out. If the red BRAKE light stays on, it indicates a problem with the main braking system, such as low fluid level detected or the parking brake is still on. If the lights stays on after the parking brake is released, check the brake fluid level in the master cylinder reservoir (see Chapter 1).

The amber ABS light indicates a problem with the ABS system, not the main or basic brake system. If the light stays on steadily, it indicates that there is a problem with the ABS system, but the main system is

1.5 The ABS modulator/motor pack (arrow) is mounted on the side of the master cylinder

still working. If you have a steady ABS light, drive to a dealer service department or other qualified repair shop for diagnosis and repair. However, if the ABS light flashes, however, there is a more serious fault in the ABS system which may have affected the regular braking system. Pull over immediately and have the vehicle towed to a dealer or other qualified repair shop for service.

Checks

Although a special electronic tester is necessary to properly diagnose the system, the home mechanic can perform a few preliminary checks before taking the vehicle to a dealer service department which is equipped with this tester:

a) *Make sure the brake calipers are in good condition.*
b) *Check the electrical connector at the controller.*
c) *Check the fuses.*
d) *Follow the wiring harness to the speed sensors and brake light switch and make sure all connections are secure and the wiring isn't damaged.*

If the above preliminary checks don't rectify the problem, the vehicle should be diagnosed by a dealer service department.

Power brake booster

The power brake booster, utilizing engine manifold vacuum and atmospheric pressure to provide assistance to the hydraulically operated brakes, is mounted on the firewall in the engine compartment.

Parking brake

The parking brake operates the rear brakes only, through cable actuation. It's activated by a lever located in between the two front seats.

Service

After completing any operation involving disassembly of any part of the brake system, always test drive the vehicle to check for proper braking performance before resuming normal driving. When testing the brakes, perform the tests on a clean, dry flat surface. Conditions other than these can lead to inaccurate test results. Test the brakes at various speeds with both light and heavy pedal pressure. The vehicle should stop evenly without pulling to one side or the other. Avoid locking the brakes because this slides the tires and diminishes braking efficiency and control of the vehicle.

Tires, vehicle load and front-end alignment are factors which also affect braking performance.

2   Disc brake pads - replacement

Refer to illustrations 2.4, 2.5, 2.6a, 2.6b and 2.8

Warning: *Disc brake pads must be replaced on both front wheels at the same time - never*

2.4 Remove the caliper mounting bolts

2.5 Hang the caliper from the strut coil spring with a piece of wire;
do not allow the caliper to hang by the flexible brake hose

2.6a Remove the outer pad from the
mounting bracket (early model shown) -
note the way the clips are installed

2.6b Unclip the inner pad from the caliper
piston (early model shown - later models
have a shim but no clip)

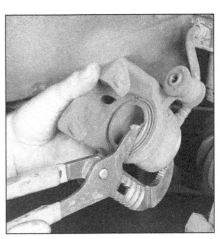

2.8 Compress the piston into the caliper
to make room for the new pads

replace the pads on only one wheel. Also, the dust created by the brake system may contain asbestos, which is harmful to your health. Never blow it out with compressed air and don't inhale any of it. An approved filtering mask should be worn when working on the brakes. Do not, under any circumstances, use petroleum based solvents to clean brake parts. Use brake cleaner or denatured alcohol only!

1   Remove the cover from the brake fluid reservoir.
2   Loosen the wheel lug nuts, raise the front of the vehicle and support it securely on jackstands.
3   Remove the front wheels. Work on one brake assembly at a time, using the assembled brake for reference if necessary.
4   Remove the caliper mounting bolts **(see illustration)**.
5   Remove the caliper from its mounting position and support it using wire or rope **(see illustration)**. Don't let it hang by the brake hose.
6   Remove the pads **(see illustrations)**.

7   Inspect the brake disc carefully as outlined in Section 4. If machining is necessary, follow the information in that Section to remove the disc.
8   Push the piston back into the bore to provide room for the new brake pads. A pair of slip joint pliers can be used to accomplish this **(see illustration)**. As the piston is depressed to the bottom of the caliper bore, the fluid in the master cylinder will rise. Make sure it does not overflow. If necessary, siphon off some of the fluid.
9   Position the pads in the caliper and install the assembly over the disc.
10   Install and tighten the caliper mounting bolts.
11   Install the wheels and lower the vehicle.
12   After the job has been completed, firmly depress the brake pedal a few times to bring the pads into contact with the disc. Check the brake fluid level and add some if necessary (see Chapter 1).
13   Check for fluid leakage and make sure the brakes operate normally before driving in traffic.

3   Disc brake caliper - removal, overhaul and installation

Refer to illustrations 3.3, 3.6, 3.8, 3.9 and 3.10

Warning: *Dust created by the brake system may contain asbestos, which is harmful to your health. Never blow it out with compressed air and don't inhale any of it. An approved filtering mask should be worn when working on the brakes. Do not, under any circumstances, use petroleum-based solvents to clean brake parts. Use brake cleaner or denatured alcohol only!*

Note: *If an overhaul is indicated (usually because of fluid leakage) explore all options before beginning the job. New and factory rebuilt calipers are available on an exchange basis, which makes this job quite easy. If it's decided to rebuild the calipers, make sure a rebuild kit is available before proceeding. Always rebuild the calipers in pairs - never rebuild just one of them.*

3.3 Unscrew the brake hose fitting bolt and detach the hose

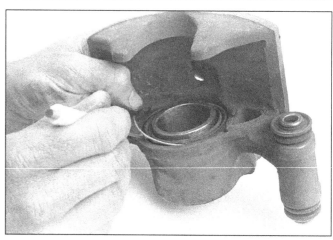

3.6 Use a screwdriver to remove the dust boot retaining ring

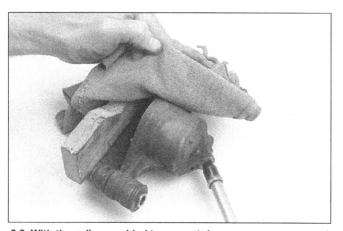

3.8 With the caliper padded to prevent damage, use compressed air to force the piston out of the bore - make sure your fingers aren't between the piston and the caliper

3.9 Use a plastic or wooden tool such as a pencil to remove the seal from the caliper bore

Removal

1   Loosen the wheel lug nuts, raise the front of the vehicle and support it securely on jackstands.

2   Remove the front wheels.

3   Remove the brake hose inlet fitting bolt and detach the hose **(see illustration)**. Have a rag handy to catch spilled fluid and wrap a plastic bag tightly around the end of the hose to prevent fluid loss and contamination.

4   Remove the mounting bolts and detach the caliper from the vehicle. Separate the pads from the caliper (see Section 2).

Overhaul

5   Clean the exterior of the caliper with brake cleaner or denatured alcohol. Never use gasoline, kerosene or petroleum-based cleaning solvents. Place the caliper on a clean workbench.

6   Carefully remove the ring retaining the dust boot to the caliper bore **(see illustration)**.

7   Carefully pry the dust boot out of the caliper bore.

8   Position a wood block or several shop rags in the caliper as a cushion, then use

compressed air to remove the piston from the caliper **(see illustration)**. Use only enough air pressure to ease the piston out of the bore. If the piston is blown out, even with the cushion in place, it may be damaged. **Warning:** *Never place your fingers in front of the piston in an attempt to catch or protect it when applying compressed air, as serious injury could occur.*

9   Using a wood or plastic tool, remove the piston seal from the groove in the caliper bore **(see illustration)**. Metal tools may cause bore damage.

10   Remove the caliper bleeder screw, then remove the sleeves and bushings from the caliper ears. Discard all rubber parts **(see illustration)**.

11   Clean the remaining parts with brake

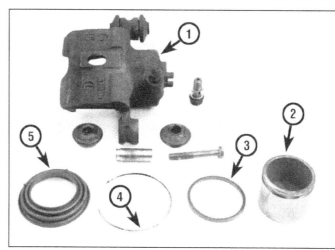

3.10 Exploded view of the caliper assembly (early model shown)

1   *Caliper body*
2   *Piston*
3   *Piston seal*
4   *Boot retaining ring*
5   *Dust boot*

4.3 The brake pads on this vehicle were obviously neglected, as they wore down to the rivets and cut deep grooves into the disc - wear this severe means the disc must be replaced

4.4a Use a dial indicator to check disc runout; if the reading exceeds the maximum allowable runout limit, the disc will have to be machined or replaced

system cleaner or denatured alcohol then blow them dry with compressed air.

12 Carefully examine the piston for nicks and burrs and loss of plating. If surface defects are present, the parts must be replaced.

13 Check the caliper bore in a similar way. Light polishing with crocus cloth is permissible to remove light corrosion and stains. Discard the mounting bolts if they're corroded or damaged.

14 When assembling, lubricate the piston bore and seal with clean brake fluid. Position the seal in the caliper bore groove.

15 Lubricate the piston with clean brake fluid, then install a new dust boot in the piston groove, with the fold toward the open end of the piston.

16 Insert the piston squarely into the caliper bore, then apply force to bottom it.

17 Position the dust boot in the cylinder groove and install the retaining ring.

18 Install the bleeder screw.

19 If the old bushings or pins will not clean up using crocus cloth, new ones should be installed. Old or new bushings or pins should be lubricated with silicone grease.

20 Install new rubber boots to the pins or bushings. Install the dust boot supports if equipped.

Installation

21 Inspect the mounting bolts for excessive corrosion.

22 Place the caliper in position over the disc and caliper mounting bracket, install the bolts and tighten them to the torque listed in this Chapter's Specifications.

23 Install the brake hose and inlet fitting bolt, using new copper washers, then tighten the bolt.

24 Be sure to bleed the brakes (see Section 10).

25 Install the wheels and lower the vehicle.

26 After the job has been completed, firmly depress the brake pedal a few times to bring the pads into contact with the disc.

4.4b Using a swirling motion, remove the glaze from the disc surface with sandpaper or emery cloth

27 Check brake operation before driving the vehicle in traffic.

4 Brake disc - inspection, removal and installation

Refer to illustrations 4.3, 4.4a, 4.4b, 4.5a, 4.5b, 4.6, 4.7, 4.9, 4.10, 4.12 and 4.13

Inspection

1 Loosen the wheel lug nuts, raise the vehicle and support it securely on jackstands. Remove the front wheels.

2 Remove the brake caliper as outlined in Section 3. It's not necessary to disconnect the brake hose. After removing the caliper bolts, suspend the caliper out of the way with a piece of wire or rope. Don't let the caliper hang by the hose and don't stretch or twist the hose.

3 Visually check the disc surface for score marks and other damage. Light scratches and shallow grooves are normal after use and

4.5a The minimum wear dimension is cast into the back side of the disc

may not always be detrimental to brake operation, but deep score marks - over 0.015-inch (0.38 mm) - require disc removal and refinishing by an automotive machine shop. Be sure to check both sides of the disc **(see illustration)**. If pulsating has been noticed during application of the brakes, suspect disc runout. Be sure to check the wheel bearings to make sure they're in good shape and properly adjusted - a bad bearing can also cause runout.

4 To check disc runout, place a dial indicator at a point about 1/2-inch from the outer edge of the disc **(see illustration)**. Set the indicator to zero and turn the disc. The indicator reading should not exceed the runout limit listed in this Chapter's Specifications. If it does, the disc should be refinished by an automotive machine shop. **Note:** *Professionals recommend resurfacing of brake discs regardless of the dial indicator reading (to produce a smooth, flat surface that will eliminate brake pedal pulsations and other undesirable symptoms related to questionable discs). At the very least, if you elect not to have the discs resurfaced, deglaze them with*

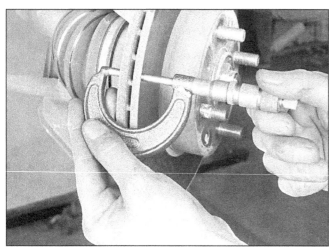

4.5b Use a micrometer to measure disc thickness

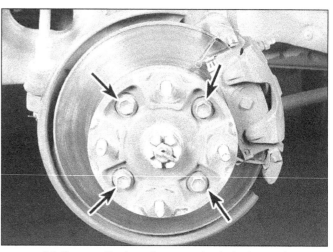

4.6 Remove the four disc mounting bolts (arrows)

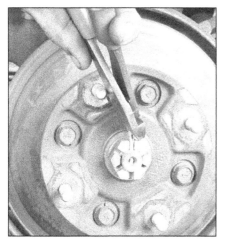

4.7 A pair of side cutters works well for removing the cotter pin.

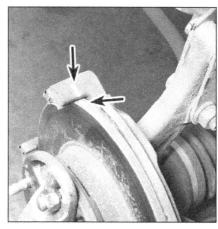

4.9 Before removing it, mark the relationship of the caliper and the caliper mount

1994 and earlier models

Removal

6   While an assistant depresses the brake pedal, loosen the bolts retaining the disc **(see illustration)**.

7   Remove the cotter pin and hub nut **(see illustration)**. If equipped with a staked hub-nut, unstake the nut **(see illustration 5.4)**.

8   Remove the caliper and pads (see Section 3).

9   Reference mark the position of the disc to the caliper mounting bracket **(see illustration)**.

10   Using a large slide hammer and proper adapter, pull off the disc and wheel hub **(see illustration)**.

11   Separate the disc from the hub.

Installation

12   Be sure the spacer on the hub is installed with the concave side to the hub **(see illustration)**.

13   Using the reference marks made during removal and the proper driver and hammer, install the disc and hub assembly to the posi-

sandpaper or emery cloth (use a swirling motion to ensure a nondirectional finish) **(see illustration)**.

5   The disc must not be machined to a thickness less than the minimum refinish thickness listed in this Chapter's Specifications. The minimum thickness is cast into the inside of the disc **(see illustration)**. The disc thickness can be checked with a micrometer **(see illustration)**.

4.10 Use a slide hammer with the proper adapter to remove the disc and hub assembly

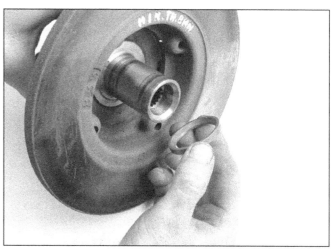

4.12 Be sure the spacer on the rear of the hub is positioned with the concave side to the hub

4.13 Use the proper drive to install the disc and hub assembly

tion it was in before removal **(see illustration)**.

14   Install the driveaxle hub nut and tighten it to the torque listed in this Chapter's Specifications. Install a new cotter pin. If necessary, turn the nut an additional amount to line up the slots in the nut with the hole in the driveaxle (don't loosen the nut to align the slots and hole). If equipped with a staked hubnut, re-stake the nut into the groove **(see illustration 5.4)**.

15   Install the caliper and brake pad assembly over the disc and position it on the steering knuckle (refer to Section 3 for the caliper installation procedure, if necessary). Tighten the caliper bolts to the torque listed in this Chapter's Specifications.

16   Install the wheel, then lower the vehicle to the ground. Tighten the lug nuts to the torque listed in the Chapter 1 Specifications.

17   Depress the brake pedal a few times to bring the brake pads into contact with the disc. Bleeding of the system will not be necessary unless the brake hose was disconnected from the caliper. Check the operation of the brakes carefully before placing the vehicle into normal service.

1995 and later models

18   Remove the two caliper mounting bracket bolts and detach the mounting bracket.

19   Remove the lug nuts you installed to hold the disc in place. There may also be two screws holding the disc to the hub; remove these screws, if installed.

20   Pull the disc off the hub and over the lug studs.

21   Installation is the reverse of removal; refer to Steps 15 through 17 of this procedure.

---

5   Drum brake shoes - replacement

Refer to illustrations 5.4, 5.5, 5.6, 5.7, 5.8, 5.12, 5.14 and 5.16

Warning: *Drum brake shoes must be replaced on both wheels at the same time - never replace the shoes on only one wheel. Also, the dust created by the brake system may contain asbestos, which is harmful to your health. Never blow it out with compressed air and don't inhale any of it. An approved filtering mask should be worn when working on the brakes. Do not, under any circumstances, use petroleum-based solvents to clean brake parts. Use brake cleaner or denatured alcohol only!*

Caution: *Whenever the brake shoes are replaced, the retractor and hold-down springs should also be replaced. Due to the continuous heating/cooling cycle that the springs are subjected to, they lose their tension over a period of time and may allow the shoes to drag on the drum and wear at a much faster rate than normal.*

Note: *Some 1995 and later models have separate drum and hub assemblies, so it is not necessary to remove the nut, as described in Step 4. On a separate-type drum there are two threaded holes on the front surface that you can thread bolts into and tighten the bolts to loosen a stuck drum.*

1   Loosen the wheel lug nuts, raise the rear of the vehicle and support it securely on jackstands. Block the front wheels to keep the vehicle from rolling.

2   Release the parking brake.

3   Remove the wheel. **Note:** *All four rear brake shoes must be replaced at the same time, but to avoid mixing up parts, work on only one brake assembly at a time.*

4   Remove the cotter pin (if equipped) and the nut retaining the brake drum **(see illustration)**. On 1995 and later models, see the **Note** above.

5   Remove the brake drums. **Note:** *If the drums cannot be pulled off with the parking brake completely released, the brake shoes will have to be retracted by removing the plug from the backing plate. Insert a screwdriver into the plug hole until its tip contacts the shoe hold-down spring* **(see illustration)**. *Push on the hold-down spring and release the parking brake lever from the hold-down spring.*

6   If the drum still will not come off, use a large slide hammer and the proper adapter to remove the drum **(see illustration)**.

5.4 Nuts that don't have a cotter pin have to be unstaked before they are loosened

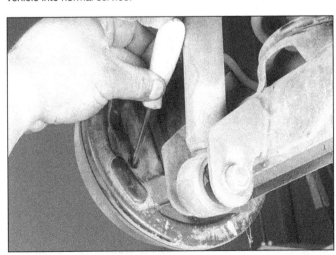

5.5 If the drum refuses to come off, try releasing the tension as shown - if it still won't come off . . .

5.6 . . . use a slide hammer to remove it

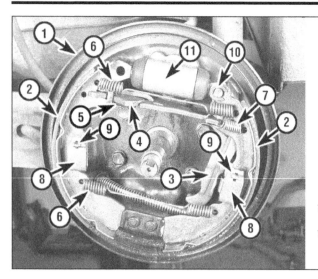

5.7 Rear brake assembly details

1   Brake backing plate
2   Brake shoe
3   Parking brake shoe lever
4   Brake strut
5   Quadrant spring
6   Shoe return spring
7   Anti-rattle spring
8   Shoe hold-down spring
9   Shoe hold-down pin
10   Parking brake lever retaining clip
11   Wheel cylinder

5.8 Use a screwdriver to pry apart the ends of the clip retaining the parking brake lever to the shoe

5.12 Lubricate the brake shoe contact area with high-temperature grease

5.14 The maximum drum diameter is cast into the inside of the rear drums

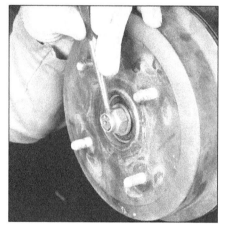

5.16 If your vehicle is not using a cotter pin on the drum retaining nut, be sure to stake the nut into the groove

7   Compress the hold-down spring while rotating the retaining pin. Remove the hold-down spring **(see illustration)**.
8   Remove the clip retaining the parking brake lever **(see illustration)**.
9   Remove the lower shoe return spring.
10   Remove the brake shoes from the backing plate and the parking brake lever.
11   Transfer the parts from the old brake shoes to the new shoes.
12   Clean the backing plate and apply high-temperature brake grease to the shoe contact areas of the backing plate **(see illustration)**.
13   Installation is the reverse of the removal procedure. Be sure to use a new clip to retain the parking brake lever.
14   Before reinstalling the drum it should be checked for cracks, score marks, deep scratches and hard spots, which will appear as small discolored areas. If the hard spots cannot be removed with fine emery cloth or if any of the other conditions listed above exist, the drum must be taken to an automotive machine shop to have it turned. **Note:** *Professionals recommend resurfacing the drums whenever a brake job is done. Resurfacing will eliminate the possibility of out-of-round*

drums. If the drums are worn so much that they can't be resurfaced without exceeding the maximum allowable diameter (cast into the drum) **(see illustration)**, *then new ones will be required. At the very least, if you elect not to have the drums resurfaced, remove the glazing from the surface with emery cloth or sandpaper, using a swirling motion.*
15   Install the brake drum.
16   Install and tighten the drum spindle nut to the torque listed in this Chapter's Specifications. Nuts that don't use cotter pins have to be staked **(see illustration)**.
17   Mount the wheel, install the lug nuts, then lower the vehicle.
18   Make a number of forward and reverse stops to adjust the brakes until satisfactory pedal action is obtained.
19   Check brake operation before driving the vehicle in traffic.

6   Rear wheel bearings - replacement

Refer to illustrations 6.1 and 6.3
1   With the rear drum removed (see Section 5), remove the bearings using a drift and

a hammer **(see illustration)**.
2   If the spacer was removed, install it with the inner lip toward the studs. Fill the cavity between the bearings with wheel bearing grease. If the bearings didn't come packed with grease from the factory, push the same wheel bearing grease into the bearings until they are thoroughly packed.

6.1 Use a drift punch and hammer to remove the wheel bearings

6.3 Use a proper driver to install the wheel bearings

3   Using the proper driver, install both the inner and outer bearings with the sealed sides facing out **(see illustration)**.

7   Wheel cylinder - removal, overhaul and installation

Refer to illustration 7.7
Note: *If an overhaul is indicated (usually because of fluid leakage or sticky operation) explore all options before beginning the job. New wheel cylinders are available, which makes this job quite easy. If it's decided to rebuild the wheel cylinder, make sure that a rebuild kit is available before proceeding. Never overhaul only one wheel cylinder - always rebuild both of them at the same time.*

Removal

1   Raise the rear of the vehicle and support it securely on jackstands. Block the front wheels to keep the vehicle from rolling.
2   Remove the brake shoe assembly (see Section 5).
3   Remove all dirt and foreign material from around the wheel cylinder.
4   Unscrew the brake line fitting, using a flare nut wrench, if available. Don't pull the brake line away from the wheel cylinder.

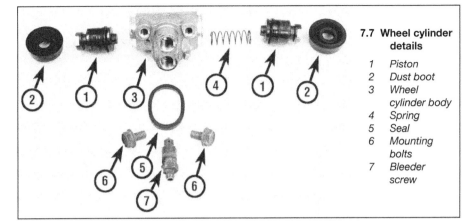

7.7 Wheel cylinder details

1   Piston
2   Dust boot
3   Wheel cylinder body
4   Spring
5   Seal
6   Mounting bolts
7   Bleeder screw

5   Remove the wheel cylinder mounting bolts.
6   Detach the wheel cylinder from the brake backing plate and place it on a clean workbench. Immediately plug the brake line to prevent fluid loss and contamination. **Note:** *If the brake shoe linings are contaminated with brake fluid, install new brake shoes.*

Overhaul

7   Remove the bleeder screw, cups, pistons, boots and spring assembly from the wheel cylinder body **(see illustration)**.
8   Clean the wheel cylinder with brake fluid, denatured alcohol or brake system cleaner. **Warning:** *Do not, under any circumstances, use petroleum-based solvents to clean brake parts!*
9   Use compressed air to remove excess fluid from the wheel cylinder and to blow out the passages.
10   Check the cylinder bore for corrosion and score marks. Crocus cloth can be used to remove light corrosion and stains, but the cylinder must be replaced with a new one if the defects cannot be removed easily, or if the bore is scored.
11   Lubricate the new cups with brake fluid.
12   Assemble the wheel cylinder components. Make sure the cup lips face in.

Installation

13   Place the wheel cylinder in position and install the bolts.

14   Connect the brake line and tighten the fitting. Install the brake shoe assembly.
15   Bleed the brakes (see Section 10).
16   Check brake operation before driving the vehicle in traffic.

8   Master cylinder - removal, overhaul and installation

Refer to illustrations 8.2, 8.6, 8.8, 8.9a and 8.9b

Removal

1   Clean around the reservoir cap and remove the brake fluid using a suction device such as a syringe.
2   Disconnect the brake lines from the master cylinder **(see illustration)**. **Note:** *Do not allow brake fluid to get on the painted surfaces.*
3   Remove the two nuts and washers mounting the master cylinder to the brake booster.
4   Remove the master cylinder.

Overhaul

5   Before attempting the overhaul of the master cylinder, obtain the proper rebuild kit, which will contain the necessary replacement parts and also any instructions which may be specific to your model.
6   Remove the fluid reservoir retaining screw **(see illustration)**.

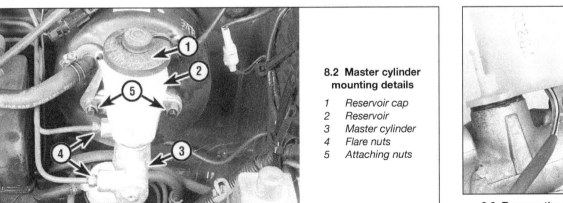

8.2 Master cylinder mounting details

1   Reservoir cap
2   Reservoir
3   Master cylinder
4   Flare nuts
5   Attaching nuts

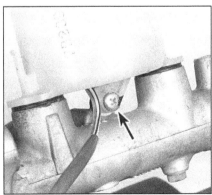

8.6 Remove the reservoir retaining screw (arrow)

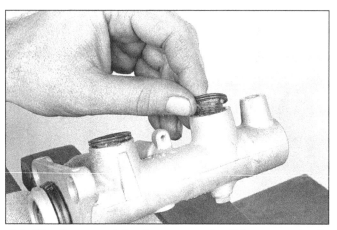

8.8 After the reservoir has been removed, pull the grommets off (if they're hardened, damaged or appear to have been leaking, replace them)

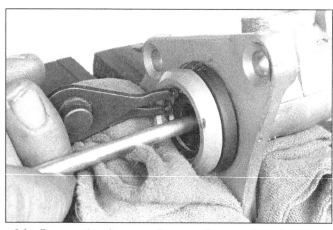

8.9a Depress the pistons and remove the snap-ring with snap-ring pliers

7   Pull up on the reservoir and detach it from the master cylinder.

8   Remove the rubber grommets from the master cylinder **(see illustration)**.

9   Remove the snap-ring from the end of the master cylinder **(see illustrations)**.

10   Remove the primary piston by inverting the cylinder and tapping it against a wood block.

11   Remove the stopper bolt for the secondary piston **(see illustration 8.9b)**.

12   Remove the secondary piston by inverting the cylinder and tapping it against a wood block.

13   Carefully inspect the bore of the master cylinder.

14   Any deep score marks or other damage will mean a new master cylinder is required. DO NOT attempt to hone the bore.

15   Replace all parts included in the rebuild kit, following any instructions in the kit.

16   Clean all reused parts with new brake fluid, brake system cleaner or denatured alcohol. Do not use any petroleum-based solvents during reassembly, lubricate all parts liberally with clean brake fluid.

17   Push the assembled components into the bore, bottoming them against the end of the master cylinder, then install the stopper bolt.

18   Install a new snap-ring, making sure it's seated properly in the groove.

19   Install the reservoir grommets and reservoir.

20   Before installing the master cylinder, it should be bench bled. Since you'll have to apply pressure to the master cylinder piston and, at the same time, control flow from the brake line outlets, the master cylinder should be mounted in a vise, with the jaws of the vise clamping on the mounting flange.

21   Insert threaded plugs into the brake line outlet holes and snug them down so no air will leak past them, but not so tight that they can't be easily loosened.

22   Fill the reservoir with brake fluid of the recommended type (see Chapter 1).

23   Remove one plug and push the piston assembly into the bore to expel the air from

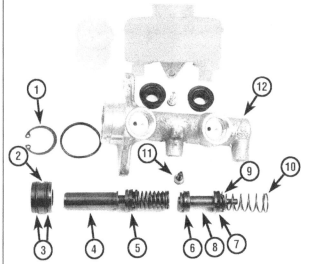

1	Snap-ring
2	Piston stopper
3	Piston stopper seals
4	Primary piston
5	Piston cup
6	Secondary piston pressure cup
7	Piston cup
8	Secondary piston
9	Secondary piston return spring seat
10	Secondary piston return spring
11	Secondary piston stopper bolt
12	Master cylinder body

8.9b Exploded view of an early-model master cylinder - some later models do not have a piston stopper

the master cylinder. A large Phillips screwdriver can be used to push on the piston assembly.

24   To prevent air from being drawn back into the master cylinder, the plug must be replaced and snugged down before releasing the pressure on the piston.

25   Repeat the procedure until only brake fluid is expelled from the brake line outlet hole. When only brake fluid is expelled, repeat the procedure at the other outlet hole and plug. Be sure to keep the master cylinder reservoir filled with brake fluid to prevent the introduction of air into the system.

26   Since high pressure isn't involved in the bench bleeding procedure, an alternative to the removal and replacement of the plugs with each stroke of the piston assembly is available. Before pushing in on the piston assembly, remove the plug as described in Step 23. Before releasing the piston, however, instead of replacing the plug, simply put your finger tightly over the hole to keep air from being drawn back into the master cylin-

der. Wait several seconds for brake fluid to be drawn from the reservoir into the bore, then depress the piston again, removing your finger as brake fluid is expelled. Be sure to put your finger back over the hole each time before releasing the piston, and when the bleeding procedure is complete for that outlet, replace the plug and tighten it before going on to the other port.

Installation

27   Installation is the reverse of removal.

28   Fill the reservoir with brake fluid.

29   Bleed the brake system (see Section 10).

---

9   Brake hoses and lines - inspection and replacement

Inspection

1   About every six months, with the vehicle raised and supported securely on jackstands,

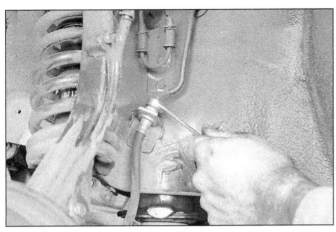

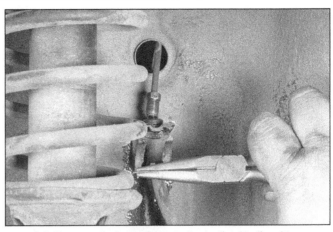

9.2 Loosen the fitting with a flare-nut wrench (if the bracket begins to bend, hold the hose fitting with another wrench)

9.3 Remove the brake hose-to-brake line U-clip with a pair of pliers

the rubber hoses which connect the steel brake lines with the front and rear brake assemblies should be inspected for cracks, chafing of the outer cover, leaks, blisters and other damage. These are important and vulnerable parts of the brake system and inspection should be complete. A light and mirror will be helpful for a thorough check. If a hose exhibits any of the above conditions, replace it with a new one.

Replacement

Front brake hose

Refer to illustrations 9.2 and 9.3

2   Using a flare-nut wrench, disconnect the brake line from the hose fitting, being careful not to bend the frame bracket or brake line **(see illustration)**.

3   Use a pair of pliers to remove the U-clip from the female fitting at the bracket, then detach the hose from the bracket **(see illustration)**.

4   Unscrew the brake hose from the caliper.

5   To install the hose, first thread it into the caliper, tightening it securely.

6   Without twisting the hose, install the female fitting in the hose bracket. It will fit the bracket in only one position.

7   Install the U-clip retaining the female fitting to the frame bracket.

8   Attach the brake line to the hose, tightening the fitting securely.

9   When the brake hose installation is complete, there should be no kinks in the hose. Make sure the hose doesn't contact any part of the suspension. Check this by turning the wheels to the extreme left and right positions. If the hose makes contact, remove it and correct the installation as necessary. Bleed the system (see Section 10).

Metal brake lines

10   When replacing brake lines be sure to use the correct parts. Don't use copper tubing for any brake system components. Purchase steel brake lines from a dealer or auto parts store.

11   Prefabricated brake line, with the tube

10.5 ABS hydraulic modulator assembly bleeder valves

A   Rear bleeder
B   Front bleeder

ends already flared and fittings installed, is available at auto parts stores and dealers.

12   When installing the new line make sure it's securely supported in the brackets and has plenty of clearance between moving or hot components.

13   After installation, check the master cylinder fluid level and add fluid as necessary. Bleed the brake system as outlined in the next Section and test the brakes carefully before driving the vehicle in traffic.

10   Brake hydraulic system - bleeding

Refer to illustrations 10.5 and 10.13

Warning: *Wear eye protection when bleeding the brake system. If the fluid comes in contact with your eyes, immediately rinse them with water and seek medical attention.*

Note: *Bleeding the hydraulic system is necessary to remove any air that manages to find its way into the system when it's been opened during removal and installation of a hose, line, caliper or master cylinder.*

ABS-equipped models only

1   Do not touch the brake pedal at any time during this preliminary procedure, which must

be performed before any bleeding operation.

2   Start the vehicle and watch the amber ABS warning light for at least ten seconds. If the warning light stays ON this long, your vehicle exhibits an ABS problem that can only be solved at a dealership or other qualified repair shop equipped with the proper scan tool.

3   If the warning light performed normally, i.e. it turned OFF after about three seconds, then turn the vehicle OFF. Repeat the procedure and if the ABS warning light turns OFF again after about three seconds, turn the vehicle OFF and you can bleed the brake system as described below for non-ABS vehicles.

4   After the bleeding process is completed, if the pedal feels soft, perform the above warning light test five times, without touching the brake pedal, then repeat the bleeding procedure.

5   **Warning:** *If an ABS-equipped model has been allowed to run dry at the master cylinder, the hydraulic modulator assembly must be bled before bleeding the calipers/wheel cylinders. Attach the bleeder hose to the rearmost bleeder valve on the modulator assembly* **(see illustration)**. *With the other end of the hose in a jar partially filled with clean brake fluid, crack the valve slowly and have an assistant depress the*

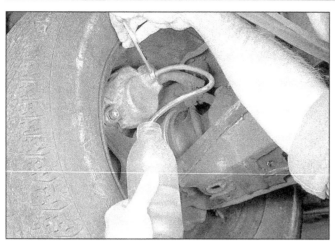

10.13 When bleeding the brakes, a hose is connected to the bleeder valve at the caliper or wheel cylinder and then submerged in brake fluid. Air will be seen as bubbles in the tube and container. All air must be expelled before moving to the next wheel

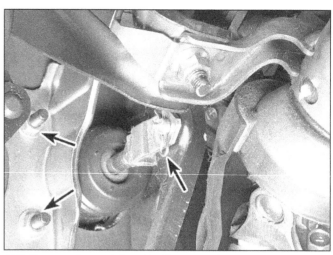

11.7 Remove the retaining clip (right arrow), pull out the clevis pin and detach the pushrod from the bake pedal; the arrows on the left point to two of the booster mounting nuts

brake pedal. Keep the pedal down until fluid flows. Close the valve and release the brake pedal. Repeat this until no air is evident as the fluid enters the jar, then repeat the procedure for the forward bleeder valve. After bleeding the hydraulic modulator assembly, bleed the rest of the braking systems as described below.

All models

Note: *This procedure applies only to ABS models which have passed the test described in Steps 1 through 3.*
6   It will probably be necessary to bleed the system at all four brakes if air has entered the system due to low fluid level, or if the brake lines have been disconnected at the master cylinder.
7   If a brake line was disconnected only at a wheel, then only that caliper or wheel cylinder must be bled.
8   If a brake line is disconnected at a fitting located between the master cylinder and any of the brakes, that part of the system served by the disconnected line must be bled.
9   Remove any residual vacuum from the brake power booster by applying the brake several times with the engine off.
10   Remove the master cylinder reservoir cover and fill the reservoir with brake fluid. Reinstall the cover. **Note:** *Check the fluid level often during the bleeding operation and add fluid as necessary to prevent the fluid level from falling low enough to allow air bubbles into the master cylinder.*
11   Have an assistant on hand, as well as a supply of new brake fluid, a clear container partially filled with clean brake fluid, a length of 3/16-inch plastic, rubber or vinyl tubing to fit over the bleeder valve and a wrench to open and close the bleeder valve.
12   Beginning at the right rear wheel, loosen the bleeder valve slightly, then tighten it to a point where it is snug but can still be loosened quickly and easily.
13   Place one end of the tubing over the

bleeder valve and submerge the other end in brake fluid in the container **(see illustration)**.
14   Have the assistant pump the brakes slowly a few times to get pressure in the system, then hold the pedal firmly depressed.
15   While the pedal is held depressed, open the bleeder valve just enough to allow a flow of fluid to leave the valve. Watch for air bubbles to exit the submerged end of the tube. When the fluid flow slows after a couple of seconds, close the valve and have your assistant release the pedal.
16   Repeat Steps 14 and 15 until no more air is seen leaving the tube, then tighten the bleeder valve and proceed to the left front wheel, the left rear wheel and the right front wheel, in that order, and perform the same procedure. Be sure to check the fluid in the master cylinder reservoir frequently.
17   Never use old brake fluid. It contains moisture which can boil, rendering the brakes useless.
18   Refill the master cylinder with fluid at the end of the operation.
19   Check the operation of the brakes. The pedal should feel solid when depressed, with no sponginess. If necessary, repeat the entire process. **Warning:** *Do not operate the vehicle if you are in doubt about the effectiveness of the brake system. On ABS-equipped models, see Step 4.*
20   Do not operate the vehicle if you have any doubts as to the effectiveness of the brake system. Seek professional advice if you can't obtain a firm pedal.

---

11   Power brake booster - check, removal and installation

Refer to illustrations 11.7, 11.14a, and 11.14b

Operating check

1   Depress the brake pedal several times with the engine off and make sure that there is no change in the pedal reserve distance.

2   Depress the pedal and start the engine. If the pedal goes down slightly, operation is normal.

Airtightness check

3   Start the engine and turn it off after one or two minutes. Depress the brake pedal several times slowly. If the pedal goes down farther the first time but gradually rises after the second or third depression, the booster is air tight.
4   Depress the brake pedal while the engine is running, then stop the engine with the pedal depressed. If there is no change in the pedal reserve travel after holding the pedal for 30 seconds, the booster is airtight.

Removal

5   Power brake booster units should not be disassembled. They require special tools not normally found in most service stations or shops. They are fairly complex and because of their critical relationship to brake performance it is best to replace a defective booster unit with a new or rebuilt one.
6   To remove the booster, first remove the brake master cylinder as described in Section 8.
7   Locate the pushrod clevis connecting the booster to the brake pedal **(see illustration)**. This is accessible from the interior in front of the driver's seat.
8   Remove the clevis pin retaining clip with pliers and pull out the pin.
9   Holding the clevis with pliers, disconnect the clevis locknut with a wrench. The clevis is now loose.
10   Disconnect the hose leading from the engine to the booster. Be careful not to damage the hose when removing it from the booster fitting.
11   Remove the four nuts and washers holding the brake booster to the firewall. You may need a light to see these, as they are up under the dash area.
12   Slide the booster straight out from the firewall until the studs clear the holes and pull

11.14a The distance from the booster mounting surface to the center line of the clevis pin hole should be as listed in this Chapter's Specifications

11.14b To adjust the length of the booster pushrod, hold the serrated portion of the rod with a pair of pliers and turn the adjusting screw in or out, as necessary, to achieve the desired setting

the booster, brackets and gaskets from the engine compartment area.

Installation

13   Installation procedures are basically the reverse of those for removal. Tighten the clevis locknut and booster mounting nuts.

14   If the power booster unit is being replaced, check the clevis rod length **(see illustration)**. Also the clearance between the master cylinder piston and the pushrod in the vacuum booster must be measured. Using a depth micrometer or vernier calipers, measure the distance from the seat (recessed area) in the master cylinder to the master cylinder mounting flange. With the engine running at idle, measure the distance from

the end of the vacuum booster pushrod to the mounting face of the booster (including gasket) where the master cylinder mounting flange seats. Subtract the two measurements to get the clearance. If the clearance is more or less than specified, turn the adjusting screw on the end of the power booster pushrod until the clearance is within the specified limit **(see illustration)**.

12 Parking brake - adjustment

Refer to illustration 12.3

1   The parking brake lever, when properly adjusted, should travel four to nine clicks when a moderate pulling force is applied. If it travels less than four clicks, there's a chance the parking brake might not be releasing completely and might be dragging on the drum. If the lever can be pulled up more than nine clicks, the parking brake may not hold adequately on an incline, allowing the car to roll.

2   To adjust the cables, from inside the vehicle gain access to the lever base.

3   Use the proper nut(s) to loosen or tighten the cables **(see illustration)**.

13 Parking brake cables - replacement

Refer to illustrations 13.4, 13.5, 13.6 and 13.7

1   Block the front wheels of the vehicle.

2   Raise and securely support the rear of the vehicle.

3   Remove the brake drum(s) and brake shoes (see Section 5).

4   Detach the parking brake cable from the brake shoe lever **(see illustration)**.

5   On Metro models, use pliers to squeeze the tangs on the cable retainer and detach the cable from the backing plate. On Sprint models, remove the cable retaining clip and detach the cable from the backing plate **(see illustration)**.

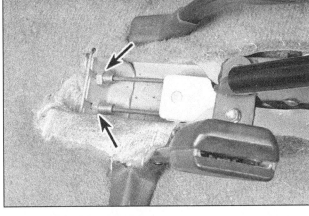

12.3 Loosen or tighten the nuts (arrows) to adjust the cables

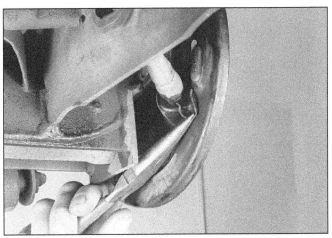

13.4 Use needle-nose pliers to pull back the parking brake cable while detaching the lever from the cable (Metro)

13.5 On Sprint models, remove the clip retaining the cable to the backing plate

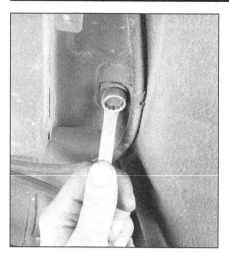

13.6 Typical cable-to-body attachment point

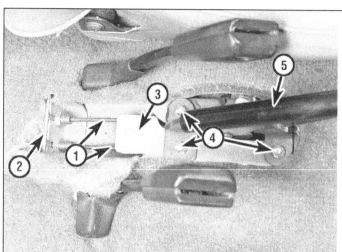

13.7 Typical parking brake lever component layout

1   Parking brake cables
2   Cable support
3   Clevis pin
4   Parking brake lever bolts
5   Parking brake lever assembly

6   On all models, detach the cable from under the vehicle at each body attaching point **(see illustration)**.

7   Detach the cable from the parking brake lever in the passenger compartment **(see illustration)**.

8   Installation is the reverse of removal. Use some silicone sealer to seal the area were the cable meets the backing plate. Adjust the cable after installation (see Section 12).

---

14   Brake light switch - removal, installation and adjustment

---

Removal and installation

1   The brake light switch is located on a bracket at the top of the brake pedal. The switch activates the brake lights at the rear of the vehicle when the pedal is depressed.

2   Disconnect the negative battery cable from the battery.

3   Disconnect the wiring harness at the brake light switch.

4   Loosen the locknut and unscrew the switch from the pedal bracket.

5   Installation is the reverse of removal.

Adjustment

6   Loosen the locknut, adjust the switch so there is a gap of 1/32 to 3/64 inch between the threaded portion and the pedal stop, then tighten the locknut.

7   Connect the wires at the switch and the battery. Make sure the rear brake lights are functioning properly.

Chapter 10
Suspension and steering systems

Contents

Specifications

Torque specifications

Ft-lbs

Metro

Front suspension

Control arm bracket nuts	92
Control arm front bracket bolts	66
Control arm rear bracket bolt	32
Knuckle-to-control arm pinch bolt	44
Stabilizer shaft link nut	20
Stabilizer shaft link-to-stabilizer shaft nut	38
Stabilizer shaft mounting bracket bolt	20
Strut-to-knuckle bolts	59
Strut-to-tower nuts	24

Rear suspension

Control rod-to-body nut	59
Control rod-to-knuckle nut	59
Stabilizer link-to-shaft nut	38
Stabilizer link-to-control arm nut	20
Stabilizer bar mounting bracket bolts	19
Strut-to-knuckle bolts	44
Strut-to-tower bolts	24
Control arm rear through bolt and nut	29
Control arm front bracket bolts	33
Control arm-to-knuckle nut and bolt	29

Steering

Steering gear mounting bolts	18
Steering shaft joint bolts	18
Steering wheel nut	24

Sprint

Control arm bolt	36 to 50
Knuckle-to-control arm pinch bolt	36 to 50
Stabilizer bar castle nut	29 to 65
Stabilizer bar mount bolt	22 to 39
Steering gear mounting bolts	15 to 22
Steering shaft joint bolts	15 to 22
Steering wheel nut	19 to 28
Strut-to-knuckle bolts	50 to 65
Strut-to-tower nut	29 to 43

1 General information

The front suspension is a MacPherson strut design. The upper end of each strut is attached to the vehicle's body strut support. The lower end of the strut is connected to the upper end of the steering knuckle. The steering knuckle is attached to a balljoint mounted on the outer end of the suspension control arm.

On Sprint models, the front control arms are connected by a stabilizer bar, which also controls fore-and-aft movement of the control arms.

On Metro models, the rear suspension also utilizes struts. The upper end of each strut is attached to the vehicle body by a strut support. The lower end of the strut is attached to a knuckle. The knuckle is located by a pair of suspension arms on each side, and a transverse mounted control rod between the body and each knuckle. A coil spring is mounted between the body and each suspension arm.

On the rear suspension of Sprint models, a solid axle is used. 1986 and earlier models use leaf springs and shock absorbers to control the suspension. 1987 and later models use coil springs, shock absorbers, trailing arms and a lateral rod used to control the suspension.

On all models, the rack and pinion steering gear is located behind the engine/transaxle assembly on the firewall and actuates the tie rods, which are attached to the steering knuckles. The steering column is designed to collapse in the event of an accident.

Frequently, when working on the suspension or steering system components, you may come across fasteners which seem impossible to loosen. These fasteners on the underside of the vehicle are continually subjected to water, road grime, mud, etc., and can become rusted or frozen, making them extremely difficult to remove. In order to unscrew these stubborn fasteners without damaging them (or other components), be sure to use lots of penetrating oil and allow it to soak in for a while. Using a wire brush to clean exposed threads will also ease removal of the nut or bolt and prevent damage to the threads. Sometimes a sharp blow with a hammer and punch will break the bond between a nut and bolt threads, but care must be taken to prevent the punch from slipping off the fastener and ruining the threads. Heating the stuck fastener and surrounding area with a torch sometimes helps too, but isn't recommended because of the obvious dangers associated with fire. Long breaker bars and extension, or "cheater", pipes will increase leverage, but never use an extension pipe on a ratchet - the ratcheting mechanism could be damaged. Sometimes tightening the nut or bolt first will help to break it loose. Fasteners that require drastic measures to remove should always be replaced with new ones.

2.4 Mounting details of the Sprint stabilizer bar (Metro models have a slightly different design)

1 Cotter pin
2 Castle nut
3 Stabilizer bar washer
4 Stabilizer bar bushing
5 Suspension control arm
6 Stabilizer bar
7 Mount bushing
8 Mounting bushing bracket
9 Mounting bracket bolts

Since most of the procedures dealt with in this Chapter involve jacking up the vehicle and working underneath it, a good pair of jackstands will be needed. A hydraulic floor jack is the preferred type of jack to lift the vehicle, and it can also be used to support certain components during various operations. **Warning:** *Never, under any circumstances, rely on a jack to support the vehicle while working on it. Whenever any of the suspension or steering fasteners are loosened or removed they must be inspected and, if necessary, replaced with new ones of the same part number or of original equipment quality and design. Torque specifications must be followed for proper reassembly and component retention. Never attempt to heat or straighten any suspension or steering components. Instead, replace any bent or damaged part with a new one.*

2 Stabilizer bar (front) - removal and installation

Refer to illustration 2.4

1 Block the rear wheels.
2 Raise the front of the vehicle and allow the control arms to hang.
3 Remove the front wheels.
4 Remove the stabilizer bracket bolts and brackets **(see illustration)**.
5 Detach the stabilizer bar from the front suspension arms.
6 Remove the stabilizer bar.
7 Installation is the reverse of removal.
8 When installing the stabilizer bar, loosely assemble all components while insuring that the stabilizer is centered before tightening components.

3 Strut assembly (front) - removal, inspection and installation

Refer to illustrations 3.3, 3.4 and 3.5

1 Raise the front of the vehicle so the front suspension hangs free.

3.3 A pair of needle-nose pliers works well for removing the clip that retains the brake line to the strut bracket

2 Remove the front wheels.
3 Remove the clip retaining the brake line to the strut bracket **(see illustration)**.
4 Remove the strut-to-knuckle nuts and bolts **(see illustration)**.
5 Working under the hood, support the strut and remove the nuts retaining the strut to the strut tower **(see illustration)**.
6 Remove the strut assembly. Be careful not to damage the outer CV joint boot during strut removal. And don't over-extend the inner CV joint (one way to prevent this from happening is to wire the top of the steering knuckle to the body).
7 Check the strut body for leaking fluid, dents, cracks and other obvious damage which would warrant repair or replacement. Check the coil spring for chips and cracks in the spring coating (this will cause premature spring failure due to corrosion).
8 Inspect the spring seat for damage, hardness and general deterioration.
9 If any undesirable conditions exist, proceed to Section 4 for the strut disassembly procedure.
10 Installation is the reverse of removal.

3.4 Remove the nuts and bolts attaching the strut to the knuckle

3.5 Two strut-to-strut tower retaining nuts (arrows) are used on Sprint models; Metro models use three

4.4 Install the spring compressor according to the tool manufacturer's instructions and compress the spring until all pressure is relieved from the upper spring seat

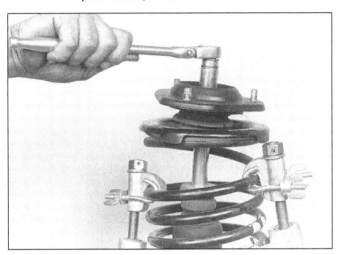

4.5 Remove the damper shaft nut

4 Strut cartridge (front) - replacement

Refer to illustrations 4.4, 4.5, 4.6 and 4.10
Warning: *Whenever any of the suspension or steering fasteners are loosened or removed they must be inspected and, if necessary, replaced with new ones of the same part number or original equipment quality and design. Torque specifications must be followed for proper reassembly and component retention.*
Note: *You'll need a spring compressor for this procedure. Spring compressors are available on a daily rental basis at most auto parts stores or equipment yards.*
1   If the struts or coil springs exhibit the telltale signs of wear (leaking fluid, loss of damping capability, chipped, sagging or cracked coil springs) explore all options before beginning any work. The strut insert assemblies are not serviceable and must be replaced if a problem develops. However, strut assemblies complete with springs may be available on an exchange basis, which eliminates much time and work. Whichever

route you choose to take, check on the cost and availability of parts before disassembling your vehicle. **Warning:** *Disassembling a strut assembly is a potentially dangerous job. Use only a high quality spring compressor and carefully follow the manufacturer's instructions furnished with the tool. After removing the coil spring from the strut assembly, set it aside in a safe, isolated area.*
2   Remove the strut and spring assembly (see Section 3).
3   Mount the strut assembly in a vise. Line the vise jaws with wood or rags to prevent damage to the unit and don't tighten the vise excessively.
4   Install the spring compressor in accordance with the manufacturer's instructions **(see illustration)**. Compress the spring until you can wiggle the strut (suspension) support.
5   Loosen the damper shaft nut with a socket wrench **(see illustration)**. **Note:** *If the shaft nut cannot be removed because the insert shaft rotates, use a pair of vise grips with a rag in the jaws for insulation clamped to the shaft to keep the shaft from rotating.*
6   Disassemble the strut assembly. **Warn-**

ing: When removing the compressed spring, it should be carefully lifted from the assembly (see illustration) and set in a safe place, such as a steel cabinet. Never place your head near the end of the spring!

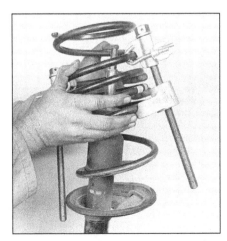

4.6 When removing the compressed spring assembly, keep the ends of the spring pointing away from your body!

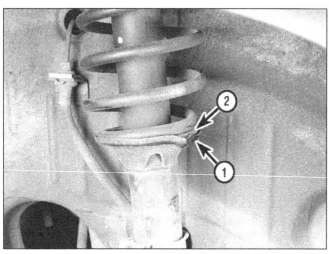

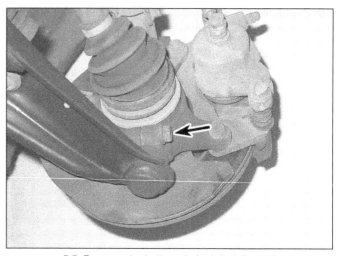

4.10 When installing the spring, make sure the end (1) fits into the recessed portion of the lower seat (2)

5.5 Remove the ball stud pinch bolt (arrow)

7   Unscrew the retaining cap for the strut cartridge.

8   Pull out the cartridge from the strut housing.

9   Check the rubber parts for damage, cracking and hardness and replace as necessary.

10   Reassembly is the reverse of disassembly. Be careful not to damage the damper shaft or the strut will leak. When installing the spring, be sure the spring ends mesh with the spring anti-rotation stops provided by the spring upper and lower seats **(see illustration)**.

5   Control arm (front) - removal, inspection and installation

Refer to illustration 5.5

Removal

Metro

1   Remove the ball stud bolt **(see illustration 5.5)**.

2   Remove the control arm rear bracket.

3   Remove the control arm front retaining nut.

4   Remove the control arm.

Sprint

5   Remove the ball stud bolt **(see illustration)**.

6   Remove the control arm-to-body mounting bolt.

7   Remove the control arm from the stabilizer bar.

Inspection

8   Check the control arm for distortion and the bushings for wear, damage and deterioration. Replace a distorted control arm with a new one. If the bushings are damaged, an arbor press is necessary to remove and install them.

7.12 Use a bearing driver to seat the bearings

Installation

9   Installation is the reverse of removal. Do not tighten the fasteners completely until the vehicle is lowered to the ground.

10   If the bushings are replaced, have the front wheel alignment checked and adjusted.

6   Balljoint (front) - replacement

On all models, the balljoint is replaced by replacing the control arm (see Section 5).

7   Steering knuckle and front wheel bearings - removal and installation

Steering knuckle

1   Raise and securely support the front of the vehicle so the front suspension hangs free.

2   Remove the front wheels.

3   Remove the brake caliper (see Chapter 9).

7.13 Installing the oil seal

4   Remove the brake disc and hub assembly (see Chapter 9).

5   Remove the ball stud bolt **(see illustration 5.5)**.

6   Remove the clip retaining the brake line to the strut bracket **(see illustration 3.3)**.

7   Remove the strut-to-knuckle bolts **(see illustration 3.4)**.

8   Remove the knuckle.

9   Installation is the reverse of removal.

Front wheel bearings

Refer to illustrations 7.12 and 7.13

10   With the steering knuckle removed, remove the bearings using a drift punch and a hammer.

11   Before installing the wheel bearings, pack them with wheel bearing grease and apply some grease to the oil seal lip. Also fill the space between the spacer and the knuckle with grease.

12   Install the inner bearing, spacer, and the outer bearing, using a suitable driver **(see illustration)**. **Note:** *Install the bearings with their sealed sides facing out. Center the spacer between the inner and outer bearings.*

13   Install the oil seal by gently tapping

8.6 If necessary, use a chisel to spread open the slit (arrow) of the knuckle to allow the removal of the strut

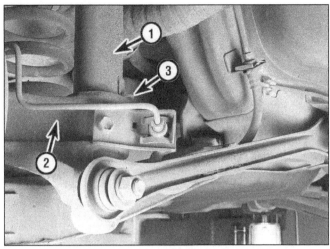

8.7 When installing be sure the strut fits properly into the knuckle

| 1 | *Strut* | 2 | *Knuckle* | 3 | *Locating tab* |

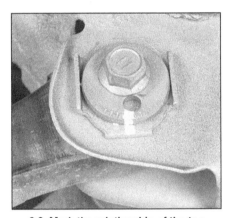

9.3 Mark the relationship of the toe adjuster to the body on the inner end of the rear suspension arms

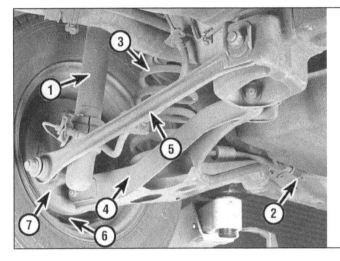

9.4 Overview of the Metro rear suspension

1   *Strut*
2   *Vehicle body*
3   *Coil spring*
4   *Suspension arm*
5   *Control rod*
6   *Brake drum*
7   *Brake backing plate*

around the edges with a blunt tool until it seats against the bearing **(see illustration)**. Be careful not to deform, tilt, or cause damage to the rubber part.
14   Reinstall the steering knuckle.

8   Rear strut assembly (Metro) - removal and installation

Refer to illustrations 8.6 and 8.7
1   Raise and securely support the rear of the vehicle.
2   Place a jack under the outer end of the rear control arm.
3   Working at the strut tower, remove the nuts retaining the upper part of the strut to the tower.
4   Compress the strut.
5   Remove the pinch bolt from the knuckle.
6   By pulling up on the lower portion of the strut, detach the strut from the knuckle. If the strut is hard to remove, open the slit of the knuckle a little **(see illustration)**. **Warning:** *During strut removal, do not lower the jack more than necessary to prevent the coil*

spring from coming out, or brake hose from being damaged.
7   Installation is the reverse of removal. Be sure the strut locating tab aligns properly with the rear knuckle **(see illustration)**.

9   Rear coil spring and suspension arms (Metro) - removal and installation

Refer to illustrations 9.3 and 9.4
1   Block the front wheels, raise and securely support the rear of the vehicle.
2   Remove the rear wheel(s).
3   Reference mark the position of the cam for rear toe adjustment **(see illustration)**.
4   Loosen the nuts mounting the suspension arm to the body **(see illustration)**.
5   Loosen the knuckle-to-suspension arm mounting nut.
6   Place a jack under the suspension arm to prevent it from lowering.
7   Remove the lower mount nut of the knuckle.
8   Raise the jack placed under the suspen-

sion arm enough to remove the knuckle-to-suspension arm mounting bolt.
9   Separate the knuckle from the suspension arm.
10   Lower the jack gradually and remove the coil spring.
11   Remove the body-to-suspension arm bolts.
12   Remove the suspension arm.
13   Installation is the reverse of removal. Be sure to place the rear toe adjustment cam in the same position as it was in before removal.

10   Rear leaf spring (1985 and 1986 Sprint) - removal and installation

1   Block the front wheels, raise and securely support the rear of the vehicle.
2   Remove the rear wheel(s).
3   Support the center of the axle with a floor jack.
4   Remove the U-bolt nuts.
5   Remove the bolt retaining the front of the leaf spring.
6   Remove the nuts and bracket retaining the rear of the leaf spring.

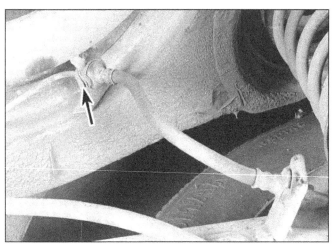

11.14 Remove the clip retaining the brake lines to the trailing arms (arrow)

11.18 Remove the four bolts retaining the backing plate (arrows)

7   Remove the leaf spring from the shackle pin.
8   Installation is the reverse of removal.

11 Rear axle (Sprint) - removal and installation

Refer to illustrations 11.14, 11.18, 11.21 and 11.24

1985 and 1986 (leaf spring equipped)

1   Raise the rear of the vehicle.
2   Remove the rear wheels.
3   Remove the brake drums.
4   Remove the clip retaining the brake line to the axle. Separate the steel brake line from the flexible hose.
5   To prevent loss of brake fluid, plug the brake hose and line.
6   Remove the nuts retaining the backing plates to the axle.
7   Remove the backing plate from the axle.
8   Remove the U-bolt nuts, the U-bolts and the jounce stops.

9   Remove the rear axle.
10   Installation is the reverse of removal. When installing the backing plate, apply sealant between the axle-to-backing plate surface. Be sure to bleed the brakes after installation (see Chapter 9).

1987 (coil spring equipped)

11   Block the front wheels, raise and support the rear of the vehicle by the body or frame.
12   Remove the rear wheels.
13   Remove the brake drums (see Chapter 9).
14   Remove the clip retaining the brake lines to the trailing arms **(see illustration)**.
15   Disconnect the brake lines from the wheel cylinders and plug the brake lines to prevent fluid leakage.
16   Remove the brake lines from the brackets on the trailing arm.
17   Remove the brake shoes and detach the parking brake cable from the shoe levers and backing plates (see Chapter 9).
18   Remove the bolts retaining the backing plate to the axle **(see illustration)**.

19   Remove the backing plates.
20   Using a floor jack, support the center of the axle.
21   Remove the bolt retaining the lateral rod-to-body **(see illustration)**.
22   Remove the lower retaining bolts for the shocks.
23   Lower the floor jack to release the spring tension.
24   Remove the trailing arm-to-body nuts and bolts and remove the axle **(see illustration)**.
25   Installation is the reverse of removal. When installing the springs be sure that the taper end of the spring faces down and meshes with the stepped part of the lower seat and the flat surface faces up.

12 Steering system - general information

All models are equipped with rack and pinion steering. The rack and pinion assembly is bolted to the firewall and operates the steering arms via tie rods. The inner ends of

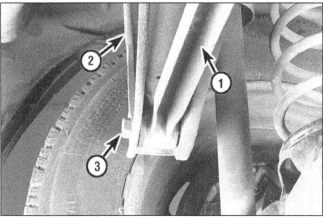

11.21 Remove the lateral rod-to-body bolt

1   *Lateral rod*   3   *Lateral rod-to-body bolt*
2   *Body*

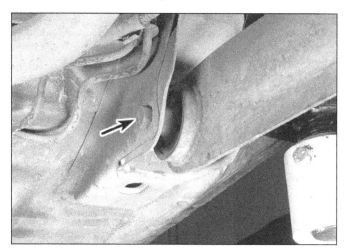

11.24 Remove the trailing arm-to-body retaining bolt and nut (arrow)

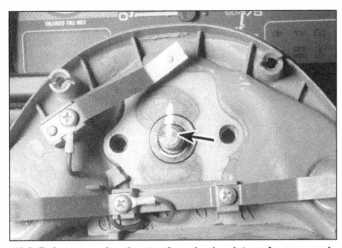

13.8 Before removing the steering wheel, paint a reference mark so the wheel can reinstalled in the same position

13.9 Use a steering wheel puller to remove the wheel

the tie rods are protected by rubber boots which should be inspected periodically for secure attachment, tears and leaking lubricant.

The steering wheel operates the steering shaft, which actuates the steering gear through universal joints. Looseness in the steering can be caused by wear in the steering shaft universal joints, the steering gear, the tie rod ends and loose retaining bolts.

13 Steering wheel - removal and installation

Refer to illustrations 13.8 and 13.9

Metro with Supplemental Inflatable Restraint (SIR) (1991 and later models)

1   On 1991 models, turn the ignition switch to the Off position; on 1992 and later models, turn the ignition switch to the lock position.
2   Remove the "SIR IG" fuse to the SIR fuse block (see Chapter 12, Section 19).
3   Remove the plastic access cover at the rear of the inflator module housing.
4   Inside the inflator module housing, detach the yellow two-way connector and the position assurance connector.
5   Detach the negative battery cable.
6   Remove the four screws from the rear of the wheel and remove the inflator module. **Warning:** *When working around or carrying an airbag module, keep the pad surface away from your face and body. When storing an airbag module, always place the padded area up and do not place anything on or near it.*
7   Remove the steering wheel retaining nut.
8   Reference mark the position of the steering wheel-to-shaft **(see illustration)**.
9   Make sure the front wheels are pointing straight ahead. While being very careful not to damage the SIR coil and turn signal/dimmer switch, use a steering wheel puller to remove the wheel **(see illustration)**. **Warn-**

ing: *Do not turn the steering column shaft while the steering wheel is removed.*
10   Remove the horn switch screws and switches from the wheel.
11   Remove the screws retaining the rear steering wheel cover and remove the cover.
12   Using the reference marks made during removal and reversing the removal procedure, install the wheel. If the coil has "R" and "L" marks, be sure to align the SIR coil as follows:

a) *Make sure the wheels are still pointing straight ahead.*
b) *Look closely at the SIR coil assembly. If the "Neutral" mark is at the alignment mark, the coil is centered and no further adjustment is necessary.*
c) *If the "R1" mark is close to the alignment mark, the coil is one rotation off, to the right of its centered state, and must be adjusted one turn counterclockwise. The "R2" mark means that the coil is off two rotations to the right, the "R3" marks means it's off three rotations, etc.*
d) *If the "L1" mark is close to the alignment mark, the coil is one rotation off, to the left of its centered state, and must be adjusted one turn clockwise. The "L2" mark means that the coil is off two rotations to the left, the "L3" mark means it's off three rotations, etc.*

If there are no "L" or "R" marks on the coil, turn the SIR coil slowly counterclockwise by hand, until the coil reaches the stop, then turn the coil clockwise 2-1/2 turns and align the arrows.

All others

13   Detach the negative cable from the battery.
14   Pull up on the steering wheel pad and remove it.
15   Remove the nut retaining the steering wheel.
16   Reference mark the position of the steering wheel to the shaft **(see illustration 13.8)**.
17   Using a steering wheel puller, remove

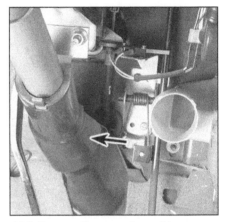

14.2 Remove the cover (arrow) to expose the steering column-to-steering shaft joint)

the wheel **(see illustration 13.9)**.
18   Using the reference marks made during removal and reversing the removal procedure, install the wheel.

14 Steering gear - removal and installation

Refer to illustrations 14.2, 14.3 and 14.8
Warning: *On airbag-equipped models, make sure the steering shaft is not turned while the steering gear is removed or you could damage the airbag system. To prevent the shaft from turning, place the ignition key in the LOCK position or thread the seat belt through the steering wheel and clip it into place. Also, before beginning work, disarm the airbag system (see Chapter 12).*
1   Move the driver's seat as far back as possible. **Caution:** *If equipped with the SIR system, lock the steering wheel in the straight ahead position and do not allow it to move during the remainder of the procedure.*
2   Pull off the front part of the floor mat on the driver's side and remove the steering shaft joint cover **(see illustration)**.

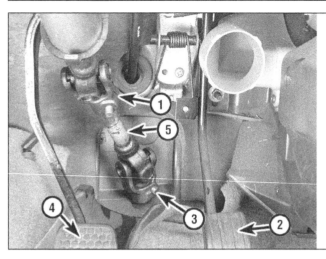

14.3 Details of the steering shaft joint

1 *Steering column-to-steering shaft joint*
2 *Accelerator pedal*
3 *Steering shaft joint lower mounting bolt*
4 *Brake pedal*
5 *Steering shaft joint*

3   Loosen the upper joint bolt for the steering shaft but don't remove it **(see illustration)**.
4   Remove the lower joint bolt for the steering shaft. Disconnect the lower joint from the pinion.
5   Raise the front of the vehicle and securely support it.
6   Remove both front wheels.
7   From both knuckles, remove the cotter pins and tie rod castle nuts.
8   Using a puller, detach the tie rod ends from the knuckles **(see illustration)**.
9   Remove the bolts mounting the steering gear to the body.
10   Installation is the reverse of removal. Be sure when connecting the steering lower joint into the steering pinion shaft, the brake discs and the steering wheel are in the straight ahead position.

15   Power steering pump - removal and installation

Removal

Refer to illustrations 15.4a and 15.4b

1   Disconnect the cable from the negative battery terminal. **Caution:** *If the stereo in your vehicle is equipped with an anti-theft system, make sure you have the correct activation code before disconnecting the battery.*
2   Using a large syringe or suction gun, suck as much fluid out of the power steering fluid reservoir as possible. Place a drain pan under the vehicle to catch any fluid that spills out when the hoses are disconnected.
3   Loosen the right front wheel lug nuts, raise the vehicle and support it securely on jackstands. Remove the right front wheel.
4   Remove the right front fender apron seal **(see illustrations)**.
5   Loosen the clamp and disconnect the fluid return hose from the pump.
6   Remove the pressure line-to-pump union bolt and separate the line from the pump. Remove the sealing washers on each side of the fitting - these should be replaced when installing the pump.

7   Loosen the adjuster and pivot bolts and remove the drivebelt.
8   Remove the pivot, adjuster and any other mounting bolts, and remove the pump.

Installation

9   Installation is the reverse of removal. Be sure to tighten the fluid line banjo bolt and the adjuster and pivot bolts to the torque listed in this Chapter's Specifications. Adjust the drivebelt tension (see Chapter 1).
10   Top up the fluid level in the reservoir (see Chapter 1) and bleed the system (see Section 21).

16   Power steering system - bleeding

1   Following any operation in which the power steering fluid lines have been disconnected, the power steering system must be bled to remove all air and obtain proper steering performance.
2   With the front wheels in the straight ahead position, check the power steering fluid level and, if low, add fluid until it reaches the Cold mark on the dipstick.
3   Start the engine and allow it to run at

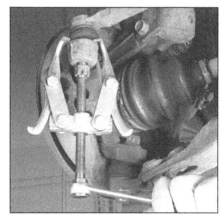

14.8 A two-jaw puller works well for separating the tie-rod end from the steering knuckle arm - note that the nut hasn't been removed (this will prevent the two components from separating violently

fast idle. Recheck the fluid level and add more if necessary to reach the Cold mark on the dipstick.
4   Bleed the system by turning the wheels from side to side, without hitting the stops. This will work the air out of the system. Keep the reservoir full of fluid as this is done.
5   When the air is worked out of the system, return the wheels to the straight ahead position and leave the vehicle running for several more minutes before shutting it off.
6   Road test the vehicle to be sure the steering system is functioning normally and noise free.
7   Recheck the fluid level to be sure it is up to the Hot mark on the dipstick while the engine is at normal operating temperature. Add fluid if necessary (see Chapter 1).

17   Tie-rod ends - removal and installation

Refer to illustration 17.2

Removal

1   Loosen the wheel lug nuts. Raise the front of the vehicle, support it securely, block the rear wheels and set the parking brake. Remove the front wheel.
2   Loosen the jam nut enough to mark the position of the tie-rod end in relation to the threads **(see illustration)**.
3   Remove the cotter pin and loosen the nut on the tie-rod end stud.
4   Disconnect the tie-rod from the steering knuckle arm with a puller **(see illustration 14.8)**. Remove the nut and separate the tie-rod.
5   Unscrew the tie-rod end from the tie-rod.

Installation

6   Thread the tie-rod end on to the marked position and insert the tie-rod stud into the steering knuckle arm. Tighten the jam nut securely.

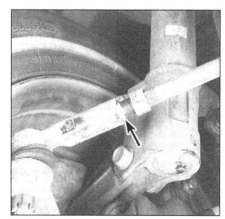

17.2 Loosen the jam nut and mark the relationship of the tie-rod end to the tie-rod with white paint

METRIC TIRE SIZES
P 185 / 80 R 13

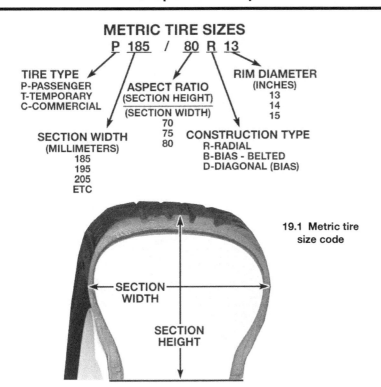

TIRE TYPE
P-PASSENGER
T-TEMPORARY
C-COMMERCIAL

ASPECT RATIO
(SECTION HEIGHT)
─────────────
(SECTION WIDTH)
70
75
80

SECTION WIDTH
(MILLIMETERS)
185
195
205
ETC

RIM DIAMETER
(INCHES)
13
14
15

CONSTRUCTION TYPE
R-RADIAL
B-BIAS - BELTED
D-DIAGONAL (BIAS)

19.1 Metric tire size code

SECTION WIDTH

SECTION HEIGHT

7   Install the castellated nut on the stud and tighten it. Install a new cotter pin.
8   Install the wheel and lug nuts. Lower the vehicle and tighten the lug nuts.
9   Have the alignment checked by a dealer service department or an alignment shop.

18 Steering gear boots - replacement

1   Loosen the lug nuts, raise the vehicle and support it securely on jackstands. Remove the wheel.
2   Remove the tie-rod end and jam nut (see Section 15).
3   Remove the steering gear boot clamps and slide the boot off.
4   Before installing the new boot, wrap the threads and serrations on the end of the steering rod with a layer of tape so the small end of the new boot isn't damaged.
5   Slide the new boot into position on the steering gear until it seats in the groove in the steering rod and install new clamps.
6   Remove the tape and install the tie-rod end (See Section 15).
7   Install the wheel and lug nuts. Lower the vehicle and tighten the lug nuts to the torque specified in Chapter 1.

19 Wheels and tires - general information

Refer to illustration 19.1

All vehicles covered by this manual are equipped with metric size fiberglass or steel belted radial tires **(see illustration)**. A label

on the drivers door jamb indicates the size tire used. Use of other size or type of tires may affect the ride and handling of the vehicle. Don't mix different types of tires, such as radials and bias belted, on the same vehicle as handling may be seriously affected. It's recommended that tires be replaced in pairs

on the same axle, but if only one tire is being replaced, be sure it's the same size, structure and tread design as the other.

Because tire pressure has a substantial effect on handling and wear, the pressure on all tires should be checked at least once a month or before any extended trips (see Chapter 1).

Wheels must be replaced if they are bent, dented, leak air, have elongated bolt holes, are heavily rusted, out of vertical symmetry or if the lug nuts won't stay tight. Wheel repairs that use welding or peening are not recommended.

Tire and wheel balance is important in the overall handling, braking and performance of the vehicle. Unbalanced wheels can adversely affect handling and ride characteristics as well as tire life. Whenever a tire is installed on a wheel, the tire and wheel should be balanced by a shop with the proper equipment.

20 Wheel alignment - general information

Refer to illustration 20.1

A wheel alignment refers to the adjustments made to the wheels so they are in proper angular relationship to the suspension and the ground. Wheels that are out of proper alignment not only affect vehicle control, but also increase tire wear. The wheel angles normally checked are camber, caster and toe-in **(see illustration)**, although the only adjustment possible is toe-in.

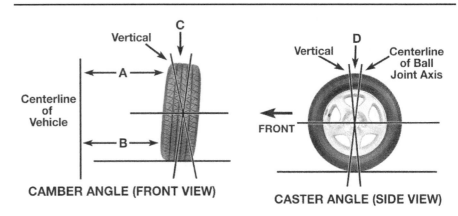

CAMBER ANGLE (FRONT VIEW)

Vertical

C

A

Centerline of Vehicle

B

Vertical

D

Centerline of Ball Joint Axis

FRONT

CASTER ANGLE (SIDE VIEW)

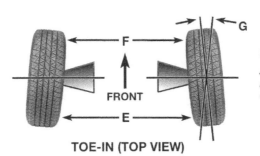

F

G

FRONT

E

TOE-IN (TOP VIEW)

20.1 Camber, caster and toe-in angles

A minus B = C (degrees camber)
E minus F = toe-in (measured in inches)
G = toe-in (expressed in degrees)

Getting the proper wheel alignment is a very exacting process, one in which complicated and expensive machines are necessary to perform the job properly. Because of this, you should have a technician with the proper equipment perform these tasks. We will, however, use this space to give you a basic idea of what is involved with wheel alignment so you can better understand the process and deal intelligently with the shop that does the work.

Toe-in is the turning in of the wheels. The purpose of a toe specification is to ensure parallel rolling of the wheels. In a vehicle with zero toe-in, the distance between the front edges of the wheels will be the same as the distance between the rear edges of the wheels. The actual amount of toe-in is normally only a fraction of an inch. On the front end, toe-in is controlled by the tie-rod end position on the tie-rod. On the rear end, it's controlled by a cam on the inner end of the suspension arm (Metro models only). Incorrect toe-in will cause the tires to wear improperly by making them scrub against the road surface.

Camber is the tilting of the wheels from the vertical when viewed from the end of the vehicle. When the wheels tilt out at the top, the camber is said to be positive (+). When the wheels tilt in at the top the camber is negative (-). The amount of tilt is measured in degrees from the vertical and this measurement is called the camber angle. This angle affects the amount of tire tread which contacts the road and compensates for changes in the suspension geometry when the vehicle is cornering or traveling over an undulating surface. Camber is not adjustable on the vehicles covered by this manual.

Caster is the tilting of the front steering axis from the vertical. A tilt toward the rear is positive caster and a tilt toward the front is negative caster. Caster is not adjustable on the vehicles covered by this manual.

Chapter 11 Body

Contents

1 General information

These models feature a "unibody" layout, using a floor pan with front and rear frame side rails which support the body components, front and rear suspension systems and other mechanical components.

Certain components are particularly vulnerable to accident damage and can be unbolted and repaired or replaced. Among these parts are the body moldings, bumpers, hood and trunk lids and all glass.

Only general body maintenance practices and body panel repair procedures within the scope of the do-it-yourselfer are included in this Chapter.

2 Body - maintenance

1   The condition of your vehicle's body is very important, because the resale value depends a great deal on it. It's much more difficult to repair a neglected or damaged body than it is to repair mechanical components. The hidden areas of the body, such as the wheel wells, the frame and the engine compartment, are equally important, although they don't require as frequent attention as the rest of the body.

2   Once a year, or every 12,000 miles, it's a

good idea to have the underside of the body steam cleaned. All traces of dirt and oil will be removed and the area can then be inspected carefully for rust, damaged brake lines, frayed electrical wires, damaged cables and other problems. The front suspension components should be greased after completion of this job.

3   At the same time, clean the engine and the engine compartment with a steam cleaner or water soluble degreaser.

4   The wheel wells should be given close attention, since undercoating can peel away and stones and dirt thrown up by the tires can cause the paint to chip and flake, allowing rust to set in. If rust is found, clean down to the bare metal and apply an anti-rust paint.

5   The body should be washed about once a week. Wet the vehicle thoroughly to soften the dirt, then wash it down with a soft sponge and plenty of clean soapy water. If the surplus dirt is not washed off very carefully, it can wear down the paint.

6   Spots of tar or asphalt thrown up from the road should be removed with a cloth soaked in solvent.

7   Once every six months, wax the body and chrome trim. If a chrome cleaner is used to remove rust from any of the vehicle's plated parts, remember that the cleaner also removes part of the chrome, so use it sparingly.

3 Vinyl trim - maintenance

Don't clean vinyl trim with detergents, caustic soap or petroleum-based cleaners. Plain soap and water works just fine, with a soft brush to clean dirt that may be ingrained. Wash the vinyl as frequently as the rest of the vehicle.

After cleaning, application of a high quality rubber and vinyl protectant will help prevent oxidation and cracks. The protectant can also be applied to weatherstripping, vacuum lines and rubber hoses, which often fail as a result of chemical degradation, and to the tires.

4 Upholstery and carpets - maintenance

1   Every three months remove the carpets or mats and clean the interior of the vehicle (more frequently if necessary). Vacuum the upholstery and carpets to remove loose dirt and dust.

2   Leather upholstery requires special care. Stains should be removed with warm water and a very mild soap solution. Use a clean, damp cloth to remove the soap, then wipe again with a dry cloth. Never use alcohol, gasoline, nail polish remover or thinner to

These photos illustrate a method of repairing simple dents. They are intended to supplement *Body repair - minor damage* in this Chapter and should not be used as the sole instructions for body repair on these vehicles.

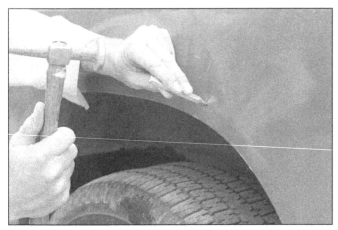

1 If you can't access the backside of the body panel to hammer out the dent, pull it out with a slide-hammer-type dent puller. In the deepest portion of the dent or along the crease line, drill or punch hole(s) at least one inch apart . . .

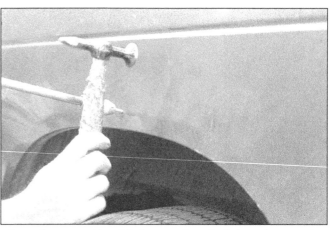

2 . . . then screw the slide-hammer into the hole and operate it. Tap with a hammer near the edge of the dent to help 'pop' the metal back to its original shape. When you're finished, the dent area should be close to its original contour and about 1/8-inch below the surface of the surrounding metal

3 Using coarse-grit sandpaper, remove the paint down to the bare metal. Hand sanding works fine, but the disc sander shown here makes the job faster. Use finer (about 320-grit) sandpaper to feather-edge the paint at least one inch around the dent area

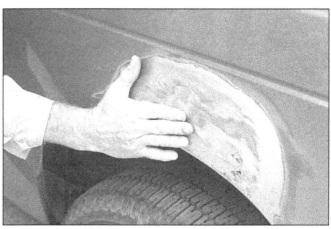

4 When the paint is removed, touch will probably be more helpful than sight for telling if the metal is straight. Hammer down the high spots or raise the low spots as necessary. Clean the repair area with wax/silicone remover

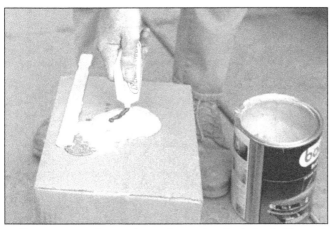

5 Following label instructions, mix up a batch of plastic filler and hardener. The ratio of filler to hardener is critical, and, if you mix it incorrectly, it will either not cure properly or cure too quickly (you won't have time to file and sand it into shape)

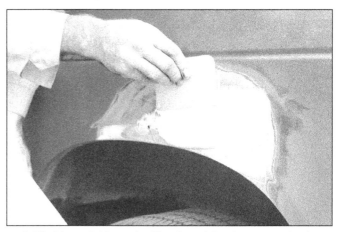

6 Working quickly so the filler doesn't harden, use a plastic applicator to press the body filler firmly into the metal, assuring it bonds completely. Work the filler until it matches the original contour and is slightly above the surrounding metal

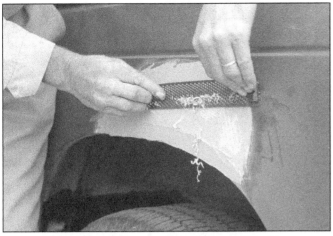

7 Let the filler harden until you can just dent it with your fingernail. Use a body file or Surform tool (shown here) to rough-shape the filler

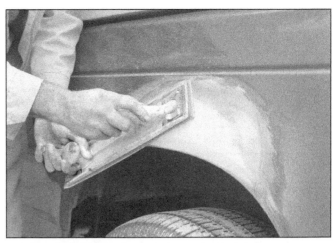

8 Use coarse-grit sandpaper and a sanding board or block to work the filler down until it's smooth and even. Work down to finer grits of sandpaper - always using a board or block - ending up with 360 or 400 grit

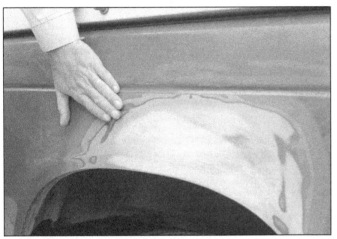

9 You shouldn't be able to feel any ridge at the transition from the filler to the bare metal or from the bare metal to the old paint. As soon as the repair is flat and uniform, remove the dust and mask off the adjacent panels or trim pieces

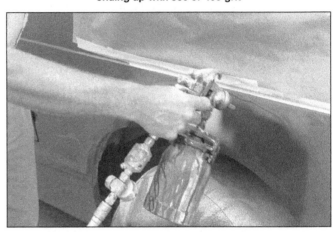

10 Apply several layers of primer to the area. Don't spray the primer on too heavy, so it sags or runs, and make sure each coat is dry before you spray on the next one. A professional-type spray gun is being used here, but aerosol spray primer is available inexpensively from auto parts stores

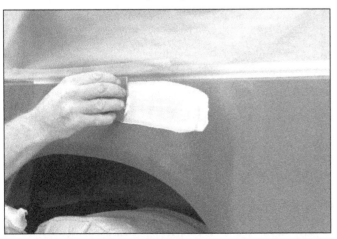

11 The primer will help reveal imperfections or scratches. Fill these with glazing compound. Follow the label instructions and sand it with 360 or 400-grit sandpaper until it's smooth. Repeat the glazing, sanding and respraying until the primer reveals a perfectly smooth surface

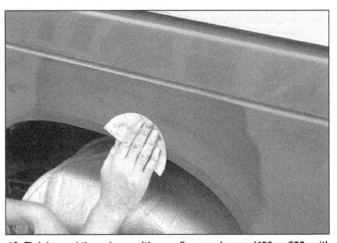

12 Finish sand the primer with very fine sandpaper (400 or 600-grit) to remove the primer overspray. Clean the area with water and allow it to dry. Use a tack rag to remove any dust, then apply the finish coat. Don't attempt to rub out or wax the repair area until the paint has dried completely (at least two weeks)

clean leather upholstery.

3   After cleaning, regularly treat leather upholstery with a leather wax. Never use car wax on leather upholstery.

4   In areas where the interior of the vehicle is subject to bright sunlight, cover leather seats with a sheet if the vehicle is to be left out for any length of time.

5   Body repair - minor damage

See photo sequence

Repair of minor scratches

1   If the scratch is superficial and does not penetrate to the metal of the body, repair is very simple. Lightly rub the scratched area with a fine rubbing compound to remove loose paint and built-up wax. Rinse the area with clean water.

2   Apply touch-up paint to the scratch, using a small brush. Continue to apply thin layers of paint until the surface of the paint in the scratch is level with the surrounding paint. Allow the new paint at least two weeks to harden, then blend it into the surrounding paint by rubbing with a very fine rubbing compound. Finally, apply a coat of wax to the scratch area.

3   If the scratch has penetrated the paint and exposed the metal of the body, causing the metal to rust, a different repair technique is required. Remove all loose rust from the bottom of the scratch with a pocket knife, then apply rust inhibiting paint to prevent the formation of rust in the future. Using a rubber or nylon applicator, coat the scratched area with glaze-type filler. If required, the filler can be mixed with thinner to provide a very thin paste, which is ideal for filling narrow scratches. Before the glaze filler in the scratch hardens, wrap a piece of smooth cotton cloth around the tip of a finger. Dip the cloth in thinner and then quickly wipe it along the surface of the scratch. This will ensure that the surface of the filler is slightly hollow. The scratch can now be painted over as described earlier in this section.

Repair of dents

4   When repairing dents, the first job is to pull the dent out until the affected area is as close as possible to its original shape. There is no point in trying to restore the original shape completely, as the metal in the damaged area will have stretched on impact and cannot be restored to its original contours. It is better to bring the level of the dent up to a point which is about 1/8-inch below the level of the surrounding metal. In cases where the dent is very shallow, it is not worth trying to pull it out at all.

5   If the back side of the dent is accessible, it can be hammered out gently from behind using a soft-face hammer. While doing this, hold a block of wood firmly against the opposite side of the metal to absorb the hammer blows and prevent the metal from being stretched.

6   If the dent is in a section of the body which has double layers, or some other factor makes it inaccessible from behind, a different technique is required. Drill several small holes through the metal inside the damaged area, particularly in the deeper sections. Screw long, self-tapping screws into the holes just enough for them to get a good grip in the metal. Now the dent can be pulled out by pulling on the protruding heads of the screws with locking pliers.

7   The next stage of repair is the removal of paint from the damaged area and from an inch or so of the surrounding metal. This is done with a wire brush or sanding disk in a drill motor, although it can be done just as effectively by hand with sandpaper. To complete the preparation for filling, score the surface of the bare metal with a screwdriver or the tang of a file, or drill small holes in the affected area. This will provide a good grip for the filler material. To complete the repair, see the subsection on filling and painting later in this Section.

Repair of rust holes or gashes

8   Remove all paint from the affected area and from an inch or so of the surrounding metal using a sanding disk or wire brush mounted in a drill motor. If these are not available, a few sheets of sandpaper will do the job just as effectively.

9   With the paint removed, you will be able to determine the severity of the corrosion and decide whether to replace the whole panel, if possible, or repair the affected area. New body panels are not as expensive as most people think and it is often quicker to install a new panel than to repair large areas of rust.

10   Remove all trim pieces from the affected area except those which will act as a guide to the original shape of the damaged body, such as headlight shells, etc. Using metal snips or a hacksaw blade, remove all loose metal and any other metal that is badly affected by rust. Hammer the edges of the hole inward to create a slight depression for the filler material.

11   Wire brush the affected area to remove the powdery rust from the surface of the metal. If the back of the rusted area is accessible, treat it with rust inhibiting paint.

12   Before filling is done, block the hole in some way. This can be done with sheet metal riveted or screwed into place, or by stuffing the hole with wire mesh.

13   Once the hole is blocked off, the affected area can be filled and painted. See the following subsection on filling and painting.

Filling and painting

14   Many types of body fillers are available, but generally speaking, body repair kits which contain filler paste and a tube of resin hardener are best for this type of repair work. A wide, flexible plastic or nylon applicator will be necessary for imparting a smooth and contoured finish to the surface of the filler material. Mix up a small amount of filler on a

clean piece of wood or cardboard (use the hardener sparingly). Follow the manufacturer's instructions on the package, otherwise the filler will set incorrectly.

15   Using the applicator, apply the filler paste to the prepared area. Draw the applicator across the surface of the filler to achieve the desired contour and to level the filler surface. As soon as a contour that approximates the original one is achieved, stop working the paste. If you continue, the paste will begin to stick to the applicator. Continue to add thin layers of paste at 20-minute intervals until the level of the filler is just above the surrounding metal.

16   Once the filler has hardened, the excess can be removed with a body file. From then on, progressively finer grades of sandpaper should be used, starting with a 180-grit paper and finishing with 600-grit wet or-dry paper. Always wrap the sandpaper around a flat rubber or wooden block, otherwise the surface of the filler will not be completely flat. During the sanding of the filler surface, the wet-or-dry paper should be periodically rinsed in water. This will ensure that a very smooth finish is produced in the final stage.

17   At this point, the repair area should be surrounded by a ring of bare metal, which in turn should be encircled by the finely feathered edge of good paint. Rinse the repair area with clean water until all of the dust produced by the sanding operation is gone.

18   Spray the entire area with a light coat of primer. This will reveal any imperfections in the surface of the filler. Repair the imperfections with fresh filler paste or glaze filler and once more smooth the surface with sandpaper. Repeat this spray-and-repair procedure until you are satisfied that the surface of the filler and the feathered edge of the paint are perfect. Rinse the area with clean water and allow it to dry completely.

19   The repair area is now ready for painting. Spray painting must be carried out in a warm, dry, windless and dust free atmosphere. These conditions can be created if you have access to a large indoor work area, but if you are forced to work in the open, you will have to pick the day very carefully. If you are working indoors, dousing the floor in the work area with water will help settle the dust which would otherwise be in the air. If the repair area is confined to one body panel, mask off the surrounding panels. This will help minimize the effects of a slight mismatch in paint color. Trim pieces such as chrome strips, door handles, etc., will also need to be masked off or removed. Use masking tape and several thicknesses of newspaper for the masking operations.

20   Before spraying, shake the paint can thoroughly, then spray a test area until the spray painting technique is mastered. Cover the repair area with a thick coat of primer. The thickness should be built up using several thin layers of primer rather than one thick one. Using 600-grit wet-or-dry sandpaper, rub down the surface of the primer until it is very smooth. While doing this, the work area

should be thoroughly rinsed with water and the wet-or-dry sandpaper periodically rinsed as well. Allow the primer to dry before spraying additional coats.

21   Spray on the top coat, again building up the thickness by using several thin layers of paint. Begin spraying in the center of the repair area and then, using a circular motion, work out until the whole repair area and about two inches of the surrounding original paint is covered. Remove all masking material 10 to 15 minutes after spraying on the final coat of paint. Allow the new paint at least two weeks to harden, then use a very fine rubbing compound to blend the edges of the new paint into the existing paint. Finally, apply a coat of wax.

6   Body repair - major damage

1   Major damage must be repaired by an auto body shop specifically equipped to perform unibody repairs. These shops have the specialized equipment required to do the job properly.
2   If the damage is extensive, the body must be checked for proper alignment or the vehicle's handling characteristics may be adversely affected and other components may wear at an accelerated rate.
3   Due to the fact that all of the major body components (hood, fenders, etc.) are separate and replaceable units, any seriously damaged components should be replaced rather than repaired. Sometimes the components can be found in a wrecking yard that specializes in used vehicle components, often at considerable savings over the cost of new parts.

7   Hinges and locks - maintenance

Once every 3000 miles, or every three

months, the hinges and latch assemblies on the doors, hood and trunk should be given a few drops of light oil or lock lubricant. The door latch strikers should also be lubricated with a thin coat of grease to reduce wear and ensure free movement. Lubricate the door and trunk locks with spray-on graphite lubricant.

8   Windshield and fixed glass - replacement

Replacement of the windshield and fixed glass requires the use of special fast-setting adhesive/caulk materials and some specialized tools. It is recommended that these operations be left to a dealer or a shop specializing in glass work.

9   Hood - removal, installation and adjustment

Refer to illustrations 9.2, 9.10a, 9.10b and 9.11
Note: *The hood is heavy and somewhat awkward to remove and install - at least two people should perform this procedure.*

Removal and installation

1   Use blankets or pads to cover the cowl area of the body and fenders. This will protect the body and paint as the hood is lifted off.
2   Draw around the bolt heads with a marking pen to ensure proper alignment during installation **(see illustration)**.
3   Disconnect any cables or wires that will interfere with removal.
4   Have an assistant support the hood. Remove the hinge-to-hood screws or bolts.
5   Lift off the hood.
6   Installation is the reverse of removal.

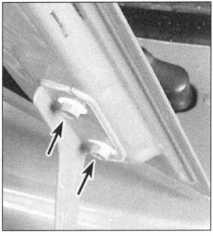

9.2 Draw alignment marks around the hood bolts to ensure proper alignment of the hood when it's reinstalled

Adjustment

7   Fore-and-aft and side-to-side adjustment of the hood is done by moving the hinge plate slot after loosening the bolts or nuts.
8   Scribe a line around the entire hinge plate so you can judge the amount of movement **(see illustration 9.2)**.
9   Loosen the bolts or nuts and move the hood into correct alignment. Move it only a little at a time. Tighten the hinge bolts or nuts and carefully lower the hood to check the position.
10   If necessary after installation, the entire hood latch assembly can be adjusted up-and-down as well as from side-to-side on the hood or radiator support so the hood closes securely, flush with the fenders. To make the adjustment, scribe a line around the hood latch mounting bolts to provide a reference point, then loosen them and reposition the latch assembly, as necessary **(see illustrations)**. Following adjustment, retighten the

9.10a On Sprint models, scribe around the latch located on the hood itself, loosen the screws, then adjust the position - the hood closed height can be adjusted by using a screwdriver to turn the locking adjustment screw (arrow)

9.10b To adjust the hood latch (arrow) horizontally or vertically, scribe around it, loosen the screws and adjust its position (tighten the screws before checking hood alignment (Metro)

9.11 To adjust the vertical height of the leading edge of the hood so that it's flush with the fenders, turn each edge cushion (arrow indicates one) clockwise (to lower the hood) or counterclockwise (to raise the hood)

mounting bolts.
11   Finally, adjust the hood bumpers on the radiator support so the hood, when closed, is flush with the fenders **(see illustration)**.
12   The hood latch assembly, as well as the hinges, should be periodically lubricated with white, lithium-base grease to prevent binding and wear.

10   Door trim panel - removal and installation

Refer to illustrations 10.2a, 10.2b, 10.2c, 10.3, 10.5 and 10.6
1   Disconnect the negative cable from the battery.
2   Remove the door handle bezel and the door pull/armrest assembly **(see illustrations)**.
3   Remove the window crank by working a cloth back-and-forth behind the handle to dislodge the retainer **(see illustration)**. With

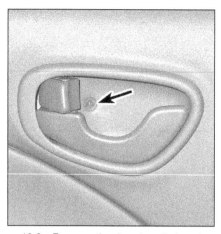

10.2a Remove the door handle bezel screw (arrow)

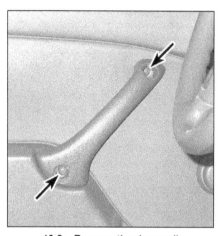

10.2c Remove the door pull screws (arrows)

the retainer removed, pull off the handle.
4   Insert a wide putty knife or a thin pry bar between the trim panel and door to disengage the retaining clips. Work around the outer edge until the panel is free.
5   Once all of the clips are disengaged,

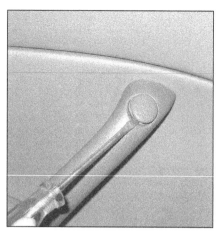

10.2b Use a screwdriver to pry off the door pull screw covers (arrow)

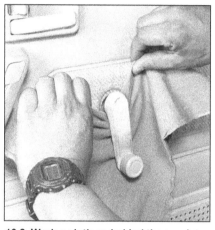

10.3 Work a cloth up behind the regulator handle and move it back-and-forth until the clip is pushed up so you can remove it

detach the trim panel, unplug any electrical connectors and remove the trim panel from the vehicle **(see illustration)**.
6   For access to the inner door, peel back the plastic water deflector, taking care not to

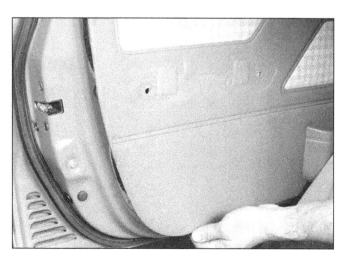

10.5 Pry under the retaining tabs until you can insert your fingers behind the panel to detach it

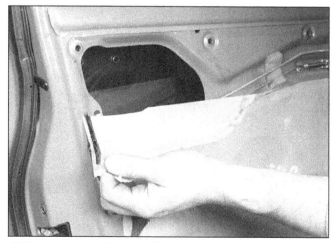

10.6 Be very careful when pulling the plastic water deflector off - don't tear it

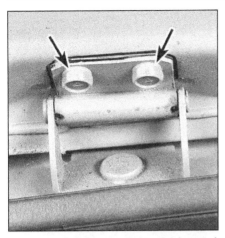

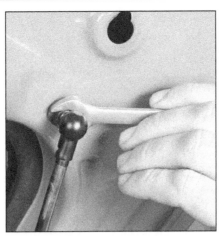

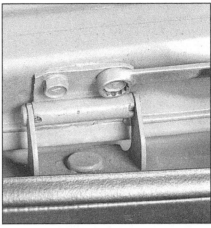

11.3 Mark around the hinge bolts (arrows) so you can return the liftgate to the same position when you install it

11.4 Use an open-end wrench to detach the support strut end

11.5 Unscrew the hinge-to-liftgate end

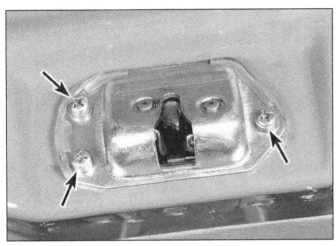

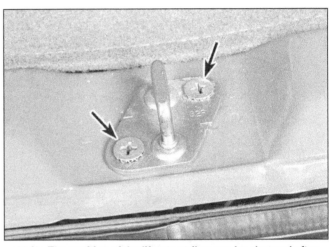

11.8a To change the engagement of the liftgate, loosen the latch mounting screws (arrows)

11.8b The position of the liftgate striker can be changed after loosening the retaining screws (arrows)

tear it **(see illustration)**. To install the trim panel, first press the water deflector into place.

7   Prior to installation of the door panel, be sure to reinstall any clips in the panel which may have come out during the removal procedure and stayed in the door.

8   Plug in any electrical connectors and place the panel in position. Press it into place until the clips are seated and install any retaining screws and armrest/door pulls. Install the manual regulator window crank or power switch assembly.

11   Liftgate - removal, installation and adjustment

Refer to illustrations 11.3, 11.4, 11.5, 11.8a, 11.8b and 11.9

1   Open the liftgate and cover the edges of the trunk compartment with pads or cloths to protect the painted surfaces when the lid is removed.

2   Disconnect any cables or electrical con-

nectors attached to the liftgate that would interfere with removal.

3   Make alignment marks around the hinge bolt mounting flanges **(see illustration)**.

4   Have an assistant support the liftgate and detach the support struts **(see illustration)**.

5   While an assistant supports the liftgate, remove the lid-to-hinge bolts on both sides and lift it off **(see illustration)**.

6   Installation is the reverse of removal. **Note:** *When reinstalling the liftgate, align the liftgate-to-hinge bolts with the marks made during removal.*

7   After installation, close the liftgate and make sure it's in proper alignment with the surrounding panels. Fore-and-aft and side-to-side adjustments of the lid are controlled by the position of the hinge bolts in the slots. To make an adjustment, loosen the hinge bolts, reposition the lid and retighten the bolts.

8   The height of the lid in relation to the surrounding body panels when closed can be changed by loosening the lock and/or striker bolts, repositioning the striker and tightening

11.9 Turn the bumpers to adjust the liftgate so it's flush with the body when closed

the bolts **(see illustrations)**.

9   Finally, adjust the bumpers on the liftgate or body so the liftgate, when closed, is flush with the body **(see illustration)**.

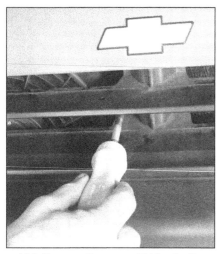

12.2 Remove the screw hidden by the grille slats using a Phillips screwdriver

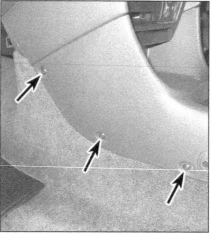

13.2 Arrows indicate the locations of the side console retaining screws - other screws are on the top and front edge

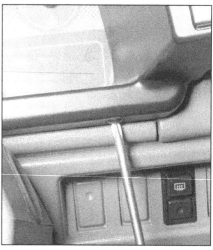

14.2a Remove the screws along the bezel lower . . .

14.2b . . . and upper edge (arrows)

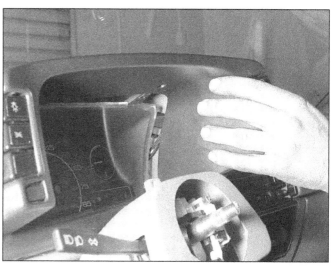

14.3 Pull the bezel straight out for access to the electrical connectors

12 Radiator grille (Sprint models only) - removal and installation

Refer to illustration 12.2

1 Remove the four screws along the top of the grille.

2 Remove the single hidden screw from the grille support **(see illustration)**.

3 Once all of the retaining screws are removed, pull the grille out and remove it.

4 Press the grille into place and install the screws.

13 Center console - removal and installation

Refer to illustration 13.2

Warning: *Later models are equipped with an airbag or Supplemental Inflatable Restraint system (SIR). When working in the vicinity of*

the airbag inflator module or the deployment sensors, the SIR system must be disabled first to avoid the possibility of accidental deployment of the airbag, which could cause personal injury (see Chapter 12).

1 Disconnect the negative cable from the battery.

2 Remove the screws, detach the console and lift it out **(see illustration)**.

3 Installation is the reverse of removal.

14 Instrument cluster bezel - removal and installation

Refer to illustrations 14.2a, 14.2b and 14.3

Warning: *Later models are equipped with an airbag or Supplemental Inflatable Restraint system (SIR). When working in the vicinity of the airbag inflator module or the deployment sensors, the SIR system must be disabled first to avoid the possibility of accidental*

deployment of the airbag, which could cause personal injury (see Chapter 12).

1 Disconnect the negative cable from the battery.

2 Remove the screws and detach the bezel **(see illustrations)**.

3 Pull the bezel out, disconnect the electrical connectors and remove the bezel and switches as an assembly **(see illustration)**.

4 Installation is the reverse of removal.

15 Door latch, lock cylinder and handle - removal and installation

1 Remove the door trim panel and water deflector (see Section 10).

Door latch

Refer to illustration 15.3

2 Reach in through the door service hole and disconnect the control links from the latch.

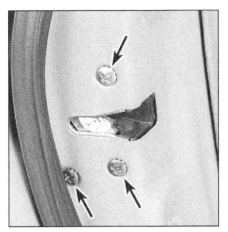

15.3 The door latch screws are located in the end of the door (arrows)

15.6 Remove the retaining clip (arrow) with a pair of pliers

15.8 Remove the two inside handle retaining nuts (arrows)

15.11 Use a screwdriver to detach the latch rod (arrow)

16.2 Use a jack (padded with rags to protect the paint) to support the door during the removal and installation procedures

3   Remove the three door lock retaining screws from the end of the door **(see illustration)**.

4   Installation is the reverse of removal. When installing the front door operating rod to the outside handle, turn the joint to obtain a clearance of 5/64-inch between the top of the slot in the operating rod and the pin on the handle lever. When placing the inside handle in position, adjust the control links to achieve the clearance between the latch inside opening lever rod and the outside opening rod of 5/64 in (2 mm).

Lock cylinder

Refer to illustration 15.6

5   Disconnect the control link from the lock cylinder.

6   Use pliers to slide the retaining clip off and remove the lock cylinder from the door **(see illustration)**.

7   Installation is the reverse of removal.

Inside handle

Refer to illustration 15.8

8   Remove the retaining screws **(see illustration)**.

9   Rotate the handle away and detach it from the door.

10   Installation is the reverse of removal.

Outside handle

Refer to illustration 15.11

11   Disconnect the control link from the handle **(see illustration)**.

12   Remove the nuts or bolts and detach the handle from the door.

13   Installation is the reverse of removal.

---

16   Door - removal, installation and adjustment

Refer to illustrations 16.2, 16.3, 16.4a and 16.4b

1   Remove the door trim panel. Disconnect any electrical connectors and push them through the door opening so they won't interfere with removal.

2   Position a jack or jackstand under the door or have an assistant on hand to support it when the hinge bolts are removed **(see illustration)**. **Note:** *If a jack or stand is used,*

16.3 Remove the door check pin by tapping it out with a hammer

place a rag between it and the door to protect the door's paint.

3   Scribe around the door bolts. Remove the door check pin **(see illustration)**.

4   Remove the hinge-to-door bolts and carefully detach the door **(see illustrations)**. Installation is the reverse of removal.

5   Following installation, make sure it's aligned properly. Adjust it if necessary as follows:

a) *Up-and-down and forward-and-backward adjustments are made by loosening the hinge-to-body bolts and moving the door, as necessary. A special offset tool may be required to reach some of the bolts.*

b) *The door lock striker can also be adjusted both up-and-down and sideways to provide a positive engagement with the locking mechanism. This is done by loosening the screws and moving the striker, as necessary.*

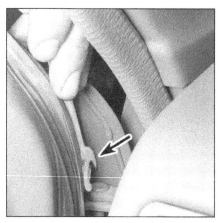

16.4a On front doors, a thin wrench will be needed to remove the hard-to-reach hinge-to-body bolts

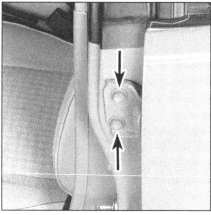

16.4b With the front door open, the rear door bolts are easily accessible (arrows)

17   Door window glass - removal and installation

1   Remove the door trim panel and water deflector (Section 10).
2   Lower the window glass.

Front door

Refer to illustration 17.3
3   Remove outer weatherstrip, then remove the glass bolts **(see illustration)**.
4   Remove the window glass by sliding it up and out of the door complete with the bottom channel.
5   If it's necessary to remove the glass, measure the distance from the end of the glass to the channel, then detach it from the channel. Lubricate the channel with soapy water and tap the new glass into place with a plastic hammer. Make sure the channel is placed the same distance from the edge of the glass as it was before.
6   Installation is the reverse of removal.

Rear door

Refer to illustrations 17.7 and 17.8
7   Remove the center guide channel and

17.3 Front door glass channel screw locations (arrows)

pull the stationary glass out of the door **(see illustration)**.
8   On Sprint models, pull the regulator arm roller out of the bottom channel and remove the glass and bottom channel as a unit **(see illustration)**.
9   On Metro models, remove the glass bolts, then remove the glass and channel as a unit.

10   If it's necessary to remove the glass, measure the distance from the end of the glass to the channel, then detach it from the channel. Lubricate the channel with soapy water and tap the new glass into place with a plastic hammer. Make sure the channel is placed the same distance from the edge of the glass as it was before.
11   Installation is the reverse of removal.

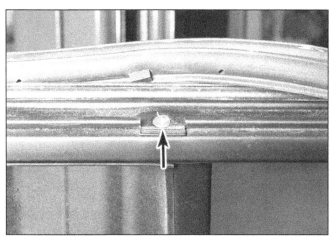

17.7 Remove the hidden screw (arrow) under the weatherstripping

17.8 Slide the glass out of the bottom of the channel

19.1a Remove the mounting screws and . . .

19.1b . . . then remove the Allen screw holding the adjuster lever

19.2 Remove the three mounting nuts (arrows) and remove the mirror from the door

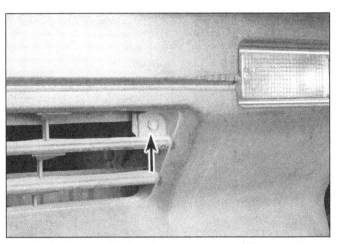

20.1a Remove the front bumper cover retaining screws

18 Door glass regulator - removal and installation

1   Remove the door trim panel and water deflector (see Section 10).
2   Remove the window assembly (see Section 17).
3   Remove the mounting bolts and detach the regulator from the door.
4   Installation is the reverse of removal.

19 Outside mirror - removal and installation

Refer to illustrations 19.1a, 19.1b and 19.2

1   Remove the screw and adjusting lever and lift off the mirror trim cover **(see illustrations)**.
2   Remove the three retaining nuts **(see illustration)** and detach the mirror. On driver side mirrors, remove the set screw, separate the cable and withdraw it along with the mirror.
3   Installation is the reverse of removal.

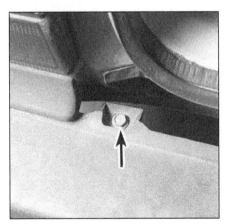

20.1b Remove the cover retaining screws on top of the bumper

20 Bumpers - removal and installation

Refer to illustrations 20.1a, 20.1b, 20.1c and 20.4

Warning: *Later models are equipped with an airbag or Supplemental Inflatable Restraint*

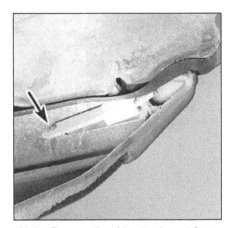

20.1c Remove the side attachment from each side of the bumper cover, from underneath (arrow)

system (SIR). When working in the vicinity of the airbag inflator module or the deployment sensors, the SIR system must be disabled first to avoid the possibility of accidental deployment of the airbag, which could cause personal injury (see Chapter 12).

1   Detach the bumper cover **(see illustrations)**.

2   Disconnect any wiring or other components that would interfere with bumper removal. **Note:** *If the vehicle is equipped with the SIR system* (see **Warning** *at the beginning of this Section), care must be taken not to damage the system sensors. If the sensors are removed, the SIR system must first be*

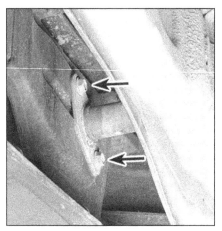

20.4 Support the bumper and remove the retaining bolts (arrows)

disarmed. When the sensors are installed, the mounting bolts must be tightened before the connectors are plugged in and the system is reactivated.
3   Support the bumper with a jack or jackstand. Alternatively, have an assistant support the bumper as the bolts are removed.
4   Remove the retaining bolts and detach the bumper **(see illustration)**.
5   Installation is the reverse of removal.
6   Tighten the retaining bolts securely.
7   Install the bumper cover and any other components that were removed.

21   Seats - removal and installation

Refer to illustrations 21.1a and 21.1b
Warning: *Some models may be equipped with an airbag or Supplemental Inflatable Restraint system (SIR). When working in the vicinity of the airbag inflator module or the deployment sensors, the SIR system must be disabled first to avoid the possibility of accidental deployment of the airbag, which could cause personal injury* (see Chapter 12).
1   Remove the retaining bolts, unplug any electrical connectors and lift the seats from

the vehicle **(see illustrations)**.
2   Installation is the reverse of removal.

22   Instrument panel - removal and installation

Warning: *Later models are equipped with an airbag or Supplemental Inflatable Restraint system (SIR). When working in the vicinity of the airbag inflator module or the deployment sensors, the SIR system must be disabled first to avoid the possibility of accidental deployment of the airbag, which could cause personal injury* (see Chapter 12).
Refer to illustrations 22.2, 22.5, 22.7, 22.10 and 22.12
1   Disconnect the negative cable from the battery.
2   Remove the instrument cluster lower covers **(see illustration)**.
3   Remove the instrument cluster (see Chapter 12).
4   Remove the center floor console (see Section 13).
5   Remove the glove box **(see illustration)**.

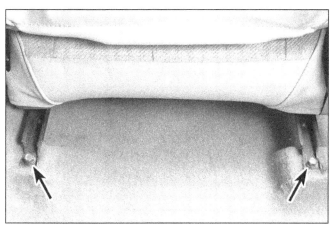

21.1a Remove the front seat retaining bolts (arrows

21.1b Lift the rear seat tab (arrow) out of the body and remove the rear seat

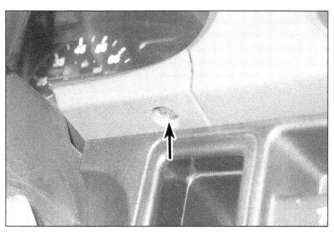

22.2 Remove the instrument cluster lower cover (arrow) on each side

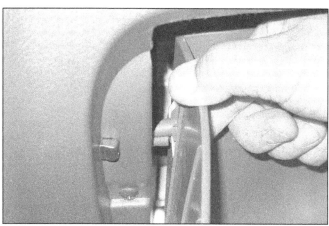

22.5 Squeeze the glove box sides in to detach it, then lower it for access, and remove the screws that attach it to the instrument panel

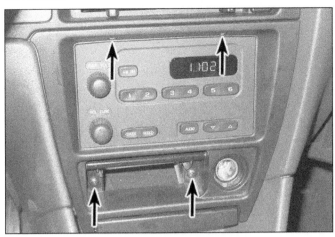

22.7 Remove the screws (arrows) and detach the center bezel and ashtray

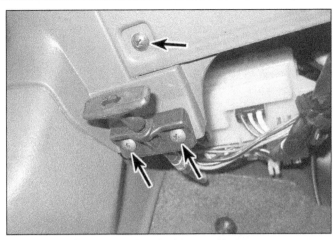

22.10 Remove the hood release and bolster screws

6   Disconnect the passenger's side air bag (if equipped) and remove it (see Chapter 12).

7   Remove the center bezel and ashtray **(see illustration)**.

8   Remove the audio unit from the center of the dashboard (see Chapter 12).

9   Remove the heater controls (see Chapter 3).

10   Remove the instrument panel cluster knee bolster cover and knee bolster, then remove the nuts/bolts securing the steering column and lower it away from the instrument panel **(see illustration)**.

11   A number of electrical connectors must be disconnected in order to remove the instrument panel. Most are designed so that they will only fit on one matching connector (male or female), but if there is any doubt, mark the connectors with masking tape and a marking pen before disconnecting them.

12   Remove all of the screws holding the instrument panel to the body **(see illustration)**. Once all are removed, lift the panel up then pull it away from the windshield and take it out through the driver's door opening.

13   Installation is the reverse of removal.

22.12 Remove the screws (arrows) holding the instrument panel to the body

Notes

Chapter 12
Chassis electrical system

Contents

1   General information

The electrical system is a 12-volt, negative ground type. Power for the lights and all electrical accessories is supplied by a lead/acid-type battery that is charged by the alternator.

This Chapter covers repair and service procedures for the various electrical components not associated with the engine. Information on the battery, alternator, distributor and starter motor can be found in Chapter 5. It should be noted that when portions of the electrical system are serviced, the negative battery cable should be disconnected from the battery to prevent electrical shorts and/or fires.

2   Electrical troubleshooting - general information

Refer to illustration 2.15

A typical electrical circuit consists of an electrical component, any switches, relays, motors, fuses, fusible links or circuit breakers related to that component and the wiring and connectors that link the component to both the battery and the chassis. To help you pinpoint an electrical circuit problem, wiring diagrams are included at the end of this Chapter.

Before tackling any troublesome electrical circuit, first study the appropriate wiring diagrams to get a complete understanding of what makes up that individual circuit. Trouble

spots, for instance, can often be narrowed down by noting if other components related to the circuit are operating properly. If several components or circuits fail at one time, chances are the problem is in a fuse or ground connection, because several circuits are often routed through the same fuse and ground connections.

Electrical problems usually stem from simple causes, such as loose or corroded connections, a blown fuse, a melted fusible link or a failed relay. Visually inspect the condition of all fuses, wires and connections in a problem circuit before troubleshooting the circuit.

If test equipment and instruments are going to be utilized, use the diagrams to plan ahead of time where you will make the neces-

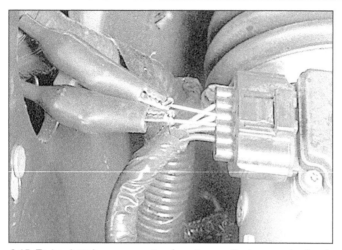

2.15 To backprobe a connector, insert a small, sharp probe (such as a straight-pin) into the back of the connector alongside the desired wire until it contacts the metal terminal inside; connect your meter leads to the probes - this allows you to test a functioning circuit

3.1a On earlier models, remove the plastic cover to access the interior fuse block located at the lower left corner of the instrument panel.

sary connections in order to accurately pinpoint the trouble spot.

The basic tools needed for electrical troubleshooting include a circuit tester or volt/ohmmeter (a 12-volt bulb with a set of test leads can also be used), a continuity tester, which includes a bulb, battery and set of test leads, and a jumper wire, preferably with a circuit breaker incorporated, which can be used to bypass electrical components **(see illustration)**. Before attempting to locate a problem with test instruments, use the wiring diagram(s) to decide where to make the connections.

Voltage checks

Voltage checks should be performed if a circuit is not functioning properly. Connect one lead of a circuit tester to either the negative battery terminal or a known good ground. Connect the other lead to a connector in the circuit being tested, preferably nearest to the battery or fuse. If the bulb of the tester lights, voltage is present, which means that the part of the circuit between the connector and the battery is problem free. Continue checking the rest of the circuit in the same fashion. When you reach a point at which no voltage is present, the problem lies between that point and the last test point with voltage. Most of the time the problem can be traced to a loose connection. **Note:** *Keep in mind that some circuits receive voltage only when the ignition key is in the Accessory or Run position.*

Finding a short

One method of finding shorts in a circuit is to remove the fuse and connect a test light or voltmeter in place of the fuse terminals. There should be no voltage present in the circuit. Move the wiring harness from side-to-side while watching the test light. If the bulb goes on, there is a short to ground somewhere in that area, probably where the insula-

tion has rubbed through. The same test can be performed on each component in the circuit, even a switch.

Ground check

Perform a ground test to check whether a component is properly grounded. Disconnect the battery and connect one lead of a self-powered test light, known as a continuity tester, to a known good ground. Connect the other lead to the wire or ground connection being tested. If the bulb goes on, the ground is good. If the bulb does not go on, the ground is not good.

Continuity check

A continuity check is done to determine if there are any breaks in a circuit - if it is passing electricity properly. With the circuit off (no power in the circuit), a self-powered continuity tester can be used to check the circuit. Connect the test leads to both ends of the circuit (or to the "power" end and a good ground), and if the test light comes on the circuit is passing current properly. If the light doesn't come on, there is a break somewhere in the circuit. The same procedure can be used to test a switch, by connecting the continuity tester to the switch terminals. With the switch turned On, the test light should come on.

Finding an open circuit

When diagnosing for possible open circuits, it is often difficult to locate them by sight because oxidation or terminal misalignment are hidden by the connectors. Merely wiggling a connector on a sensor or in the wiring harness may correct the open circuit condition. Remember this when an open circuit is indicated when troubleshooting a circuit. Intermittent problems may also be caused by oxidized or loose connections.

Electrical troubleshooting is simple if

you keep in mind that all electrical circuits are basically electricity running from the battery, through the wires, switches, relays, fuses and fusible links to each electrical component (light bulb, motor, etc.) and to ground, from which it is passed back to the battery. Any electrical problem is an interruption in the flow of electricity to and from the battery.

Connectors

Most electrical connections on these vehicles are made with multi-wire plastic connectors. The mating halves of many connectors are secured with locking clips molded into the plastic connector shells. The mating halves of large connectors, such as some of those under the instrument panel, are held together by a bolt through the center of the connector.

To separate a connector with locking clips, use a small screwdriver to pry the clips apart carefully, then separate the connector halves. Pull only on the shell, never pull on the wiring harness as you may damage the individual wires and terminals inside the connectors. Look at the connector closely before trying to separate the halves. Often the locking clips are engaged in a way that is not immediately clear. Additionally, many connectors have more than one set of clips.

Each pair of connector terminals has a male half and a female half. When you look at the end view of a connector in a diagram, be sure to understand whether the view shows the harness side or the component side of the connector. Connector halves are mirror images of each other, and a terminal shown on the right side end view of one half will be on the left side end view of the other half. When inserting a test probe into a female terminal, be careful not to distort the terminal opening. Doing so can lead to a poor connection and corrosion at that terminal later.

It is often necessary to take circuit voltage measurements with a connector con-

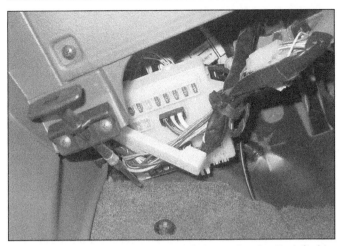

3.1b On later models the fuse block is located under the left side of the instrument panel

3.1c The power distribution blocks are located in the engine compartment adjacent to the battery - they contain cartridge type fusible links, miniaturized fuses, and relays

nected. Whenever possible, carefully insert a straight pin into the rear of the connector shell to contact the terminal inside, then attach your meter lead to the pin. Shoving the meter's probe into the connector may spread or damage the connector. This kind of connection is called "backprobing" **(see illustration)**.

3 Fuses - general information

Refer to illustrations 3.1a, 3.1b, 3.1c and 3.3

1   The electrical circuits of the vehicle are protected by a combination of fuses and circuit breakers. The fuse and relay blocks are located under the instrument panel and on the left side of the engine compartment **(see illustrations)**.

2   Each of the fuses is designed to protect a specific circuit (or circuits), and the various circuits are identified on the fuse panel cover.

3   Miniaturized fuses are employed in the fuse block. These compact fuses, with blade terminal design, allow fingertip removal and replacement. If an electrical component fails, always check the fuse first. To check the fuses, turn the ignition key to the On position and, using a test light, probe each exposed terminal of each fuse. If the test light glows on both terminals of a fuse, the fuse is good.

If power is available on one side of the fuse but not the other, the fuse is blown. When removed, a blown fuse is easily identified through the clear plastic body. Visually inspect the element for evidence of damage **(see illustration)**.

4   Be sure to replace blown fuses with the correct type. Fuses of different ratings are physically interchangeable, but only fuses of the proper rating should be used. Replacing a fuse with one of a higher or lower value than specified is not recommended. Each electrical circuit needs a specific amount of protection. The amperage value of each fuse is molded into the fuse body.

5   If the replacement fuse immediately fails, don't replace it again until the cause of the problem is isolated and corrected. In most cases, the cause will be a short circuit in the wiring caused by a broken or deteriorated wire.

4 Fusible links - general information

Some circuits are protected by fusible links. The links are used in circuits that are not normally fused, such as the ignition circuit. They look like large fuses, and are normally located in the engine compartment fuse block.

The fusible links on these models are similar to fuses in that they can be visually checked to determine if they are melted.

To replace a fusible link, first disconnect the negative battery cable. Unplug the burned out link and replace it with a new one (available from your dealer). Always determine the cause for the overload that melted the fusible link before installing a new one.

5 Circuit breakers - general information

Circuit breakers protect components such as moonroof motors, power window motors and airbag inflator resistors.

On some models the circuit breaker resets itself automatically, so an electrical overload in a circuit-breaker-protected system will cause the circuit to fail momentarily, then come back on. If the circuit does not come back on, check it immediately. Once the condition is corrected, the circuit breaker will resume its normal function. Some circuit breakers must be reset manually.

6 Relays - general information

General information

1   Many electrical accessories in the vehicle, such as the fuel injection system, horns, starter, cooling fans and fog lamps use relays to transmit the electrical signal to the component. Relays use a low-current circuit (the control circuit) to open and close a high-current circuit (the power circuit). If the relay is defective, that component will not operate properly. The various relays are mounted in engine compartment fuse box and various other locations throughout the vehicle. If a faulty relay is suspected, it can be removed and tested using the procedure below or by a dealer service department or a repair shop. Defective relays must be replaced as a unit.

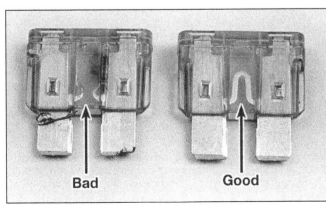

3.3 The fuses can easily be checked visually to see if they are blown (the fuse on the left is blown)

Bad   Good

7   Turn signal/hazard flasher - Check and replacement

Warning: *Later models covered by this manual are equipped with Supplemental Inflatable Restraints (SIR), more commonly known as airbags. Always disable the airbag system before working in the vicinity of the airbag system components to avoid the possibility of accidental deployment of the airbag, which could cause personal injury* (see Section 22).

Check

1   Turn signal and hazard flashers are controlled from a single electronic flasher unit that is located in the wiring harness behind the passenger compartment fuse block (Section 3). It can be located by listening for the clicks when the signals are on.
2   When the flasher unit is functioning properly, an audible click can be heard during its operation. If the turn signals fail on one side or the other and the flasher unit does not make its characteristic clicking sound, a faulty turn signal bulb is indicated.
3   If both turn signals fail to blink, the problem may be due to a blown fuse, a faulty flasher unit, a broken switch or a loose or open connection. If a quick check of the fuse box indicates that the turn signal fuse has blown, check the wiring for a short before installing a new fuse.

Replacement

4   To replace the flasher, unplug it from the harness.
5   Make sure that the replacement unit is identical to the original. Compare the old one to the new one before installing it.
6   Installation is the reverse of removal.

8   Instrument panel switches - replacement

Warning: *Later models covered by this manual are equipped with Supplemental Inflatable Restraints (SIR), more commonly known as airbags. Always disable the airbag system before working in the vicinity of the airbag system components to avoid the possibility of accidental deployment of the airbag, which could cause personal injury* (see Section 22).

1   Disconnect the negative cable at the battery.

Headlight and windshield and wiper/washer switches

Refer to illustrations 8.2
2   On 1991 and earlier models the headlight and wiper/washer switches are located in the instrument cluster **(see illustration)**.
3   Remove the instrument cluster bezel (see Chapter 11) and unplug the electrical connectors.
4   Remove the screws and separate the switches from back of the cluster.

Dash light rheostat

Refer to illustration 8.6

1991 and earlier models
5   Pull off the knob, remove the retaining nut and trim plate, push the rheostat back through the opening then disconnect the electrical connector and remove the rheostat from under the instrument panel **(see illustration)**.

1992 and later models
6   Remove the steering cover and lower instrument panel lower cover, reach up behind the rheostat and disconnect the electrical connector. Use a small screwdriver to press the retaining clips in, then pull the rheostat out of the front of the instrument panel.

Rear defogger switch
7   Reach up behind the instrument panel and unplug the switch electrical connector. Squeeze the clips on the sides of the switch together, push the switch out of the instrument panel and withdraw it.

Rear wiper/washer switch
8   Reach up behind the instrument panel, unplug the electrical connector, then squeeze the clips on the sides of the switch together. Push the switch out of the instrument panel.

9   Combination switch - replacement

Refer to illustration 9.4
Warning: *Later models covered by this manual are equipped with Supplemental Inflatable Restraints (SIR), more commonly known as airbags. Always disable the airbag system before working in the vicinity of the airbag system components to avoid the possibility of accidental deployment of the airbag, which could cause personal injury* (see Section 22).
1   The combination switch is located on the top of the steering column.

8.6 On early models, pull off the knob, remove the nut (arrow) and push the rheostat back through the hole in the dash to remove it

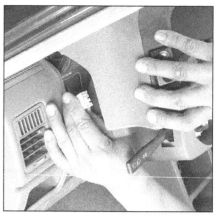

8.2 On 1991 and earlier models, remove the electrical connector from the back of the switch, then remove the cluster bezel

2   Disconnect the negative battery cable.
3   Position the front wheels and the steering wheel in the straight ahead position. Remove the steering wheel (see Chapter 10). Remove the knee bolster and steering column covers (see Chapter 11).
4   Remove the switch retaining screws, disconnect the electrical connectors, then detach the combination switch from the steering column **(see illustration)**.
5   Installation is the reverse of removal.

10   Ignition switch and key lock cylinder - replacement

Refer to illustrations 10.6 and 10.7
Warning: *Later models covered by this manual are equipped with Supplemental Inflatable Restraints (SIR), more commonly known as airbags. Always disable the airbag system before working in the vicinity of the airbag system components to avoid the possibility of accidental deployment of the airbag, which could cause personal injury* (see Section 22).
1   Disconnect the negative battery cable.
2   The ignition switch and key lock cylinder

9.4 To remove the combination switch, remove the screws (arrows) and unplug the electrical connector

10.6 Remove the screw (arrow) and detach the ignition switch from the key lock cylinder

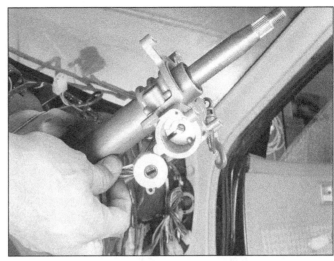

10.7 Insert the lock cylinder tab in the ignition switch notch

are incorporated into one unit on these models. If the ignition switch is faulty however, the ignition switch assembly can be replaced separately.

3   Remove the steering column covers and the lower instrument panel cover (see Chapter 11).

4   Remove the steering wheel.

5   Trace the wiring harness from the switch to the connector under the instrument cluster and disconnect it.

6   Remove the screw and detach the ignition switch **(see illustration)**.

7   Installation is the reverse of removal, making sure to align the slot in the switch with the tab on the lock cylinder **(see illustration)**.

---

11   Radio - removal and installation

Refer to illustrations 11.2a and 11.2b

Warning: *Later models covered by this manual are equipped with Supplemental Inflatable Restraints (SIR), more commonly known as airbags. Always disable the airbag system before working in the vicinity of the airbag system components to avoid the possibility of accidental deployment of the airbag, which could cause personal injury (see Section 22).*

1   Disconnect the negative battery cable.

Sprint

2   Pull off the control knobs, then remove the nuts and radio faceplate **(see illustrations)**.

3   Remove the ashtray.

4   Pull the radio out, reach behind it and unplug the electrical connector and antenna lead, then remove the radio.

5   Installation is the reverse of removal.

Metro

1990 and earlier

6   Removal of the radio on these models requires the use of removal fork tool, available at auto parts stores. You can also fabri-

cate your own tool from pieces of coat hanger.

7   Reach up behind the radio and remove the retaining bolt, then unplug the electrical connector and antenna lead.

8   Insert the fork tools into the holes at each side of the faceplate, release the radio and pull it straight out to remove it.

9   Install by pressing the radio into place until it locks. The remainder of installation is the reverse of removal

10   Remove the mounting screws and pull the radio out of the instrument panel **(see illustration)**.

11   Disconnect the electrical connectors and the antenna lead and remove the radio from the vehicle **(see illustration)**.

12   Installation is the reverse of removal.

1991 and later models

Refer to illustrations 11.14 and 11.15

13   Remove the center bezel and the glove box from the instrument panel for access (see Chapter 11).

11.2a On Sprint models, pull off the knobs and use a small socket to remove the nuts then . . .

11.2b . . . lift off the radio face plate

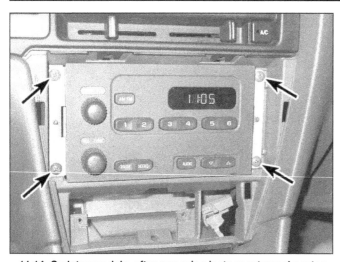

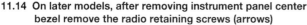

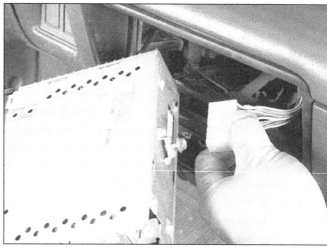

11.14 On later models, after removing instrument panel center bezel remove the radio retaining screws (arrows)

11.15 Pull the radio out and disconnect the electrical connectors

14   Remove the attaching screws **(see illustration)**.

15   Working through the glove box opening, remove the radio-to-mounting bracket screw, and push the radio out of the instrument panel, then unplug the electrical connector and antenna lead **(see illustration)**.

16   Installation is the reverse of removal.

12   Headlights - removal and installation

Sealed beam type

1   Remove the radiator grill (see Chapter 11).

2   Remove the headlight retainer screws, taking care not to disturb the adjustment screws.

3   Remove the retainer and pull the headlight out for access and unplug the connector.

4   Remove the headlight.

5   Plug in the connector securely, place the headlight in position and install the retainer and screws. Tighten the screws securely.

6   Install the radiator grille.

Halogen bulb-type

Refer to illustrations 12.8a, 12.8b and 12.9

Warning: *Halogen gas-filled bulbs are under pressure and may shatter if the surface is scratched or the bulb is dropped. Wear eye protection and handle the bulbs carefully, grasping only the base whenever possible. Do not touch the surface of the bulb with your fingers because the oil from your skin could cause it to overheat and fail prematurely. If you do touch the bulb surface, clean it with rubbing alcohol.*

7   Open the hood.

8   Rotate the knurled plastic collar counterclockwise, then withdraw the halogen bulb **(see illustrations)**.

9   Disconnect the electrical connector **(see illustration)**.

10   Without touching the glass with your bare fingers, insert the new bulb into the socket assembly, place the bulb holder in position and lock it in the place by rotating the knurled plastic collar clockwise.

11   Reinstall the electrical connector and test the headlight operation.

13   Headlights - adjustment

Refer to illustrations 13.1a, 13.1b, 13.1c and 13.2

Note: *The headlights must be aimed correctly. If adjusted incorrectly they could blind the driver of an oncoming vehicle and cause a serious accident or seriously reduce your ability to see the road. The headlights should be checked for proper aim every 12 months and any time a new headlight is installed or front end body work is performed. It should be emphasized that the following procedure is only an interim step which will provide temporary adjustment until the headlights can be adjusted by a properly equipped shop.*

1   The headlights on earlier models have two adjusting screws. The vertical screw is accessible from the top while the horizontal screw is located on the side of the housing **(see illustration)**. Early model adjusters are turned using a Phillips screwdriver. Later models have a horizontal and vertical adjuster with a leveling bubble. Use a flat tip screwdriver to adjust the horizontal adjusting gear (1 in illustration 13.1b) until the 0 (2 in

12.8a Unscrew the collar and withdraw the bulb holder from the housing

12.8b Later model headlight bulb holder retaining collar (arrow)

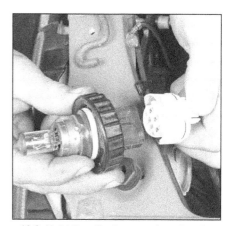

12.9 Hold the clip down and unplug the connector from the bulb holder

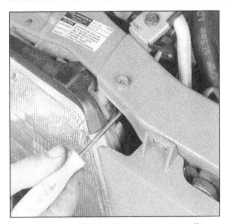

13.1a Headlight adjustment on earlier models is made with a Phillips screwdriver at the top (vertical) or the side (horizontal)

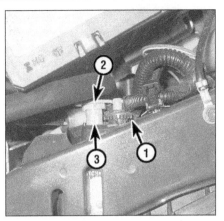

13.1b Later model headlight horizontal adjustment details

1   Turn the horizontal adjuster with a screwdriver
2   The 0 must be aligned with the adjuster centerline
3   Centerline

Note: *It may be easier to position the tape on the wall with the vehicle parked only a few inches away.*

4   Adjustment should be made with the vehicle parked 25 feet from the wall, sitting level, the gas tank half-full and no unusually heavy load in the vehicle.

5   Starting with the low beam adjustment, position the high intensity zone so it is two inches below the horizontal line and two inches to the right of the headlight vertical

13.1c To make the vertical adjustment on later models, turn the adjustment screw (4) until the bubble (5) is centered

line. Adjustment is made by turning the top adjusting screw clockwise to raise the beam and counterclockwise to lower the beam. The adjusting screw on the side should be used in the same manner to move the beam left or right.

6   With the high beams on, the high intensity zone should be vertically centered with the exact center just below the horizontal line. **Note:** *It may not be possible to position the headlight aim exactly for both high and low beams. If a compromise must be made, keep in mind that the low beams are the most used and have the greatest effect on safety.*

7   Have the headlights adjusted by a properly equipped headlight aiming facility at the earliest opportunity.

14 Headlight housing - replacement

Refer to illustration 14.2

1   Unplug the electrical connectors, and remove the halogen bulbs (see Section 12).
2   Remove the headlight housing mounting bolts and remove the housing **(see illustration)**.
3   Installation is the reverse of removal. After you're done, adjust the headlights (see Section 13).

the **illustration)** lines up with the centerline (3 in the illustration). After making the horizontal adjustment, turn the vertical adjusting screw located at the side of the housing to center the bubble (4 and 5 in the **illustration 13.1c)**.
2   The following adjustment method requires a level area and a blank wall. Position the vehicle with the headlights aimed squarely at the wall. Place masking tape vertically on the wall in reference to the vehicle centerline and the centerlines of both headlights **(see illustration)**.
3   Position a horizontal tape line in reference to the centerline of all the headlights.

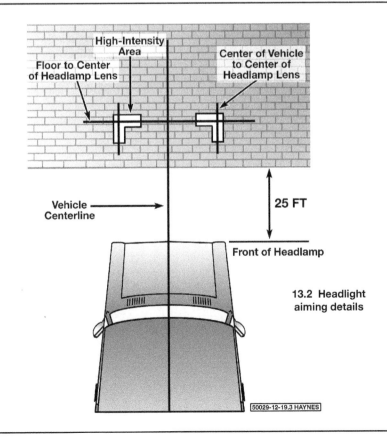

13.2 Headlight aiming details

50029-12-19.3 HAYNES

14.2 Remove screws and lift the headlight housing out of the body

15.1 The center high-mounted stop light cover is retained by screws.

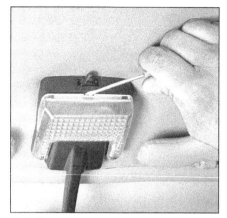

15.2 The dome light bulb can be replaced after prying off the lens

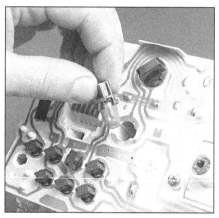

15.4 After removing the instrument cluster, turn it over and replace the bulbs by rotating them and pulling straight out

15 Bulb replacement

Refer to illustrations 15.1, 15.2, 15.4 and 15.5

Warning: *Bulbs remain hot for up to twenty minutes after they're turned off. Be sure bulbs are off and cool before you touch them.*

1   The lenses of many lights are in place by

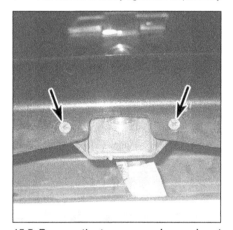

15.5 Remove the two screws (arrows) and lower the license plate light housing

screws, which makes it a simple procedure to gain access to the bulbs **(see illustration)**.
2   On some lights the lenses are held in place by clips. The lenses can be removed either by unsnapping them or by using a small screwdriver to pry then out of the holder **(see illustration)**.
3   Several types of bulbs are used. Some, like the taillights, are removed by pushing them in and turning them counterclockwise.
4   To gain access to the instrument panel lights, the instrument panel will have to be removed first **(see illustration)**.
5   Remove the two screws and lower the license plate light housing for access to the bulb **(see illustration)**.

16 Wiper motor - replacement

Refer to illustrations 16.2a, 16.2b, 16.3a, 16.3b and 16. 5

1   Disconnect the cable from the negative battery terminal.
2   Disconnect the electrical connector from the wiper motor, remove the retaining bolts and remove the motor **(see illustrations)**.

3   Remove the nut and disconnect the wiper linkage from the motor **(see illustrations)**.
4   Remove the wiper motor and mount as an assembly **(see illustration)**.
5   Installation is the reverse of removal.

17 Instrument cluster - removal and installation

Warning: *Later models covered by this manual are equipped with Supplemental Inflatable Restraints (SIR), more commonly known as airbags. Always disable the airbag system before working in the vicinity of the airbag system components to avoid the possibility of accidental deployment of the airbag, which could cause personal injury (see Section 22).*

1   Disconnect the negative cable from the battery.

1985 through 1991 models
Refer to illustrations 17.3, 17.5a and 17.5b

2   Remove the instrument cluster bezel (see Chapter 11).
3   Remove the instrument cluster screws,

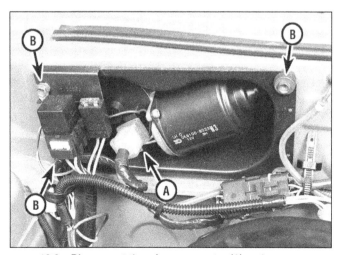

16.2a Disconnect the wiper connector (A) and remove the mounting bolts (B) (early models)

16.2b Later model wiper motor bolt and connector locations (arrows)

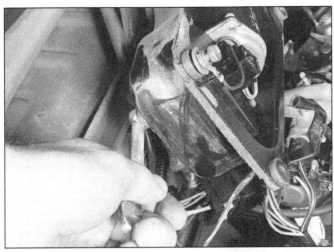

16.3a Reach behind the motor, use a wrench to remove the nut from the shaft, then detach the wiper linkage (early model)

16.3b Later model wiper motor-to-linkage retaining nut

16.5 Lift the wiper motor and mount off the firewall

17.3 Instrument cluster screw locations (arrows)

pull out the cluster and unplug the electrical connectors **(see illustration)**.

4   Detach the speedometer cable from the transaxle and pull it up to provide sufficient slack to allow cluster removal.

5   Pull the cluster out of the instrument panel and detach the electrical connector and speedometer cable **(see illustrations)**.

6   Installation is the reverse of removal. Be sure to connect the positive cable to the battery first, then the negative cable.

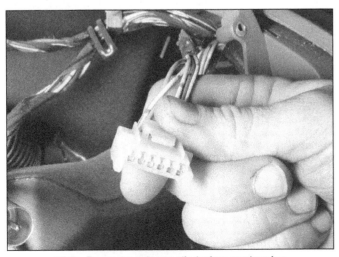

17.5a Press the release clip before unplugging the electrical connector

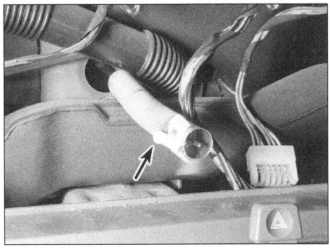

17.5b Press the speedometer cable clip in to release the cable from the cluster

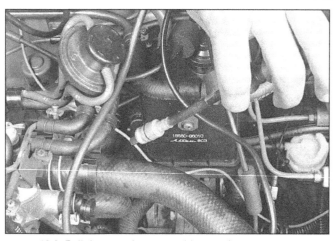

18.2 Pull the speedometer cable out of the transaxle

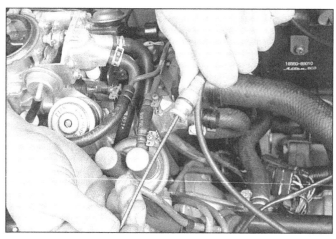

18.3 Withdraw the drive cable from the speedometer housing

1992 and later models

7   Disable the airbag (if equipped) (see Section 22).
8   Remove the steering column cover.
9   Lower the steering column.
10   Remove the instrument cluster bezel (see Chapter 11).
11   Unplug the electrical connectors from the switches mounted on the cluster trim bezel (some models).
12   Disconnect the speedometer cable at the transaxle.
13   Remove the retaining screws, pull the instrument cluster assembly from the dash and disconnect the speedometer cable and all electrical connectors from the back of the cluster. Remove the cluster.
14   Installation is the reverse of removal. Be sure to connect the positive cable to the battery first, then the negative cable.

18   Speedometer cable replacement

Refer to illustrations 18.2 and 18.3
1   Disconnect the negative cable from the battery
2   Disconnect the speedometer cable from the transaxle. On Sprint models this is accomplished by removing a bolt in the

19.2 Horn mounting bolt (A) and electrical connectors (B)

transaxle and then withdrawing the cable end **(see illustration)**. On Metro models, remove the bolt and detach the speedometer drive gear housing from the transaxle.
3   Pull the drive cable out of the speedometer cable housing **(see illustration)**. If you are also replacing the cable housing, disconnect the housing at the instrument cluster (see the previous Section).
4   Insert the new drive cable into the housing, making sure it seats securely in the instrument cluster, then connect the cable to the transaxle.

19   Horn - replacement

Refer to illustration 19.1
1   To access the horn the left bumper cover must be removed.
2   Disconnect the electrical connectors and remove the bracket bolt **(see illustration)**.
3   Installation is the reverse of removal.

20   Daytime Running Lights (DRL) - general information

The Daytime Running Lights (DRL) system used on some models illuminates the headlights whenever the engine is running. The only exception is with the engine running and the parking brake engaged. Once the parking brake is released, the lights will remain on as long as the ignition switch is on, even if the parking brake is later applied.
The DRL system supplies reduced power to the headlights so they won't be too bright for daytime use, while prolonging headlight life.

21   Rear window defogger - check and repair

1   The rear window defogger consists of a number of horizontal heating elements baked onto the inside surface of the glass. Power is

supplied through a large fuse from the power distribution box in the engine compartment. The heater is controlled by the instrument panel switch.
2   Small breaks in the element can be repaired without removing the rear window.

Check

Refer to illustrations 21.5, 21.6 and 21.8
3   Turn the ignition switch and defogger switches to the On position.
4   Using a voltmeter, place the positive probe against the defogger grid positive terminal and the negative probe against the ground terminal. If battery voltage is not indicated, check the fuse, defogger switch, defogger relay and related wiring. If voltage is indicated, but all or part of the defogger doesn't heat, proceed with the following tests.
5   When measuring voltage during the next two tests, wrap a piece of aluminum foil around the tip of the voltmeter positive probe and press the foil against the heating element with your finger **(see illustration)**. Place the negative probe on the defogger grid ground terminal.
6   Check the voltage at the center of each heating element **(see illustration)**. If the voltage is 5 to 6 volts, the element is okay (there is no break). If the voltage is 0 volts, the element is broken between the center of the element and the positive end. If the voltage is 10 to 12 volts the element is broken between the center of the element and the ground side. Check each heating element.
7   If none of the elements are broken, connect the negative probe to a good chassis ground. The voltage reading should stay the same, if it doesn't the ground connection is bad.
8   To find the break, place the voltmeter negative probe against the defogger ground terminal. Place the voltmeter positive probe with the foil strip against the heating element at the positive side and slide it toward the negative side. The point at which the voltmeter deflects from several volts to zero is the point where the heating element is broken **(see illustration)**.

21.5 When measuring the voltage at the rear window defogger grid, wrap a piece of aluminum foil around the positive probe of the voltmeter and press the foil against the wire with your finger

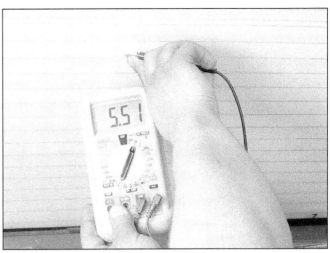

21.6 To determine if a wire has broken, check the voltage at the center of each wire. If the voltage is 5 to 6 volts, the wire is unbroken; if the voltage is 10 to 12 volts, the wire is broken between the center of the wire and the ground side; if the voltage is 0 volts, the wire is broken between the center of the wire and the power side

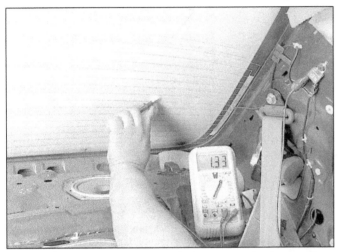

21.8 To find the break, place the voltmeter negative lead against the defogger ground terminal, place the voltmeter positive lead with the foil strip against the wire at the positive terminal end and slide it toward the negative terminal end - the point at which the voltmeter deflects from several volts to zero volts is the point at which the wire is broken

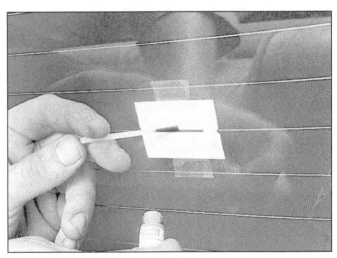

21.14 To use a defogger repair kit, apply masking tape to the inside of the window at the damaged area, then brush on the special conductive coating

Repair

Refer to illustration 21.14

9   Repair the break in the element using a repair kit specifically for this purpose, such as Dupont paste No. 4817 (or equivalent). The kit includes conductive plastic epoxy.

10   Before repairing a break, turn off the system and allow it to cool for a few minutes.

11   Lightly buff the element area with fine steel wool; then clean it thoroughly with rubbing alcohol.

12   Use masking tape to mask off the area being repaired.

13   Thoroughly mix the epoxy, following the kit instructions.

14   Apply the epoxy material to the slit in the masking tape, overlapping the undamaged area about 3/4-inch on either end **(see illustration)**.

15   Allow the repair to cure for 24 hours before removing the tape and using the system.

---

22   Airbag system - general information, removal and installation

---

Refer to illustration 22.13

General information

1   Later models are equipped with a Supplemental Inflatable Restraint System (SIR), more commonly known as airbags. There are two airbags (on later models), one for the driver and for the front seat passenger. The SIR system is designed to protect the driver (and on later models, the passenger as well) from serious injury in the event of a head-on or frontal collision. All models have a sensing/diagnostic control unit, located under the center console. **Warning:** *If your vehicle is ever involved in a flood, or the interior carpeting is ever soaked for any reason, disconnect the battery and do not start the vehicle until the airbag system can be checked by your dealer. If the SIR system is subjected to flooding, the airbags could go off upon starting the vehicle, even without an accident taking place.*

Airbag modules

2   The airbag modules consist of a housing incorporating the cushion (airbag) and inflator unit. The inflator assembly is mounted on the back of the housing over a hole through which gas is expelled, inflating the bag almost instantaneously when an electrical

signal is sent from the system. The specially wound wire on the driver's side is called a clockspring. The clockspring is a flat, ribbon-like electrically conductive tape that is wound many times so that it can transmit an electrical signal regardless of steering wheel position.

Sensing/diagnostic control unit and sensors

3   The sensing/diagnostic control unit contains an on-board microprocessor which monitors the operation of the system, and also contains a crash sensor. It checks this system every time the engine is started, causing the "Airbag" light to flash seven times then go off, if the system is operating properly. If there is a fault in the system, the light will go on and continue, either illuminated steadily or blinking, and the unit will store fault codes indicating the nature of the fault.

Operation

4   For the airbag(s) to deploy, the accelerometer in the diagnostic control unit is activated. The control unit then compares this force to a value stored in its memory.The control unit determines if the force is in excess and the safing sensor must be activated. When this condition occurs, the circuit to the airbag inflator is closed and the airbag inflates. If the battery is destroyed by the impact, or is too low to power the inflator, a back-up power unit provides power.

Self-diagnosis system

5   A self-diagnosis circuit in the control unit displays a light when the ignition switch is turned to the On position. If the system is operating normally, the light should go out after a few seconds. If the light doesn't come on, or doesn't go out after a few seconds, or if it comes on while you're driving the vehicle, there's a malfunction in the SIR system. Have it inspected and repaired as soon as possible. Do not attempt to troubleshoot or service the SIR system yourself. Even a small mistake could cause the SIR system to malfunction when you need it.

Servicing components near the SIR system

6   Nevertheless, there are times when you need to remove the steering wheel, radio or

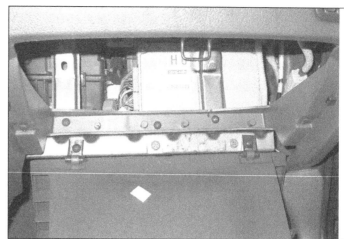

22.13 Passenger side airbag details

service other components on or near the instrument panel. At these times, you'll be working around components and wire harnesses for the SIR system. The SIR wiring harnesses are easy to identify: They're all bright yellow. Do not unplug the connectors for these wires. And do not use electrical test equipment on yellow wires; it could cause the airbag(s) to deploy. *ALWAYS DISABLE THE SIR SYSTEM BEFORE WORKING NEAR THE SIR SYSTEM COMPONENTS OR RELATED WIRING.*

Disabling the SIR system

Warning: *Any time you are working in the vicinity of airbag wiring or components, DISABLE THE SIR SYSTEM.*

7   Turn the steering wheel so that the front wheels are pointing straight ahead.
8   Turn the ignition switch to Off (1991 models) or Lock (1992 and later models). Disconnect the battery negative cable, then disconnect the positive cable and wait two minutes.
9   Remove the SIR IG from the fuse block.

Drivers side airbag

10   Remove the access panel below the instrument panel and disconnect the electrical connector leading up the column to the drivers airbag.

Passenger's side airbag

11   Remove the glove box (see Chapter 11).
12   Disconnect the passenger's side airbag electrical connector.
13   Remove the screws and detach the

airbag from the instrument panel **(see illustration)**. **Caution:** *The airbag assembly is heavier than it looks. Use both hands when removing it from the dash.*
14   Installation is the reverse of removal.

Enabling the SIR system

15   After you've disabled the airbag and performed the necessary service, reconnect the electrical connectors to the airbags. Reinstall the glove box.
16   Turn the ignition switch to the Off position.
17   Reattach the positive battery cable first and then the negative cable.

23   Wiring diagrams - general information

Since it isn't possible to include all wiring diagrams for every year and model covered by this manual, the following diagrams are those that are typical and most commonly needed.

Prior to troubleshooting any circuits, check the fuse and circuit breakers (if equipped) to make sure they're in good condition. Make sure the battery is properly charged and check the cable connections (see Chapter 1).

When checking a circuit, make sure that all connectors are clean, with no broken or loose terminals. When unplugging a connector, do not pull on the wires. Pull only on the connector housings themselves.

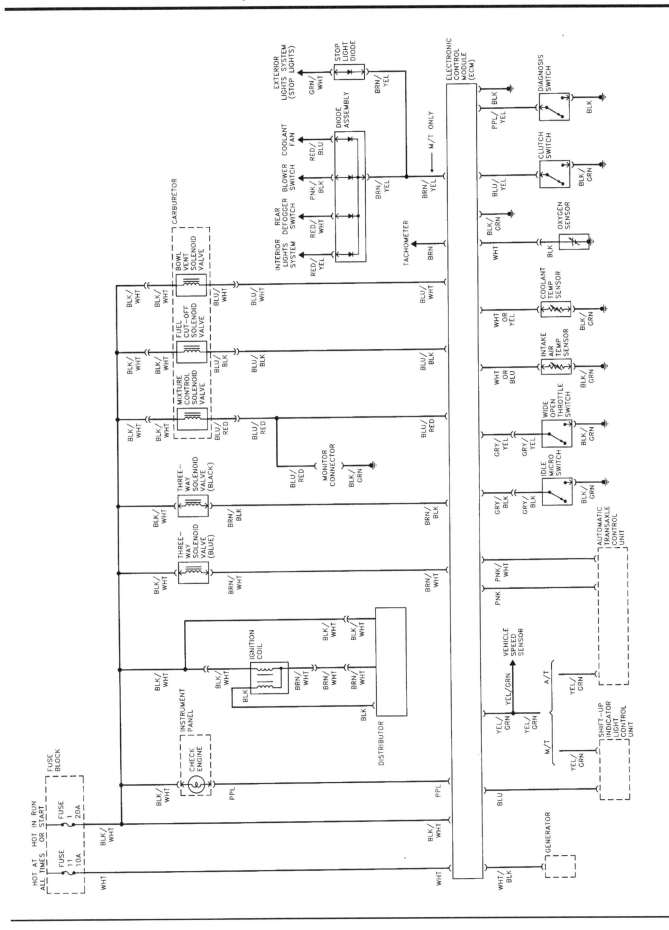

Engine control system - carbureted models

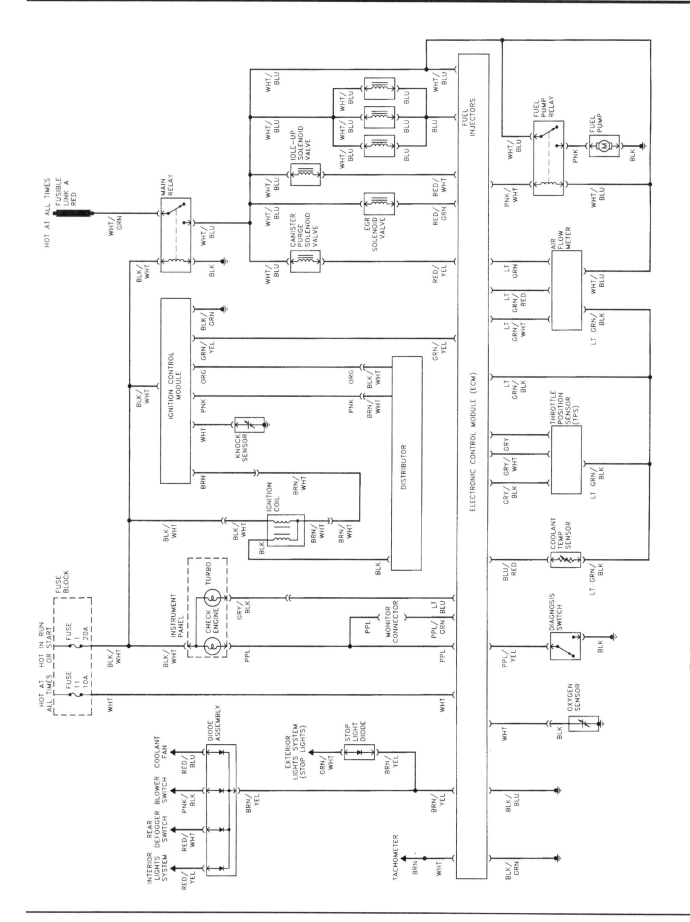

Engine control system - early multi-port fuel-injected models (later models similar)

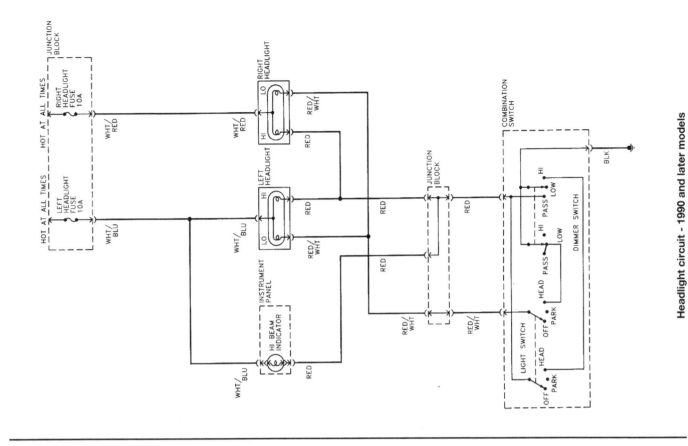

Headlight circuit - 1990 and later models

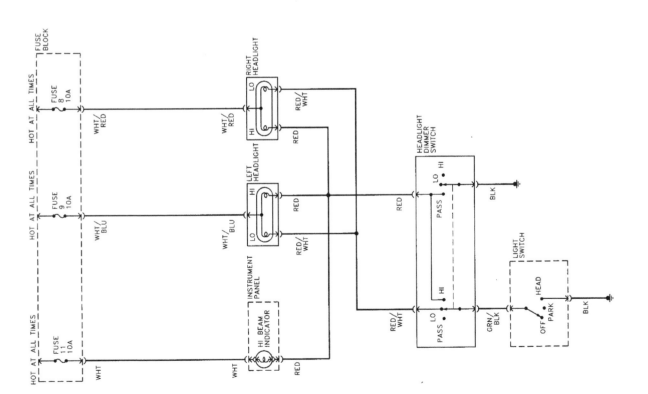

Headlight circuit - 1989 and earlier models

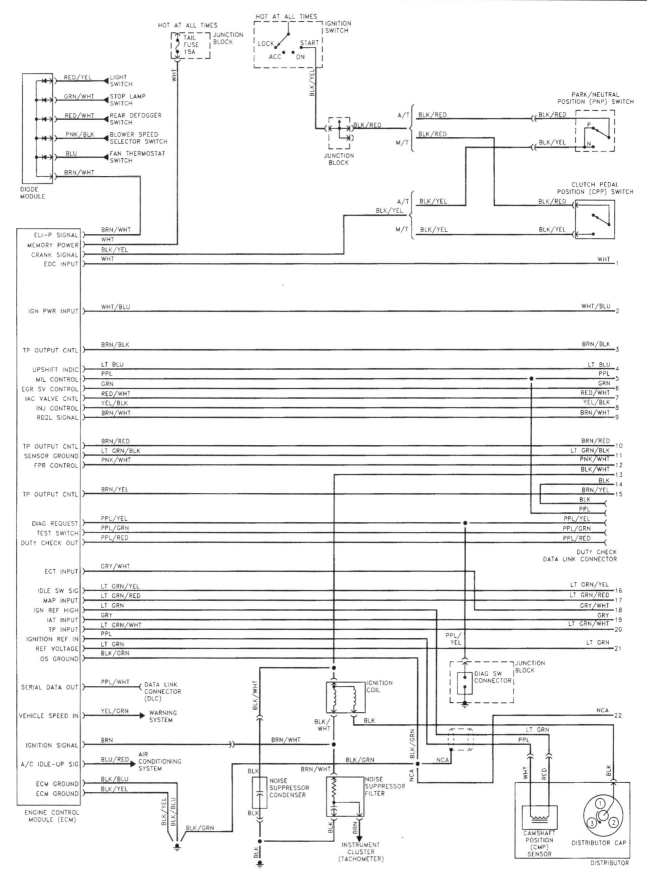

Engine control system - typical TBI fuel-injected models (1 of 2)

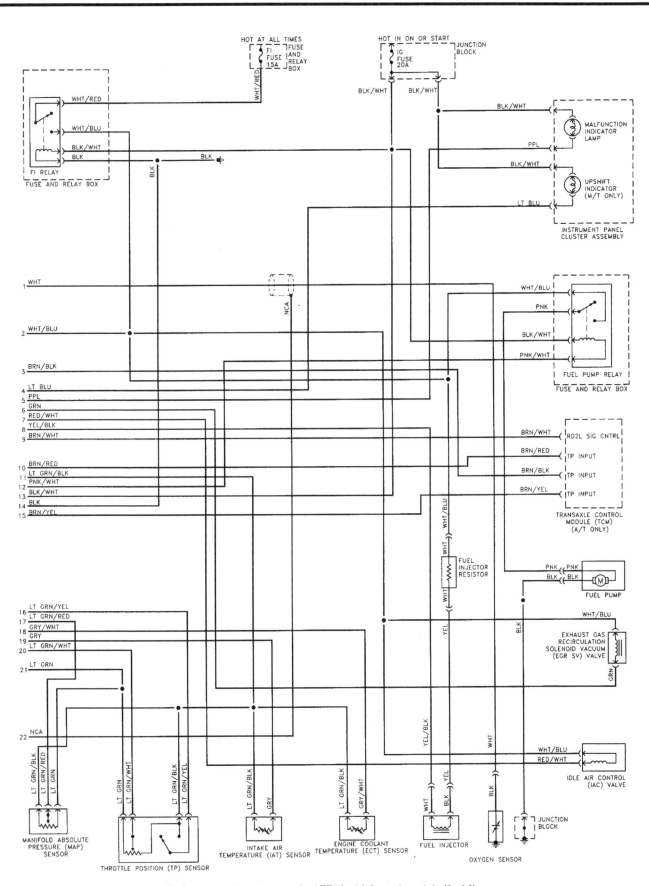

Engine control system - typical TBI fuel-injected models (2 of 2)

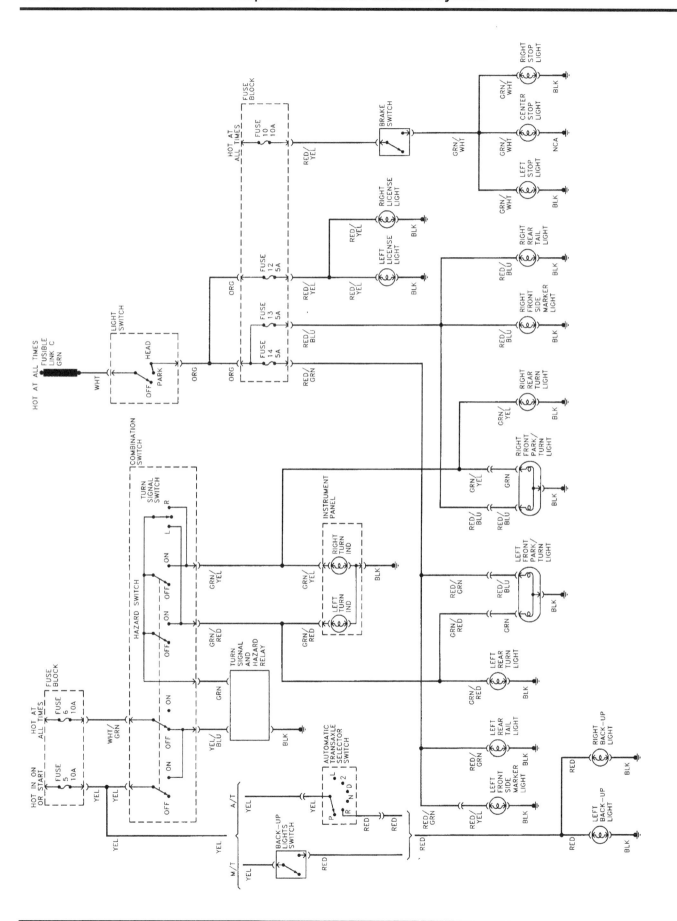

Exterior lighting circuit (except headlights) - 1989 and earlier models

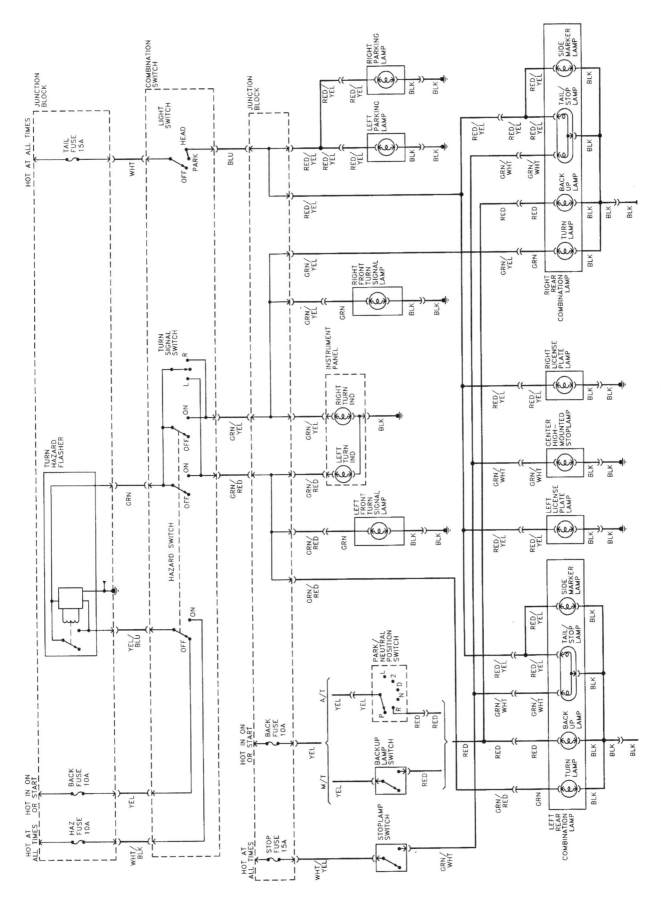

Exterior lighting circuit (except headlights) - 1990 and later models

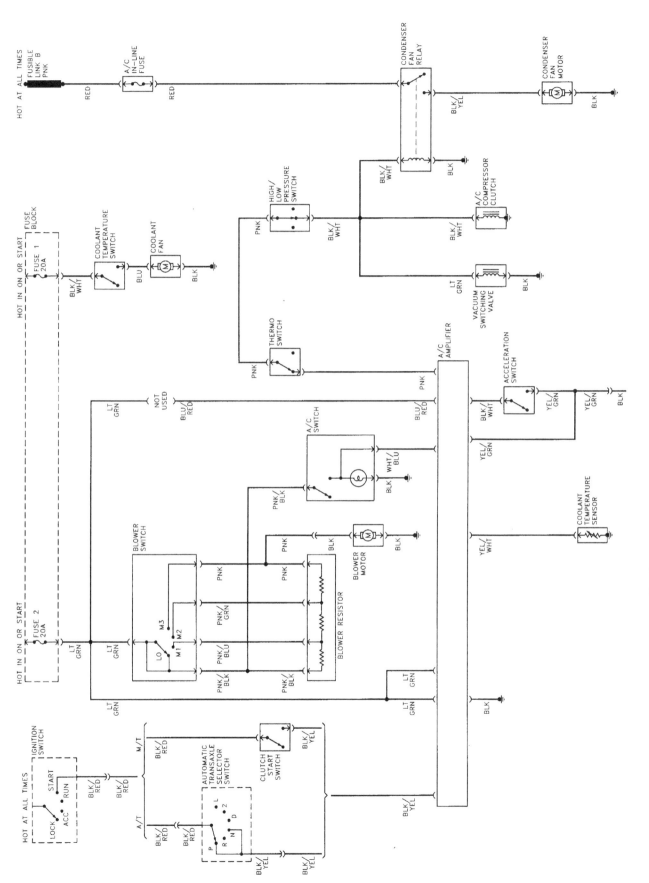

Heating and air conditioning system - 1985 and 1986 models

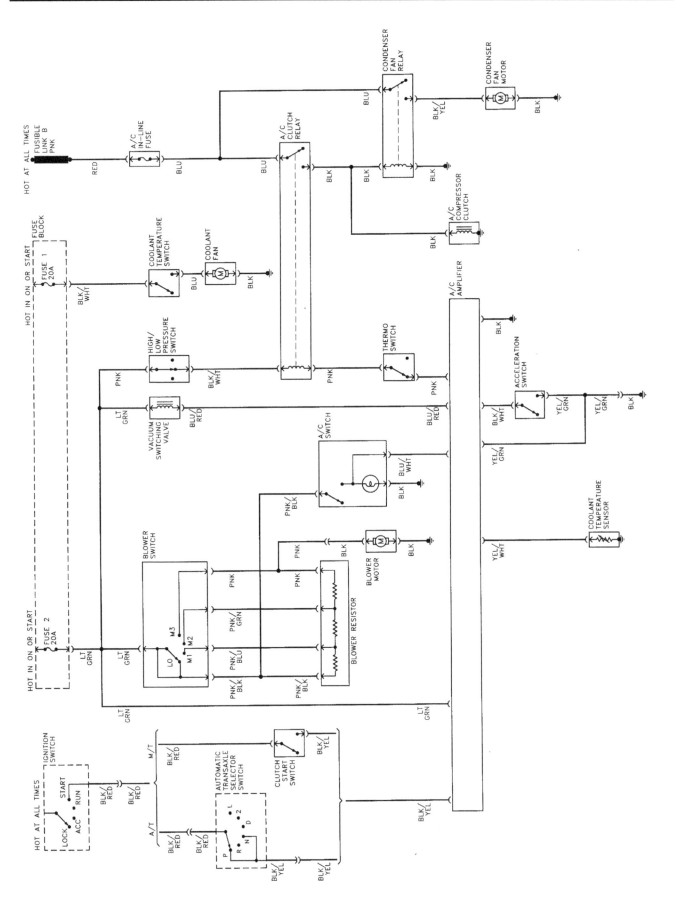

Heating and air conditioning system - 1987 through 1989 models

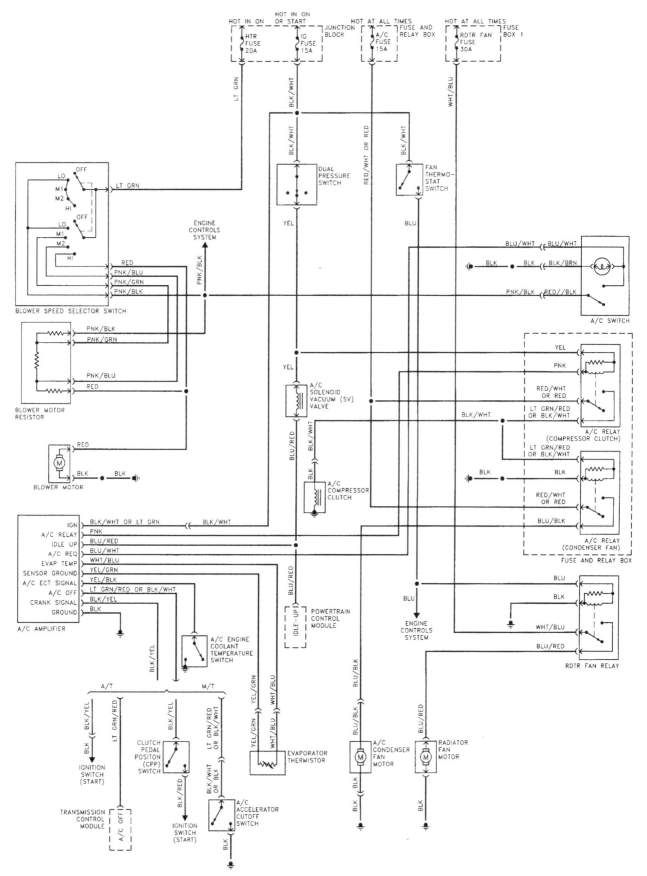

Heating and air conditioning system - 1990 and later models

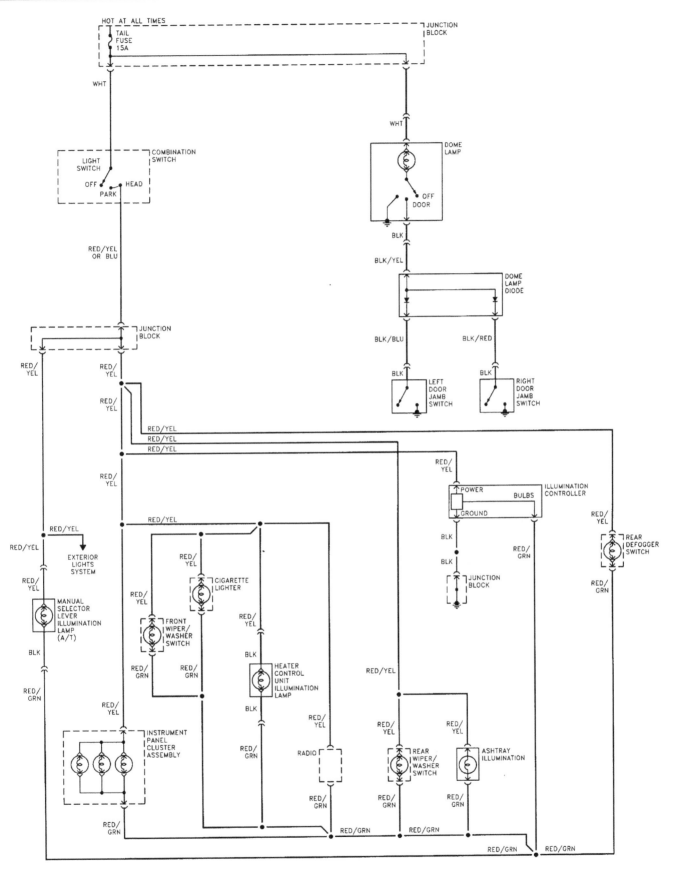

Interior lighting circuit - 1990 and later models

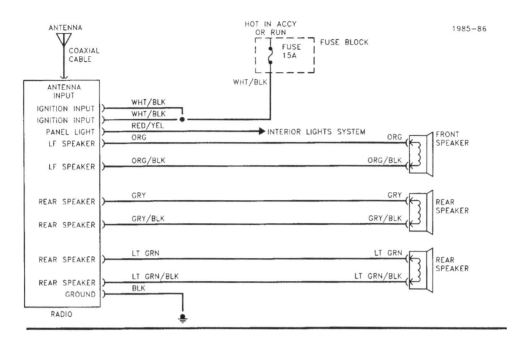

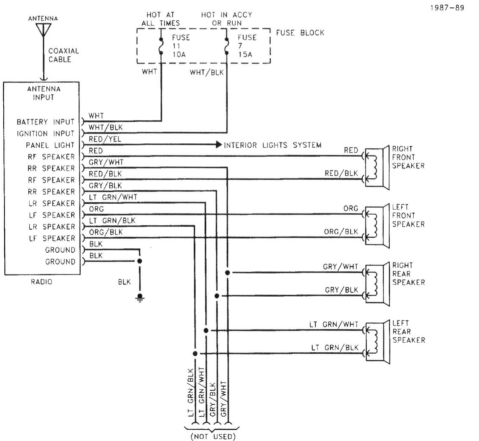

Radio circuit - 1985 and 1986 models (top); 1987 through 1989 models (bottom)

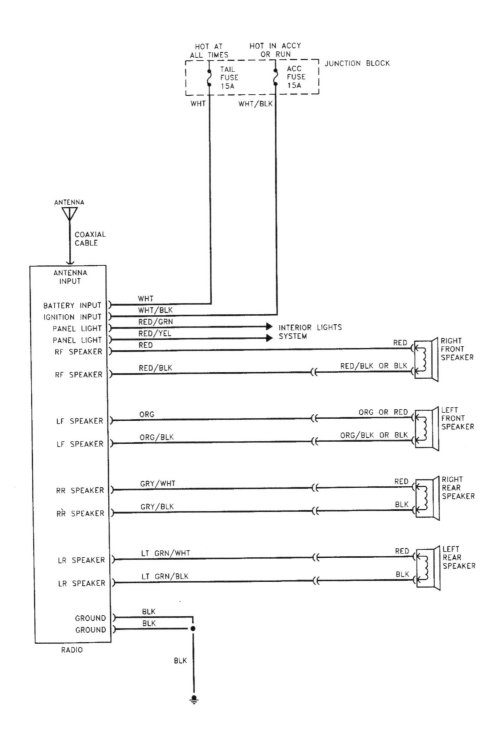

Radio circuit - 1990 and later models

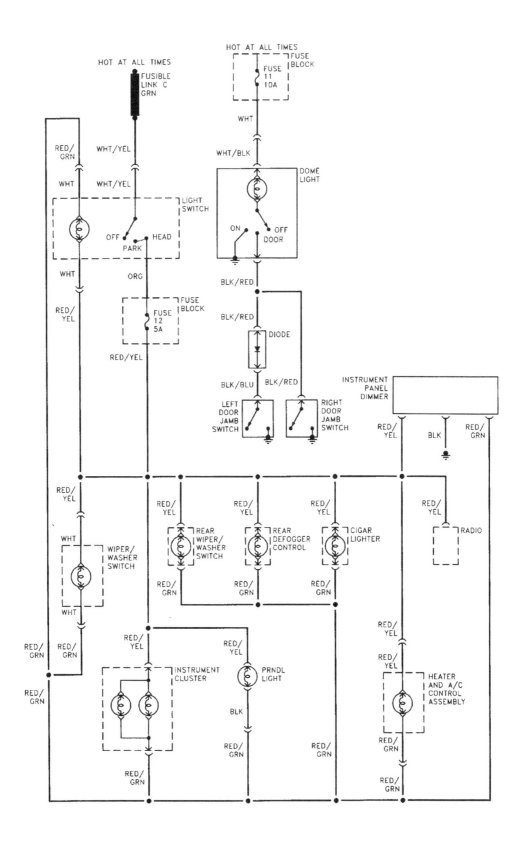

Interior lighting circuit - 1989 and earlier models

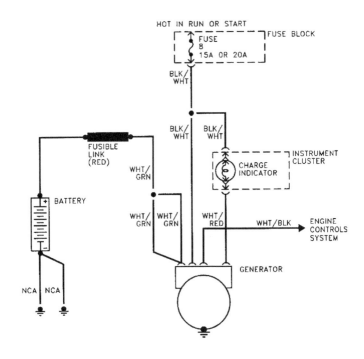

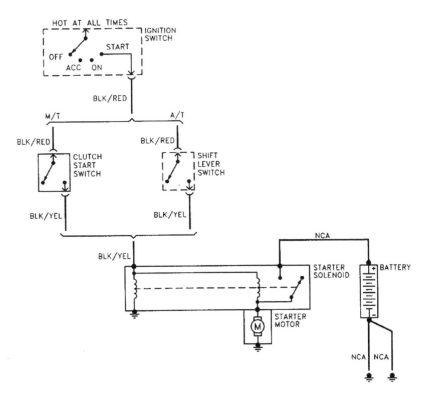

Starting and charging systems - 1989 and earlier Sprint models

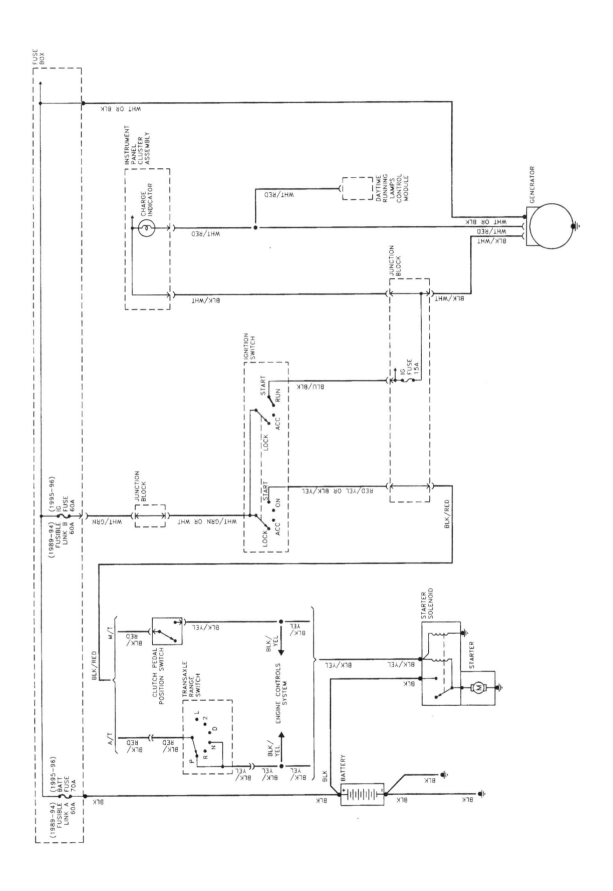

Starting and charging systems - 1989 and later **Metro** models

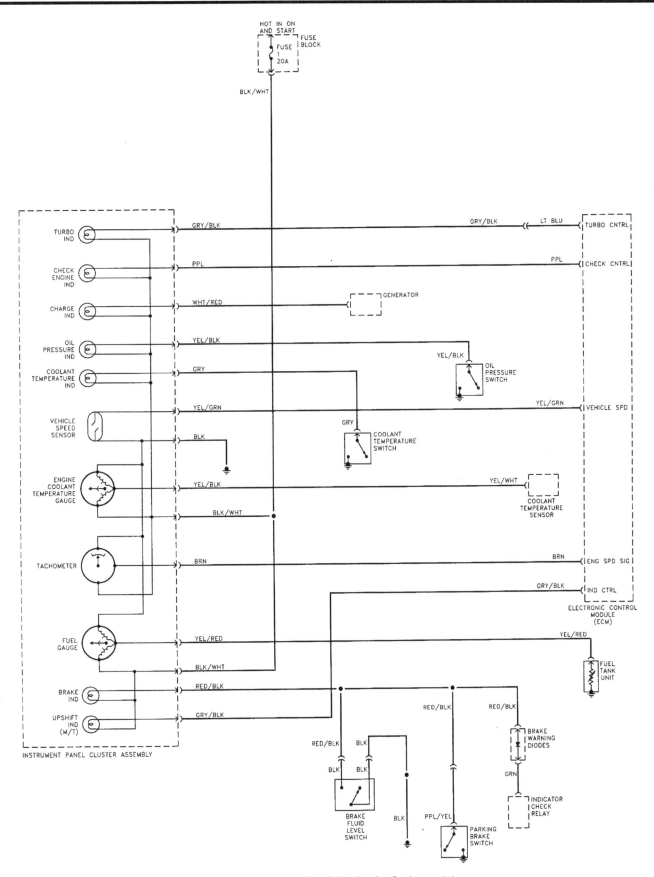

Instrument panel warning light circuit - Sprint models

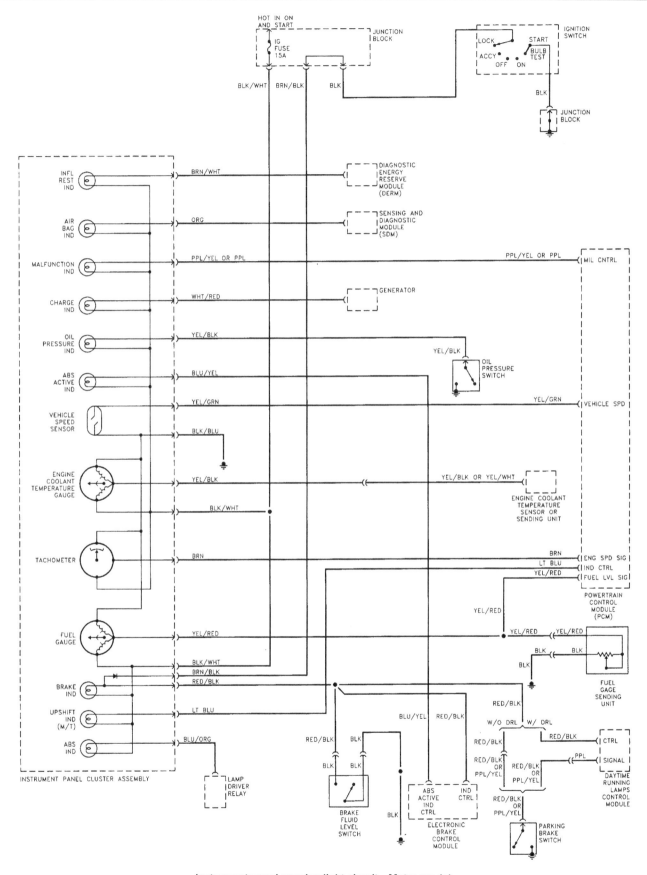

Instrument panel warning light circuit - Metro model

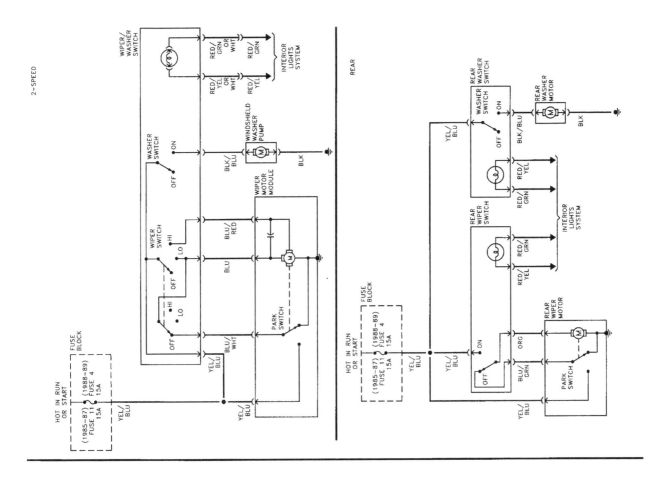

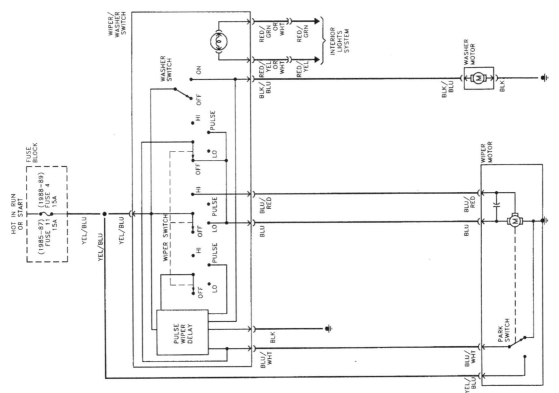

Wiper/washer systems - 1989 and earlier models

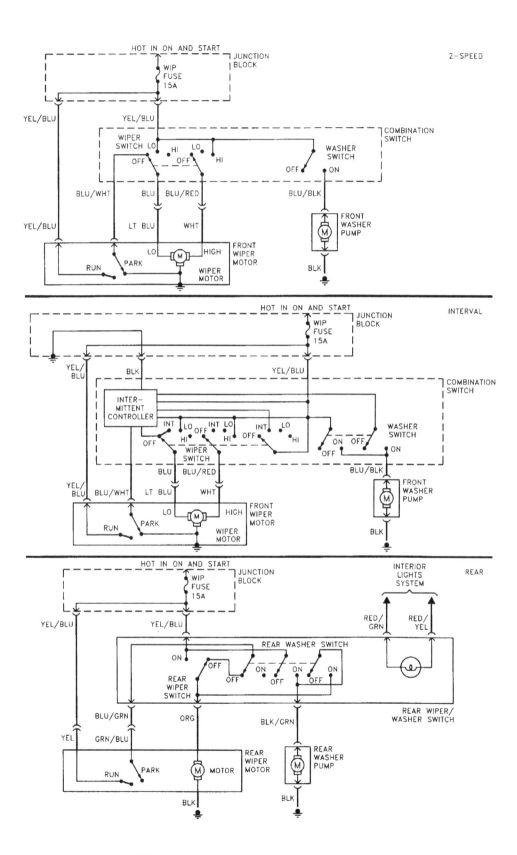

Wiper/washer systems - 1990 and later models

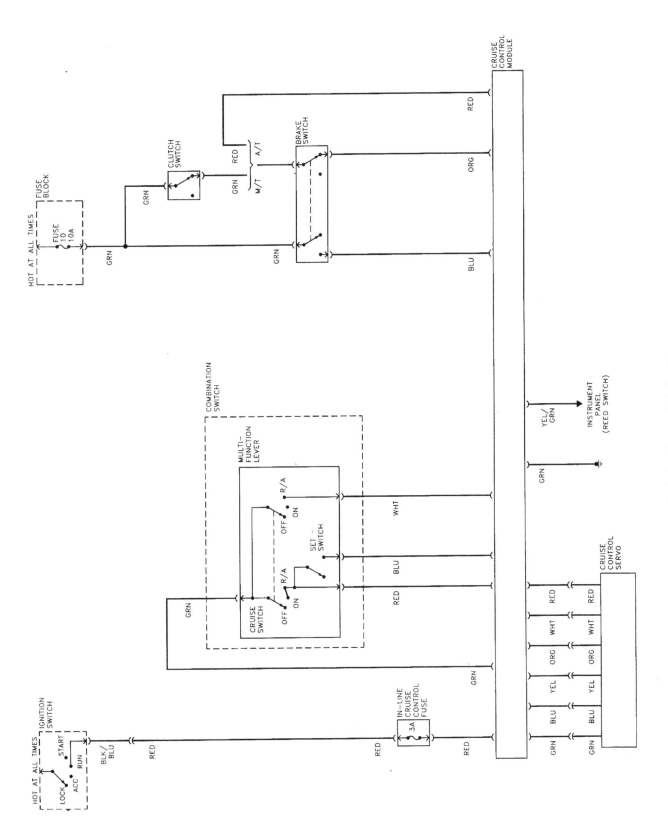

Typical cruise control circuit

Notes

Index

Haynes Automotive Manuals

NOTE: If you do not see a listing for your vehicle, please visit **haynes.com** *for the latest product information and check out our* ***Online Manuals!***

ACURA

12020 **Integra** '86 thru '89 **& Legend** '86 thru '90
12021 **Integra** '90 thru '93 **& Legend** '91 thru '95
Integra '94 thru '00 - *see HONDA Civic (42025)*
MDX '01 thru '07 - *see HONDA Pilot (42037)*
12050 **Acura TL** all models '99 thru '08

AMC

14020 **Mid-size models** '70 thru '83
14025 **(Renault) Alliance & Encore** '83 thru '87

AUDI

15020 **4000** all models '80 thru '87
15025 **5000** all models '77 thru '83
15026 **5000** all models '84 thru '88
Audi A4 '96 thru '01 - *see VW Passat (96023)*
15030 **Audi A4** '02 thru '08

AUSTIN-HEALEY

Sprite - *see MG Midget (66015)*

BMW

18020 **3/5 Series** '82 thru '92
18021 **3-Series** incl. Z3 models '92 thru '98
18022 **3-Series** incl. Z4 models '99 thru '05
18023 **3-Series** '06 thru '14
18025 **320i all 4-cylinder models** '75 thru '83
18050 **1500 thru 2002** except Turbo '59 thru '77

BUICK

19010 **Buick Century** '97 thru '05
Century (front-wheel drive) - *see GM (38005)*
19020 **Buick, Oldsmobile & Pontiac Full-size (Front-wheel drive)** '85 thru '05
Buick Electra, LeSabre and Park Avenue; **Oldsmobile** Delta 88 Royale, Ninety Eight and Regency; **Pontiac** Bonneville
19025 **Buick, Oldsmobile & Pontiac Full-size (Rear wheel drive)** '70 thru '90
Buick Estate, Electra, LeSabre, Limited, **Oldsmobile** Custom Cruiser, Delta 88, Ninety-eight, **Pontiac** Bonneville, Catalina, Grandville, Parisienne
19027 **Buick LaCrosse** '05 thru '13
Enclave - *see GENERAL MOTORS (38001)*
Rainier - *see CHEVROLET (24072)*
Regal - *see GENERAL MOTORS (38010)*
Riviera - *see GENERAL MOTORS (38030, 38031)*
Roadmaster - *see CHEVROLET (24046)*
Skyhawk - *see GENERAL MOTORS (38015)*
Skylark - *see GENERAL MOTORS (38020, 38025)*
Somerset - *see GENERAL MOTORS (38025)*

CADILLAC

21015 **CTS & CTS-V** '03 thru '14
21030 **Cadillac Rear Wheel Drive** '70 thru '93
Cimarron - *see GENERAL MOTORS (38015)*
DeVille - *see GENERAL MOTORS (38031 & 38032)*
Eldorado - *see GENERAL MOTORS (38030)*
Fleetwood - *see GENERAL MOTORS (38031)*
Seville - *see GM (38030, 38031 & 38032)*

CHEVROLET

10305 **Chevrolet Engine Overhaul Manual**
24010 **Astro & GMC Safari Mini-vans** '85 thru '05
24013 **Aveo** '04 thru '11
24015 **Camaro V8** all models '70 thru '81
24016 **Camaro** all models '82 thru '92
24017 **Camaro & Firebird** '93 thru '02
Cavalier - *see GENERAL MOTORS (38016)*
Celebrity - *see GENERAL MOTORS (38005)*
24018 **Camaro** '10 thru '15
24020 **Chevelle, Malibu & El Camino** '69 thru '87
Cobalt - *see GENERAL MOTORS (38017)*
24024 **Chevette & Pontiac T1000** '76 thru '87
Citation - *see GENERAL MOTORS (38020)*
24027 **Colorado & GMC Canyon** '04 thru '12
24032 **Corsica & Beretta** all models '87 thru '96
24040 **Corvette** all V8 models '68 thru '82
24041 **Corvette** all models '84 thru '96
24042 **Corvette** all models '97 thru '13
24044 **Cruze** '11 thru '19
24045 **Full-size Sedans** Caprice, Impala, Biscayne, Bel Air & Wagons '69 thru '90
24046 **Impala SS & Caprice and Buick Roadmaster** '91 thru '96
Impala '00 thru '05 - *see LUMINA (24048)*
24047 **Impala & Monte Carlo** all models '06 thru '11
Lumina '90 thru '94 - *see GM (38010)*
24048 **Lumina & Monte Carlo** '95 thru '05
Lumina APV - *see GM (38035)*
24050 **Luv Pick-up** all 2WD & 4WD '72 thru '82
24051 **Malibu** '13 thru '19
24055 **Monte Carlo** all models '70 thru '88
Monte Carlo '95 thru '01 - *see LUMINA (24048)*
24059 **Nova** all V8 models '69 thru '79
24060 **Nova and Geo Prizm** '85 thru '92
24064 **Pick-ups** '67 thru '87 - Chevrolet & GMC
24065 **Pick-ups** '88 thru '98 - Chevrolet & GMC
24066 **Pick-ups** '99 thru '06 - Chevrolet & GMC
24067 **Chevrolet Silverado & GMC Sierra** '07 thru '14
24068 **Chevrolet Silverado & GMC Sierra** '14 thru '19
24070 **S-10 & S-15 Pick-ups** '82 thru '93, **Blazer & Jimmy** '83 thru '94,
24071 **S-10 & Sonoma Pick-ups** '94 thru '04, including **Blazer, Jimmy & Hombre**
24072 **Chevrolet TrailBlazer, GMC Envoy & Oldsmobile Bravada** '02 thru '09
24075 **Sprint** '85 thru '88 **& Geo Metro** '89 thru '01
24080 **Vans - Chevrolet & GMC** '68 thru '96
24081 **Chevrolet Express & GMC Savana** Full-size Vans '96 thru '19

CHRYSLER

10310 **Chrysler Engine Overhaul Manual**
25015 **Chrysler Cirrus, Dodge Stratus, Plymouth Breeze** '95 thru '00
25020 **Full-size Front-Wheel Drive** '88 thru '93
K-Cars - *see DODGE Aries (30008)*
Laser - *see DODGE Daytona (30030)*
25025 **Chrysler LHS, Concorde, New Yorker, Dodge** Intrepid, **Eagle** Vision, '93 thru '97
25026 **Chrysler LHS, Concorde, 300M, Dodge** Intrepid, '98 thru '04
25027 **Chrysler 300** '05 thru '18, **Dodge Charger** '06 thru '18, **Magnum** '05 thru '08 **& Challenger** '08 thru '18
25030 **Chrysler & Plymouth Mid-size** front wheel drive '82 thru '95
Rear-wheel Drive - *see Dodge (30050)*
25035 **PT Cruiser** all models '01 thru '10
25040 **Chrysler Sebring** '95 thru '06, **Dodge Stratus** '01 thru '06 **& Dodge Avenger** '95 thru '00
25041 **Chrysler Sebring** '07 thru '10, **200** '11 thru '17 **Dodge Avenger** '08 thru '14

DATSUN

28005 **200SX** all models '80 thru '83
28012 **240Z, 260Z & 280Z** Coupe '70 thru '78
28014 **280ZX** Coupe & 2+2 '79 thru '83
300ZX - *see NISSAN (72010)*
28018 **510 & PL521 Pick-up** '68 thru '73
28020 **510** all models '78 thru '81
28022 **620 Series Pick-up** all models '73 thru '79
720 Series Pick-up - *see NISSAN (72030)*

DODGE

400 & 600 - *see CHRYSLER (25030)*
30008 **Aries & Plymouth Reliant** '81 thru '89
30010 **Caravan & Plymouth Voyager** '84 thru '95
30011 **Caravan & Plymouth Voyager** '96 thru '02
30012 **Challenger & Plymouth Sapporro** '78 thru '83
30013 **Caravan, Chrysler Voyager & Town & Country** '03 thru '07
30014 **Grand Caravan & Chrysler Town & Country** '08 thru '18
30016 **Colt & Plymouth Champ** '78 thru '87
30020 **Dakota Pick-ups** all models '87 thru '96
30021 **Durango** '98 & '99 **& Dakota** '97 thru '99
30022 **Durango** '00 thru '03 **& Dakota** '00 thru '04
30023 **Durango** '04 thru '09 **& Dakota** '05 thru '11
30025 **Dart, Demon, Plymouth Barracuda, Duster & Valiant** 6-cylinder models '67 thru '76
30030 **Daytona & Chrysler Laser** '84 thru '89
Intrepid - *see CHRYSLER (25025, 25026)*
30034 **Neon** all models '95 thru '99
30035 **Omni & Plymouth Horizon** '78 thru '90
30036 **Dodge & Plymouth Neon** '00 thru '05
30040 **Pick-ups** full-size models '74 thru '93
30042 **Pick-ups** full-size models '94 thru '08
30043 **Pick-ups** full-size models '09 thru '18
30045 **Ram 50/D50 Pick-ups & Raider and Plymouth Arrow Pick-ups** '79 thru '93
30050 **Dodge/Plymouth/Chrysler** RWD '71 thru '89
30055 **Shadow & Plymouth Sundance** '87 thru '94
30060 **Spirit & Plymouth Acclaim** '89 thru '95
30065 **Vans - Dodge & Plymouth** '71 thru '03

EAGLE

Talon - *see MITSUBISHI (68030, 68031)*
Vision - *see CHRYSLER (25025)*

FIAT

34010 **124 Sport Coupe & Spider** '68 thru '78
34025 **X1/9** all models '74 thru '80

FORD

10320 **Ford Engine Overhaul Manual**
10355 **Ford Automatic Transmission Overhaul**
11500 **Mustang '64-1/2 thru '70 Restoration Guide**
36004 **Aerostar Mini-vans** all models '86 thru '97
36006 **Contour & Mercury Mystique** '95 thru '00
36008 **Courier Pick-up** all models '72 thru '82
36012 **Crown Victoria & Mercury Grand Marquis** '88 thru '11
36014 **Edge** '07 thru '19 **& Lincoln MKX** '07 thru '18
36016 **Escort & Mercury Lynx** all models '81 thru '90
36020 **Escort & Mercury Tracer** '91 thru '02
36022 **Escape** '01 thru '17, **Mazda Tribute** '01 thru '11, **& Mercury Mariner** '05 thru '11
36024 **Explorer & Mazda Navajo** '91 thru '01
36025 **Explorer & Mercury Mountaineer** '02 thru '10
36026 **Explorer** '11 thru '17
36028 **Fairmont & Mercury Zephyr** '78 thru '83
36030 **Festiva & Aspire** '88 thru '97
36032 **Fiesta** all models '77 thru '80
36034 **Focus** all models '00 thru '11
36035 **Focus** '12 thru '14
36045 **Fusion** '06 thru '14 **& Mercury Milan** '06 thru '11
36048 **Mustang V8** all models '64-1/2 thru '73
36049 **Mustang II** 4-cylinder, V6 & V8 models '74 thru '78
36050 **Mustang & Mercury Capri** '79 thru '93
36051 **Mustang** all models '94 thru '04
36052 **Mustang** '05 thru '14
36054 **Pick-ups & Bronco** '73 thru '79
36058 **Pick-ups & Bronco** '80 thru '96
36059 **F-150** '97 thru '03, **Expedition** '97 thru '17, **F-250** '97 thru '99, **F-150 Heritage** '04 **& Lincoln Navigator** '98 thru '17
36060 **Super Duty Pick-ups & Excursion** '99 thru '10
36061 **F-150** full-size '04 thru '14
36062 **Pinto & Mercury Bobcat** '75 thru '80
36063 **F-150** full-size '15 thru '17
36064 **Super Duty Pick-ups** '11 thru '16
36066 **Probe** all models '89 thru '92
Probe '93 thru '97 - *see MAZDA 626 (61042)*
36070 **Ranger & Bronco II** gas models '83 thru '92
36071 **Ranger** '93 thru '11 **& Mazda Pick-ups** '94 thru '09
36074 **Taurus & Mercury Sable** '86 thru '95
36075 **Taurus & Mercury Sable** '96 thru '07
36076 **Taurus** '08 thru '14, **Five Hundred** '05 thru '07, **Mercury Montego** '05 thru '07 **& Sable** '08 thru '09
36078 **Tempo & Mercury Topaz** '84 thru '94
36082 **Thunderbird & Mercury Cougar** '83 thru '88
36086 **Thunderbird & Mercury Cougar** '89 thru '97
36090 **Vans** all V8 Econoline models '69 thru '91
36094 **Vans** full size '92 thru '14
36097 **Windstar** '95 thru '03, **Freestar & Mercury Monterey Mini-van** '04 thru '07

GENERAL MOTORS

10360 **GM Automatic Transmission Overhaul**
38001 **GMC Acadia** '07 thru '16, **Buick Enclave** '08 thru '17, **Saturn Outlook** '07 thru '10 **& Chevrolet Traverse** '09 thru '17
38005 **Buick Century, Chevrolet Celebrity, Oldsmobile Cutlass Ciera & Pontiac 6000** all models '82 thru '96
38010 **Buick Regal** '88 thru '04, **Chevrolet Lumina** '88 thru '04, **Oldsmobile Cutlass Supreme** '88 thru '97 **& Pontiac Grand Prix** '88 thru '07
38015 **Buick Skyhawk, Cadillac Cimarron, Chevrolet Cavalier, Oldsmobile Firenza, Pontiac J-2000 & Sunbird** '82 thru '94
38016 **Chevrolet Cavalier & Pontiac Sunfire** '95 thru '05
38017 **Chevrolet Cobalt** '05 thru '10, **HHR** '06 thru '11, **Pontiac G5** '07 thru '09, **Pursuit** '05 thru '06 **& Saturn ION** '03 thru '07
38020 **Buick Skylark, Chevrolet Citation, Oldsmobile Omega, Pontiac Phoenix** '80 thru '85
38025 **Buick Skylark** '86 thru '98, **Somerset** '85 thru '87, **Oldsmobile Achieva** '92 thru '98, **Calais** '85 thru '91, **& Pontiac Grand Am** all models '85 thru '98
38026 **Chevrolet Malibu** '97 thru '03, **Classic** '04 thru '05, **Oldsmobile Alero** '99 thru '03, **Cutlass** '97 thru '00, **& Pontiac Grand Am** '99 thru '03
38027 **Chevrolet Malibu** '04 thru '12, **Pontiac G6** '05 thru '10 **& Saturn Aura** '07 thru '10
38030 **Cadillac Eldorado, Seville, Oldsmobile Toronado & Buick Riviera** '71 thru '85
38031 **Cadillac Eldorado, Seville, DeVille, Fleetwood, Oldsmobile Toronado & Buick Riviera** '86 thru '93
38032 **Cadillac DeVille** '94 thru '05, **Seville** '92 thru '04 **& Cadillac DTS** '06 thru '10
38035 **Chevrolet Lumina APV, Oldsmobile Silhouette & Pontiac Trans Sport** all models '90 thru '96
38036 **Chevrolet Venture** '97 thru '05, **Oldsmobile Silhouette** '97 thru '04, **Pontiac Trans Sport** '97 thru '98 **& Montana** '99 thru '05
38040 **Chevrolet Equinox** '05 thru '17, **GMC Terrain** '10 thru '17 **& Pontiac Torrent** '06 thru '09

GEO

Metro - *see CHEVROLET Sprint (24075)*
Prizm - *'85 thru '92 see CHEVY (24060), '93 thru '02 see TOYOTA Corolla (92036)*
40030 **Storm** all models '90 thru '93
Tracker - *see SUZUKI Samurai (90010)*

(Continued on other side)

Haynes Automotive Manuals (continued)

NOTE: *If you do not see a listing for your vehicle, please visit* **haynes.com** *for the latest product information and check out our* **Online Manuals!**

GMC
Acadia - *see GENERAL MOTORS (38001)*
Pick-ups - *see CHEVROLET (24027, 24068)*
Vans - *see CHEVROLET (24081)*

HONDA
42010 **Accord CVCC** all models '76 thru '83
42011 **Accord** all models '84 thru '89
42012 **Accord** all models '90 thru '93
42013 **Accord** all models '94 thru '97
42014 **Accord** all models '98 thru '02
42015 **Accord** '03 thru '12 & **Crosstour** '10 thru '14
42016 **Accord** '13 thru '17
42020 **Civic 1200** all models '73 thru '79
42021 **Civic 1300 & 1500 CVCC** '80 thru '83
42022 **Civic 1500 CVCC** all models '75 thru '79
42023 **Civic** all models '84 thru '91
42024 **Civic & del Sol** '92 thru '95
42025 **Civic** '96 thru '00, **CR-V** '97 thru '01
    & **Acura Integra** '94 thru '00
42026 **Civic** '01 thru '11 & **CR-V** '02 thru '11
42027 **Civic** '12 thru '15 & **CR-V** '12 thru '16
42030 **Fit** '07 thru '13
42035 **Odyssey** all models '99 thru '10
    Passport - *see ISUZU Rodeo (47017)*
42037 **Honda Pilot** '03 thru '08, **Ridgeline** '06 thru '14
    & **Acura MDX** '01 thru '07
42040 **Prelude CVCC** all models '79 thru '89

HYUNDAI
43010 **Elantra** all models '96 thru '19
43015 **Excel & Accent** all models '86 thru '13
43050 **Santa Fe** all models '01 thru '12
43055 **Sonata** all models '99 thru '14

INFINITI
G35 '03 thru '08 - *see NISSAN 350Z (72011)*

ISUZU
Hombre - *see CHEVROLET S-10 (24071)*
47017 **Rodeo** '91 thru '02, **Amigo** '89 thru '94 & '98 thru '02
    & **Honda Passport** '95 thru '02
47020 **Trooper** '84 thru '91 & **Pick-up** '81 thru '93

JAGUAR
49010 **XJ6** all 6-cylinder models '68 thru '86
49011 **XJ6** all models '88 thru '94
49015 **XJ12 & XJS** all 12-cylinder models '72 thru '85

JEEP
50010 **Cherokee, Comanche & Wagoneer Limited**
    all models '84 thru '01
50011 **Cherokee** '14 thru '19
50020 **CJ** all models '49 thru '86
50025 **Grand Cherokee** all models '93 thru '04
50026 **Grand Cherokee** '05 thru '19
    & **Dodge Durango** '11 thru '19
50029 **Grand Wagoneer & Pick-up** '72 thru '91
    Grand Wagoneer '84 thru '91, Cherokee &
    Wagoneer '72 thru '83, Pick-up '72 thru '88
50030 **Wrangler** all models '87 thru '17
50035 **Liberty** '02 thru '12 & **Dodge Nitro** '07 thru '11
50050 **Patriot & Compass** '07 thru '17

KIA
54050 **Optima** '01 thru '10
54060 **Sedona** '02 thru '14
54070 **Sephia** '94 thru '01, **Spectra** '00 thru '09,
    Sportage '05 thru '20
54077 **Sorento** '03 thru '13

LEXUS
ES 300/330 - *see TOYOTA Camry (92007, 92008)*
ES 350 - *see TOYOTA Camry (92009)*
RX 300/330/350 - *see TOYOTA Highlander (92095)*

LINCOLN
MKX - *see FORD (36014)*
Navigator - *see FORD Pick-up (36059)*
59010 **Rear-Wheel Drive Continental** '70 thru '87,
    Mark Series '70 thru '92 & **Town Car** '81 thru '10

MAZDA
61010 **GLC (rear-wheel drive)** '77 thru '83
61011 **GLC (front-wheel drive)** '81 thru '85
61012 **Mazda3** '04 thru '11
61015 **323 & Protogé** '90 thru '03
61016 **MX-5 Miata** '90 thru '14
61020 **MPV** all models '89 thru '98
    Navajo - *see Ford Explorer (36024)*
61030 **Pick-ups** '72 thru '93
    Pick-ups '94 thru '09 - *see Ford Ranger (36071)*
61035 **RX-7** all models '79 thru '85
61036 **RX-7** all models '86 thru '91
61040 **626 (rear-wheel drive)** all models '79 thru '82
61041 **626 & MX-6 (front-wheel drive)** '83 thru '92
61042 **626** '93 thru '01 & **MX-6/Ford Probe** '93 thru '02
61043 **Mazda6** '03 thru '13

MERCEDES-BENZ
63012 **123 Series Diesel** '76 thru '85
63015 **190 Series** 4-cylinder gas models '84 thru '88
63020 **230/250/280** 6-cylinder SOHC models '68 thru '72
63025 **280 123 Series** gas models '77 thru '81
63030 **350 & 450** all models '71 thru '80
63040 **C-Class:** C230/C240/C280/C320/C350 '01 thru '07

MERCURY
64200 **Villager & Nissan Quest** '93 thru '01
    All other titles, see FORD Listing.

MG
66010 **MGB** Roadster & GT Coupe '62 thru '80
66015 **MG Midget, Austin Healey Sprite** '58 thru '80

MINI
67020 **Mini** '02 thru '13

MITSUBISHI
68020 **Cordia, Tredia, Galant, Precis & Mirage** '83 thru '93
68030 **Eclipse, Eagle Talon & Plymouth Laser** '90 thru '94
68031 **Eclipse** '95 thru '05 & **Eagle Talon** '95 thru '98
68035 **Galant** '94 thru '12
68040 **Pick-up** '83 thru '96 & **Montero** '83 thru '93

NISSAN
72010 **300ZX** all models including Turbo '84 thru '89
72011 **350Z & Infiniti G35** all models '03 thru '08
72015 **Altima** all models '93 thru '06
72016 **Altima** '07 thru '12
72020 **Maxima** all models '85 thru '92
72021 **Maxima** all models '93 thru '08
72025 **Murano** '03 thru '14
72030 **Pick-ups** '80 thru '97 & **Pathfinder** '87 thru '95
72031 **Frontier** '98 thru '04, **Xterra** '00 thru '04,
    & **Pathfinder** '96 thru '04
72032 **Frontier & Xterra** '05 thru '14
72037 **Pathfinder** '05 thru '14
72040 **Pulsar** all models '83 thru '86
72042 **Roque** all models '08 thru '20
72050 **Sentra** all models '82 thru '94
72051 **Sentra & 200SX** all models '95 thru '06
72060 **Stanza** all models '82 thru '90
72070 **Titan pick-ups** '04 thru '10, **Armada** '05 thru '10
    & **Pathfinder Armada** '04
72080 **Versa** all models '07 thru '19

OLDSMOBILE
73015 **Cutlass** V6 & V8 gas models '74 thru '88
    For other OLDSMOBILE titles, see BUICK,
    CHEVROLET or GENERAL MOTORS listings.

PLYMOUTH
For PLYMOUTH titles, see DODGE listing.

PONTIAC
79008 **Fiero** all models '84 thru '88
79018 **Firebird** V8 models except Turbo '70 thru '81
79019 **Firebird** all models '82 thru '92
79025 **G6** all models '05 thru '09
79040 **Mid-size Rear-wheel Drive** '70 thru '87
    Vibe '03 thru '10 - *see TOYOTA Corolla (92037)*
    For other PONTIAC titles, see BUICK,
    CHEVROLET or GENERAL MOTORS listings.

PORSCHE
80020 **911** Coupe & Targa models '65 thru '89
80025 **914** all 4-cylinder models '69 thru '76
80030 **924** all models including Turbo '76 thru '82
80035 **944** all models including Turbo '83 thru '89

RENAULT
Alliance & Encore - *see AMC (14025)*

SAAB
84010 **900** all models including Turbo '79 thru '88

SATURN
87010 **Saturn** all S-series models '91 thru '02
    Saturn Ion '03 thru '07- *see GM (38017)*
    Saturn Outlook - *see GM (38001)*
87020 **Saturn L-series** all models '00 thru '04
87040 **Saturn VUE** '02 thru '09

SUBARU
89002 **1100, 1300, 1400 & 1600** '71 thru '79
89003 **1600 & 1800** 2WD & 4WD '80 thru '94
89080 **Impreza** '02 thru '11, **WRX** '02 thru '14,
    & **WRX STI** '04 thru '14
89100 **Legacy** all models '90 thru '99
89101 **Legacy & Forester** '00 thru '09
89102 **Legacy** '10 thru '16 & **Forester** '12 thru '16

SUZUKI
90010 **Samurai/Sidekick & Geo Tracker** '86 thru '01

TOYOTA
92005 **Camry** all models '83 thru '91
92006 **Camry** '92 thru '96 & **Avalon** '95 thru '96
92007 **Camry, Avalon, Solara, Lexus ES 300** '97 thru '01

92008 **Camry, Avalon, Lexus ES 300/330** '02 thru '06
    & **Solara** '02 thru '08
92009 **Camry, Avalon & Lexus ES 350** '07 thru '17
92015 **Celica Rear-wheel Drive** '71 thru '85
92020 **Celica Front-wheel Drive** '86 thru '99
92025 **Celica Supra** all models '79 thru '92
92030 **Corolla** all models '75 thru '79
92032 **Corolla** all rear-wheel drive models '80 thru '87
92035 **Corolla** all front-wheel drive models '84 thru '92
92036 **Corolla & Geo/Chevrolet Prizm** '93 thru '02
92037 **Corolla** '03 thru '19, **Matrix** '03 thru '14,
    & **Pontiac Vibe** '03 thru '10
92040 **Corolla Tercel** all models '80 thru '82
92045 **Corona** all models '74 thru '82
92050 **Cressida** all models '78 thru '82
92055 **Land Cruiser** FJ40, 43, 45, 55 '68 thru '82
92056 **Land Cruiser** FJ60, 62, 80, FZJ80 '80 thru '96
92060 **Matrix** '03 thru '11 & **Pontiac Vibe** '03 thru '10
92065 **MR2** all models '85 thru '87
92070 **Pick-up** all models '69 thru '78
92075 **Pick-up** all models '79 thru '95
92076 **Tacoma** '95 thru '04, **4Runner** '96 thru '02
    & **T100** '93 thru '08
92077 **Tacoma** all models '05 thru '18
92078 **Tundra** '00 thru '06 & **Sequoia** '01 thru '07
92079 **4Runner** all models '03 thru '09
92080 **Previa** all models '91 thru '95
92081 **Prius** all models '01 thru '12
92082 **RAV4** all models '96 thru '12
92085 **Tercel** all models '87 thru '94
92090 **Sienna** all models '98 thru '10
92095 **Highlander** '01 thru '19
    & **Lexus RX330/330/350** '99 thru '19
92179 **Tundra** '07 thru '19 & **Sequoia** '08 thru '19

TRIUMPH
94007 **Spitfire** all models '62 thru '81
94010 **TR7** all models '75 thru '81

VW
96008 **Beetle & Karmann Ghia** '54 thru '79
96009 **New Beetle** '98 thru '10
96016 **Rabbit, Jetta, Scirocco & Pick-up**
    gas models '75 thru '92 & Convertible '80 thru '92
96017 **Golf, GTI & Jetta** '93 thru '98, **Cabrio** '95 thru '02
96018 **Golf, GTI, Jetta** '99 thru '05
96019 **Jetta, Rabbit, GLI, GTI & Golf** '05 thru '11
96020 **Rabbit, Jetta & Pick-up** diesel '77 thru '84
96021 **Jetta** '11 thru '18 & **Golf** '15 thru '19
96023 **Passat** '98 thru '05 & **Audi A4** '96 thru '01
96030 **Transporter 1600** all models '68 thru '79
96035 **Transporter 1700, 1800 & 2000** '72 thru '79
96040 **Type 3 1500 & 1600** all models '63 thru '73
96045 **Vanagon Air-Cooled** all models '80 thru '83

VOLVO
97010 **120, 130 Series & 1800 Sports** '61 thru '73
97015 **140 Series** all models '66 thru '74
97020 **240 Series** all models '76 thru '93
97040 **740 & 760 Series** all models '82 thru '88
97050 **850 Series** all models '93 thru '97

TECHBOOK MANUALS
10205 **Automotive Computer Codes**
10206 **OBD-II & Electronic Engine Management**
10210 **Automotive Emissions Control Manual**
10215 **Fuel Injection Manual** '78 thru '85
10225 **Holley Carburetor Manual**
10230 **Rochester Carburetor Manual**
10305 **Chevrolet Engine Overhaul Manual**
10320 **Ford Engine Overhaul Manual**
10330 **GM and Ford Diesel Engine Repair Manual**
10331 **Duramax Diesel Engines** '01 thru '19
10332 **Cummins Diesel Engine Performance Manual**
10333 **GM, Ford & Chrysler Engine Performance Manual**
10334 **GM Engine Performance Manual**
10340 **Small Engine Repair Manual,** 5 HP & Less
10341 **Small Engine Repair Manual,** 5.5 thru 20 HP
10345 **Suspension, Steering & Driveline Manual**
10355 **Ford Automatic Transmission Overhaul**
10360 **GM Automatic Transmission Overhaul**
10405 **Automotive Body Repair & Painting**
10410 **Automotive Brake Manual**
10411 **Automotive Anti-lock Brake (ABS) Systems**
10420 **Automotive Electrical Manual**
10425 **Automotive Heating & Air Conditioning**
10435 **Automotive Tools Manual**
10445 **Welding Manual**
10450 **ATV Basics**

Over a 100 Haynes
motorcycle manuals
also available

10/22
